# METHODS IN MOLECULAR BIOLOGY™

*Series Editor*
**John M. Walker**
**School of Life Sciences**
**University of Hertfordshire**
**Hatfield, Hertfordshire, AL10 9AB, UK**

For further volumes:
http://www.springer.com/series/7651

# Mitochondrial Disorders

## Biochemical and Molecular Analysis

Edited by

## Lee-Jun C. Wong

*Mitochondrial Diagnostic Laboratory, Department of Molecular and Human Genetics,*
*Baylor College of Medicine, Houston, TX, USA*

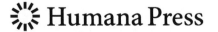 Humana Press

*Editor*
Lee-Jun C. Wong, Ph.D. FACMG
Clinical Molecular Genetics and Clinical Biochemical Genetics
Professor, Department of Molecular and Human Genetics
Director, Mitochondrial Diagnostic Laboratory
Baylor College of Medicine
One Baylor Plaza, NAB 2015
Houston, Texas 77030, USA
ljwong@bcm.edu

ISSN 1064-3745          e-ISSN 1940-6029
ISBN 978-1-61779-503-9          e-ISBN 978-1-61779-504-6
DOI 10.1007/978-1-61779-504-6
Springer New York Dordrecht Heidelberg London

Library of Congress Control Number: 2011943084

Printed on acid-free paper

Humana Press is part of Springer Science+Business Media (www.springer.com)

# Preface

A major function of mitochondria is the production of energy molecule ATP, by the way of electron transport chain and respiration, in a process called oxidative phosphorylation (OXPHOS). In order to carry out OXPHOS, the assembly of fully functional mitochondria requires the participation of approximately 1,500 genes encoded by both the mitochondrial and nuclear genomes. Thus, molecular defects in either of the two genomes may cause mitochondrial dysfunction, giving rise to either Mendelian or Matrilineal disorders. Each cell may contain hundreds to thousands of copies of the mitochondrial genome. Depending on the specific genetic defect, the distribution of the affected tissues, and the proportion of mutant to wild-type mitochondrial DNA (mtDNA) (termed heteroplasmy), the clinical manifestations of the disease are remarkably variable and heterogeneous. Therefore, for any given patient, establishing a diagnosis of a mitochondrial disorder can be very difficult. It requires an evaluation of the family pedigree, in conjunction with a thorough assessment of the clinical, histopathological, imaging, biochemical, and molecular features of the case. Given the breadth and complexity of the problem, these studies are usually provided by several different clinical specialties and/or laboratories; each focused on one or more particular areas. The laboratory and clinical methodologies used may vary widely, and to date there has been no systematic presentation of the numerous protocols that are applied to the assessment of these clinically and genetically heterogeneous mitochondrial disorders. It is the main objective of this volume of *Methods in Molecular Biology* to provide such a collection of protocols.

This volume is divided into three parts. The first part is the nonprotocol section that contains three chapters describing the complexity of these dual genome disorders. Chapter 1 provides an overview of the mitochondrial syndromes caused by common point mutations or deletions of the mtDNA, leading to the concepts and methods of analyzing mutation heteroplasmy, tissue distribution, and the mtDNA content. Molecular defects in a group of nuclear genes responsible for mtDNA biogenesis and the maintenance of mtDNA integrity may cause mtDNA defects secondary to nuclear gene mutations. Chapter 2 focuses on mitochondrial disorders caused by molecular defects in nuclear genes. The strategies used to distinguish nuclear and mitochondrial etiologies of the disease, and approaches to pinpoint an appropriate class of nuclear genes for further sequence analysis are described. The third chapter presents useful diagnostic algorithms. Throughout these chapters, the rationale for the application of the necessary diagnostic method included in this volume is described.

The second part of this volume is devoted to biochemical protocols that are used to study mitochondrial disorders. These include methods for mitochondrial functional studies such as the assays of electron transport chain complex activities, the measurement of ATP synthesis, oxygen consumption, and pyruvate dehydrogenase (Chapters 4–7); the analysis of thymidine phosphorylase activity and measurements of unbalanced dNTP concentrations (Chapters 8 and 9); assessment of CoQ by two different methods (Chapters 10 and 11); morphological and histochemical methods to evaluate mitochondrial dysfunction (Chapter 12); blue native gel analysis of higher-order respiratory chain complexes and mitochondrial protein translation (Chapters 13 and 14); and tools and novel technologies used to study mitochondrial function and gene expression such as cybrids, fluorescence-activated cell sorting, and gene expression arrays (Chapters 15–17).

The third part of this volume focuses on the DNA-based approaches used to identify molecular defects. This part includes screening of the known common mtDNA point mutations and large deletions (Chapter 18); sequence analysis of both nuclear and mitochondrial genomes (Chapter 19); the utility of oligonucleotide array comparative genome hybridization to evaluate genomic deletions and copy number changes (Chapter 20); quantitative analysis of mutant heteroplasmy and mtDNA depletions (Chapters 21 and 22); and, finally, the interpretation of variants identified by sequencing (Chapter 23).

There are a number of procedures that can be used to evaluate mitochondrial disorders, such as electron microscopy and immunofluorescence methods, that are not provided in this volume. Furthermore, a novel one-step comprehensive molecular analysis by the enrichment of all ~1,500 target genes followed by deep sequencing is being currently developed. However, due to the limitations of space, a detailed exploration of these topics is not included.

I am grateful to all contributing authors whose input made this volume, *Mitochondrial Disorders: Biochemical and Molecular Analysis*, possible. I particularly appreciate the patience of the authors who submitted their chapters on time.

*Houston, TX, USA*                                                      *Lee-Jun C. Wong*

# Contents

# Contributors

KIMBERLY A. CHAPMAN • *Department of Genetics, Children's National Medical Center, Washington, DC, USA; Division of Human Genetics, Department of Pediatrics, The Children's Hospital of Philadelphia, Philadelphia, PA, USA*

MEGAN E. CORNWELL • *Medical Genetics Laboratories, Department of Molecular and Human Genetics, Baylor College of Medicine, Houston, TX, USA*

WILLIAM J. CRAIGEN • *Department of Molecular and Human Genetics, Baylor College of Medicine, Houston, TX, USA*

STEPHEN DINGLEY • *Division of Human Genetics, Department of Pediatrics, The Children's Hospital of Philadelphia, Philadelphia, PA, USA*

BEATRIZ DORADO • *Department of Neurology, H. Houston Merritt Clinical Research Center, Columbia University Medical Center, New York, NY, USA*

MARNI J. FALK • *Division of Human Genetics, Department of Pediatrics, The Children's Hospital of Philadelphia and University of Pennsylvania School of Medicine, Philadelphia, PA, USA*

KRISTEN C. FLOYD • *Medical Genetics Laboratories, Department of Molecular and Human Genetics, Baylor College of Medicine, Houston, TX, USA*

ANN E. FRAZIER • *Murdoch Children's Research Institute, Parkville, VIC, Australia*

BRETT H. GRAHAM • *Department of Molecular and Human Genetics, Baylor College of Medicine, Houston, TX, USA*

GEORGE GRAHAME • *Center for Inherited Disorders of Energy Metabolism, University Hospitals Case Medical Center, Case Western Reserve University, Cleveland, OH, USA*

MANUELA M. GRAZINA • *Laboratory of Biochemical Genetics (CNC/UC), Faculty of Medicine, University of Coimbra, Coimbra, Portugal*

SI HOUN HAHN • *Seattle Children's Hospital Research Institute, Seattle, WA, USA; Department of Pediatrics, University of Washington School of Medicine, Seattle, WA, USA*

MICHELLE C. HALBERG • *Medical Genetics Laboratories, Department of Molecular and Human Genetics, Baylor College of Medicine, Houston, TX, USA*

MICHIO HIRANO • *H. Houston Merritt Clinical Research Center, Department of Neurology, Columbia University Medical Center, New York, NY, USA*

BENNY ABRAHAM KAIPPARETTU • *Department of Molecular and Human Genetics, Baylor College of Medicine, Houston, TX, USA*

SANDRA KERFOOT • *Seattle Children's Hospital Research Institute, Seattle, WA, USA*

DOUGLAS KERR • *Center for Inherited Disorders of Energy Metabolism, University Hospitals Case Medical Center, Case Western Reserve University, Cleveland, OH, USA*

MEGAN L. LANDSVERK • *Medical Genetics Laboratories, Department of Molecular and Human Genetics, Baylor College of Medicine, Houston, TX, USA*

SCOT C. LEARY • *Department of Biochemistry, University of Saskatchewan, Saskatoon, SK, Canada*

ZHIHONG LI • *Department of Molecular and Human Genetics, Baylor College of Medicine, Houston, TX, USA*

LUIS C. LÓPEZ • *Instituto de Biotecnologia, Centro de Investigacion Biomedica, Parque Technologico de Ciencias de la Salud, Universidad de Granada, Armilla, Granada, Spain*

YEWEI MA • *Department of Molecular and Human Genetics, Baylor College of Medicine, Houston, TX, USA*

RAMON MARTÍ • *Laboratori de Patologia Mitocondrial, Institut de Recerca Hospital Universitari Vall D'Hebron, Universitat Autonoma de Barcelona, Barcelona, Spain; Biomedical Network Research Centre on Rare Diseases (CIBERER), Instituto de Salud Carlos III, Barcelona, Spain*

MICHAEL V. MILES • *Division of Pathology and Laboratory Medicine, Departments of Pediatrics and Pathology & Laboratory Medicine, Cincinnati Children's Hospital Medical Center and University of Cincinnati College of Medicine, Cincinnati, OH, USA*

GHUNWA NAKOUZI • *Center for Inherited Disorders of Energy Metabolism, University Hospitals Case Medical Center, Case Western Reserve University, Cleveland, OH, USA*

MEAGAN E. PALCULICT • *Medical Genetics Laboratories, Department of Molecular and Human Genetics, Baylor College of Medicine, Houston, TX, USA*

ERZSEBET POLYAK • *Division of Human Genetics, Department of Pediatrics, The Children's Hospital of Philadelphia, Philadelphia, PA, USA*

MRUDULA RAKHADE • *Mitochondrial Diagnostic Laboratory, Medical Genetics Laboratories, Department of Molecular and Human Genetics, Baylor College of Medicine, Houston, TX, USA*

FLORIN SASARMAN • *Montreal Neurological Institute and Department of Human Genetics, McGill University, Montreal, QC, Canada*

FERNANDO SCAGLIA • *Department of Molecular and Human Genetics, Baylor College of Medicine, Houston, TX, USA*

ERIC A. SHOUBRIDGE • *Department of Human Genetics, Montreal Neurological Institute, McGill University, Montreal, QC, Canada*

PETER H. TANG • *Division of Pathology and Laboratory Medicine, Cincinnati Children's Hospital Medical Center, Cincinnati, OH, USA*

SHA TANG • *Medical Genetics Laboratories, Department of Molecular and Human Genetics, Baylor College of Medicine, Houston, TX, USA*

KURENAI TANJI • *Neuromuscular Pathology Laboratory, Division of Neuropathology, Department of Pathology and Cell Biology, Columbia University, New York, NY, USA*

DAVID R. THORBURN • *Murdoch Childrens Research Institute and Victorian Clinical Genetics Services Pathology, Royal Children's Hospital, Melbourne, VIC, Australia; Department of Paediatrics, University of Melbourne, Melbourne, VIC, Australia*

VALERIA VASTA • *Seattle Children's Hospital Research Institute, Seattle, WA, USA*

VICTOR VENEGAS • *Department of Molecular and Human Genetics, Baylor College of Medicine, Houston, TX, USA*

SAJNA ANTONY VITHAYATHIL • *Department of Molecular and Human Genetics, Baylor College of Medicine, Houston, TX, USA*

JING WANG • *Medical Genetics Laboratories, Department of Molecular and Human Genetics, Baylor College of Medicine, Houston, TX, USA*

VICTOR WEI ZHANG • *Mitochondrial Diagnostic Laboratory, Medical Genetics Laboratories, Department of Molecular and Human Genetics, Baylor College of Medicine, Houston, TX, USA*

ZHE ZHANG • *Center for Biomedical Informatics, The Children's Hospital of Philadelphia, Philadelphia, PA, USA*

# Part I

# Mitochondrial Disorder: A Complex Disease of the Two Genomes

# Chapter 1

# Mitochondrial DNA Mutations: An Overview of Clinical and Molecular Aspects

## William J. Craigen

## Abstract

Mutations that arise in mitochondrial DNA (mtDNA) may be sporadic, maternally inherited, or Mendelian in character and include mtDNA rearrangements such as deletions, inversions or duplications, point mutations, or copy number depletion. Primary mtDNA mutations occur sporadically or exhibit maternal inheritance and arise due in large part to the high mutation rate of mtDNA. mtDNA mutations may also occur because of defects in the biogenesis or maintenance of mtDNA, reflecting the contribution of nuclear-encoded genes to these processes, and in this case exhibit Mendelian inheritance. Whether maternally inherited, sporadic, or Mendelian, mtDNA mutations can exhibit a complex and broad spectrum of disease manifestations due to the central role mitochondria play in a variety of cellular functions. In addition, because there exist hundreds to thousands of copies of mtDNA in each cell, the proportion of mutant mtDNA molecules can have a profound effect on the cellular and clinical phenotype. This chapter reviews the classification of mtDNA mutations and the clinical features that determine the diagnosis of a primary mtDNA disorder.

**Key words:** Mitochondrial DNA mutations, Electron transport chain, Heteroplasmy, MtDNA deletion, MtDNA depletion

## 1. Introduction

Mitochondria are essential organelles that are present in virtually all eukaryotic cells and are the modern day remnants of the ancient evolutionary symbiotic marriage of a protobacterium and progenitor eukaryote. Historically, mitochondria have been viewed as simply a source of cellular energy, yet mitochondria perform crucial roles in a number of metabolic and developmental processes, including ATP production via the oxidative phosphorylation (OXPHOS) pathway, modulating apoptosis or programmed cell death, providing a means to buffer and regulate calcium homeostasis, and participating

Lee-Jun C. Wong (ed.), *Mitochondrial Disorders: Biochemical and Molecular Analysis*, Methods in Molecular Biology, vol. 837,
DOI 10.1007/978-1-61779-504-6_1, © Springer Science+Business Media, LLC 2012

in cell cycle regulation through "retrograde signaling" (1, 2). Increasingly, signal transduction pathways are recognized to converge on mitochondria in previously unrecognized ways, including STAT3, AKT, PKA, and PKC signaling cascades (3–7), although defining the functional significance of these pathways is an ongoing challenge. The complexity and centrality of mitochondrial functions means that mitochondria participate directly or indirectly in an enormous variety of diseases, not just rare monogenic multisystem disorders but also common multifactorial disorders such as diabetes, Alzheimer disease, and Parkinson disease. Furthermore, progressive mitochondrial dysfunction has been implicated in the normal aging process (8).

The term mitochondrial disorder generally refers to diseases that are caused by disturbances in the OXPHOS system, and given the dual genomes nature of the mitochondrial electron transport chain (ETC), where 13 protein proteins are encoded by mitochondrial DNA (mtDNA) and the remainder by nuclear genes, there is tremendous genetic, biochemical, and clinical complexity to this heterogeneous group of often multisystem and fatal diseases.

A functional ETC leads to the coordinated transport of electrons and protons, resulting in the production of ATP. The ETC is embedded in the mitochondrial inner membrane and consists of almost 90 proteins assembled into 5 multiprotein enzyme complexes (complexes I–V) that can be assayed biochemically using enzyme assays and functionally by measuring oxygen consumption, ATP synthesis, or mitochondrial inner membrane electrochemical potential. Other biophysical approaches such as evaluating the integrity of the multiprotein complexes via blue native gel electrophoresis are increasingly employed for diagnostic purposes. Based upon biochemical and molecular studies performed at major referral centers, around two thirds of ETC defects consist of isolated enzyme deficiencies, while one third of cases are due to multiple enzyme complexes (9). Because of the dual genetic systems encoding components of ETC and the need for a parallel system for the synthesis of proteins within mitochondria (translation), in addition to mechanisms required for the biosynthesis and maintenance of mtDNA and the biogenesis of the organelle itself, there are remarkably diverse causes for mitochondrial disorders. Isolated OXPHOS deficiencies are generally caused by mutations in genes encoding subunits of the OXPHOS system, whether nuclear or mtDNA-encoded, or in genes encoding proteins required for the assembly of specific OXPHOS enzyme complexes, whereas combined deficiencies in the ETC complexes may reflect the consequence of mutations in mtDNA-encoded transfer RNAs or ribosomal RNAs, or due to arrangements or depletion of mtDNA (10). Both heritable and sporadic (new mutation) forms of mtDNA mutations occur, and mutations can be observed in either a mosaic composition within an individual (heteroplasmy) or in a uniform state (homoplasmy), with the severity of pathogenicity influencing the degree to which the proportion of mutant mtDNA molecules is tolerated. This chapter focuses on disorders caused by primary mutations of mtDNA, while disorders where mtDNA mutations arise as a consequence of defects in nuclear-encoded genes necessary for the replication and maintenance of mtDNA are discussed in the following chapter.

## 2. History

The recognition of cytoplasmic inheritance dates to botanists of the nineteenth century. However, the identification of mtDNA was not made until the early 1960s when Schatz reported its isolation from yeast (11) and Nass observed DNA fibers within mitochondria by electron microscopy (12). It was not until 1988 that heteroplasmic deletions of mtDNA in patients with mitochondrial myopathies were detected (13). Similar large deletions were subsequently uncovered in patients with Kearns-Sayre syndrome (14, 15), a multisystem sporadic disorder form of chronic progressive external ophthalmoplegia (CPEO). Subsequently, an mtDNA point mutation leading to a missense substitution of a histidine for arginine in subunit 4 of NADH dehydrogenase (complex I) was uncovered as the basis for Leber's hereditary optic neuropathy (LHON) (16). Soon thereafter, additional point mutations in mtDNA-encoded tRNA genes were found to cause the mitochondrial syndromes myoclonic epilepsy with ragged-red fibers (MERRF) (17) and mitochondrial encephalomyopathy, lactic acidosis, and stroke-like episodes (MELAS) (18). Finally, the first example of a mitochondrial ribosomal RNA mutation associated with nonsyndromic hearing loss and antibiotic-induced hearing loss was described in 1993 (19). Thus, mutations in each functional class of genes found in mtDNA; protein-coding genes, tRNAs, and rRNAs, can be a cause of mitochondrial disease.

## 3. mtDNA Structure

The human mtDNA genome is composed of 16,569 base pairs (bp), encoding at total of 37 genes in a remarkably compact form. There are 13 protein-coding genes, 22 tRNA genes, and 2 ribosomal genes, with the overall organization of the genome shown in Fig. 1. The protein-coding genes contribute to the ETC complexes I, III, IV, and V, with complex II being exclusively nuclear encoded. Preservation of complex II (succinate dehydrogenase) activity can suggest the

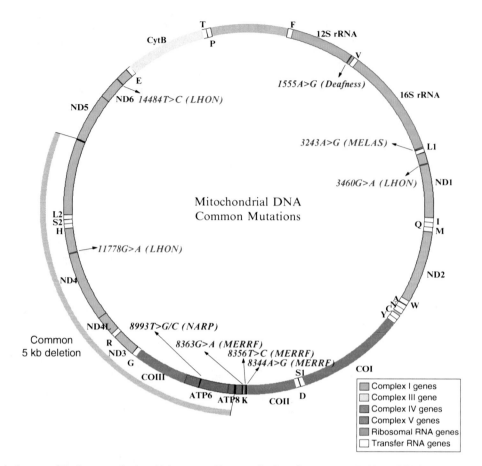

Fig. 1. A diagram of the human mitochondrial genome. The organization of genes encoded in mtDNA is shown, along with the positions of the mutations referred to in the text.

presence of an mtDNA-mediated disease, whether maternally inherited or Mendelian. Of the 13 protein-coding genes, 7 contribute to complex I (ND1, ND2, ND3, ND4, ND4L, ND5, and ND6), 1 is a component of complex III (cytochrome b), 3 proteins form the core of complex IV (cytochrome c oxidase; COX I, COX II, and COX III), and 2 proteins are part of complex V (ATPase 6 and ATPase 8). Based upon the exclusively maternal inheritance of mtDNA, recombination between parental genomes would not be a source of mtDNA variation, but, rather, variation reflects both historical population lineages and a high mutation rate (20). DNA sequence variation at a population level has been categorized into "haplogroups." This variation has been used to reconstruct historic population movements and has a variety of practical applications such as forensics (21). Population variation also has effects on the pathogenicity of particular mutations via still poorly understood interactions between the mutation and the genetic "background" (22) and may predispose individuals to more common disorders (23).

## 4. mtDNA Mutations

Given the wide clinical variability and a lack of simple, definitive testing, the prevalence of mitochondrial disorders is difficult to accurately measure (24). However, estimates from laboratory referral centers and population screening have been reported (25, 26), and the overall frequency of ETC disorders has been estimated to be approximately 1:5,000–8,000, including both primary mtDNA disorders and Mendelian diseases. Studies of adult populations, where primary mtDNA disorders are more common, suggest a prevalence of about 1:10,000 for mtDNA mutations (27, 28). Employing a small number of specific mutations as a screen in newborns, Elliott and colleagues demonstrated a remarkably high rate of 1:200 newborns harboring an mtDNA mutation, with a correspondingly high rate of new mutation (29). While it is likely that the majority of these individuals will remain asymptomatic, these high rates of detection reflect the propensity for mutation in the mitochondrial genome.

Factors that both define and influence the inheritance and development of mtDNA disease include maternal transmission, the degree of heteroplasmy and the attendant threshold at which a tissue experiences dysfunction, and the mitotic segregation of the mutation.

The mammalian oocyte contains over a 100,000 mitochondria, while sperm do not contribute to the zygote mitochondrial population. A single exception to this biological truism has been reported, although it was identified only in the setting of a mitochondrial disorder (30, 31). During subsequent embryonic development, there is a gradual "dilution" in the number of mitochondria per cell until mitochondrial biogenesis begins. In order to try to explain the intergenerational changes in the degree of heteroplasmy that can be observed, it has been debated whether the embryonic reduction in mtDNA copy number in primordial germ cells leads to shifts in heteroplasmy (32), whether it occurs postnatally in primordial germ cells (33), or whether other factors such as the preferential replication of a subpopulation of mtDNA in germ cells drive the rapid shifts in heteroplasmy that can be seen (34). A final answer remains to be conclusively determined.

Each somatic cell contains hundreds to thousands of copies of mtDNA that during cell division distribute randomly among daughter cells. In normal tissues, all mtDNA molecules are thought to be identical. While some deleterious mutations are mild enough to be tolerated in all mtDNA molecules, such as those causing LHON, and thus are referred to as homoplasmic, many deleterious mtDNA mutations impair mitochondrial functions to a degree that is not compatible with cell survival. The relative proportion of mutant to normal mtDNA genomes can vary among different tissues, and similarly, different tissues exhibit varying sensitivity to a disruption in mitochondrial function. This latter concept is referred to as the threshold effect.

The segregation of heteroplasmic mtDNA to daughter cells, referred to as mitotic segregation, also influences the development of mitochondrial dysfunction. For pathologic mtDNA variants, the exact mechanisms influencing the pattern of segregation are poorly understood but may reflect the survival of the resulting daughter cells, the relative replication efficiency of the two genomes, interactions of the mtDNA genomes with nucleoid proteins that package the mtDNA, or other mtDNA modifications. However, some insights have been gleaned from studying the segregation patterns of apparently neutral mtDNA sequence variants in model systems such as the mouse, and these studies clearly reveal that mtDNA segregation varies with age, is at least partially under the control of nuclear genes (35), and depends on the tissue identity in which it occurs (36). Recently, using heteroplasmic mouse strains, a nuclear gene that influences mtDNA segregation in leukocytes was identified to be *Gimap3*, a mitochondrial outer membrane GTPase protein of unknown function (37).

With these concepts in mind, a brief review of the types of mtDNA mutations is presented, categorized either as mtDNA rearrangements or point mutations.

## 5. mtDNA Rearrangements

mtDNA deletions, duplications, and other more complex rearrangements are observed in disease states. In addition, multiple deletions and mtDNA depletion can be observed in the context of a Mendelian disorder (see Chapter 2). Patients harboring primary mtDNA deletions (in contrast to those patients in whom the deletion is a manifestation of a Mendelian disorder of mtDNA integrity) generally exhibit one of three sporadic conditions. First, Pearson syndrome is an often fatal disorder of infancy or early childhood that is characterized by sideroblastic anemia and exocrine pancreas insufficiency and may be complicated by gastrointestinal problems and growth failure (38). Kearns-Sayre syndrome is a multisystem disorder characterized by impaired eye movements (chronic progressive external ophthalmoplegia (CPEO)), pigmentary retinopathy, and a cardiac conduction defect. The signs and symptoms arise before 20 years of age. Other clinical problems may include endocrinopathies such as diabetes mellitus, hypoparathyroidism, and short stature, progressive neurologic impairments such as ataxia or dementia. Laboratory abnormalities are common, including lactic acidosis, elevated cerebrospinal fluid (CSF) protein, and scattered cytochrome oxidase–negative RRF in skeletal muscle biopsies. Finally, isolated CPEO with or without proximal muscle weakness is the mildest clinical syndrome associated with mtDNA deletions (39). Patients with CPEO but without other symptoms of Kearns-Sayre syndrome often develop neuromuscular symptoms as they age, and conversely Pearson syndrome patients who survive infancy may develop Kearns-Sayre syndrome at a later age (40). In young patients with multisystem disease, mtDNA deletion testing may be abnormal in blood samples since the deletion is more likely to be a de novo germ line or early embryonic event, whereas in older patients, the mtDNA deletion is more likely a somatic event in the affected tissue. Thus, in patients with a delayed onset of disease, the deletion is typically not detectable in blood specimens, and it is necessary to use skeletal muscle for mtDNA deletion testing (41).

At the molecular level, approximately 60% of mtDNA deletions occur in a region of the mtDNA genome that is flanked by short direct repeat sequences, one of which is usually lost during the deletion process, and these have been referred to as class I deletions (42). Such repeats are thought to play a role in the formation of mtDNA deletions. Approximately 30% of mtDNA deletions are flanked by imperfect repeats containing a few mismatches (class II deletions), and about 10% have no repeats at the deletion flanking regions (43). The most common mtDNA deletion, which is present in approximately one third of patients, is a 5-kb deletion (m.8470–m.13447) that is flanked by a 13-bp class I direct repeat (42). While it has been speculated that defects in mtDNA replication due to misalignment of direct repeats may cause mtDNA deletions (44), an alternative mechanism involving the repair of mtDNA damage has recently been proposed (45). A recent report summarizing the molecular and clinical characteristics in 67 patients of varying age reported that the deletion breakpoints found in the youngest patients have significantly lower breakpoint homology relative to the older patients, with fewer class I breakpoints and an almost threefold decreased incidence of the common 5-kb mtDNA deletion relative to older patients, as well as increased heterogeneity in the breakpoint distribution. The severity of disease appears not to be affected by the size of the mtDNA deletion or the particular genes deleted (46). These findings suggest that the molecular events responsible for mtDNA deletions in young patients may differ from those found in older patients.

## 6. mtDNA Point Mutations

Over 200 pathogenic point mutations have been identified in mtDNA from patients with a wide variety of disorders (http://mitomap.org/MITOMAP), many of which are maternally inherited and involve multiple organ systems but on occasion can be sporadic and tissue specific. These can impair mtDNA-encoded proteins, tRNAs, or rRNAs and potentially interfere with replication, transcription, or RNA processing. Examples of some of these mechanisms of disease are provided by four of the most common point mutations and their associated clinical syndromes.

Mitochondrial tRNAs are structurally distinct from other tRNAs; they are shorter than bacterial or eukaryotic cytoplasmic tRNAs and lack a variety of conserved nucleotides that are involved in the prototypic tertiary interactions that create the canonical L-shape of tRNAs, possibly resulting in a weaker tertiary structure. In addition, in comparison to cytosolic tRNAs, posttranscriptional base modification appears to be more important for the proper tertiary structure and function of mitochondrial tRNAs (47). A pathogenic tRNA mutation leads to a combined OXPHOS defect, in part through a decreased overall rate of mitochondrial protein synthesis. Depending on which tRNA is mutated; there will be varying effects on the individual ETC complexes based upon the percentage of the corresponding amino acid in the different ETC complex subunits. The pathogenic mechanisms leading to defective translation caused by a tRNA mutation are numerous, including impaired transcription termination, impaired tRNA maturation, defective posttranscriptional modification of the tRNA, effects on tRNA folding and stability, reduced aminoacylation, decreased binding to the translation factor mtEFTu or the mitochondrial ribosome, and altered codon decoding (48).

The tRNA$^{Leu(UUR)}$ gene (*MT-TL1*) is particularly rife with pathogenic mutations, with nearly 30 different mutations to date, but mutations have now been detected in all 22 tRNA genes. The prototypic tRNA mutation is the 3243A>G mutation in tRNALeu$^{(UUR)}$. This mutation causes a variety of clinical disorders, the best known being MELAS (mitochondrial encephalomyopathy, lactic acidosis, and stroke-like episodes) syndrome, which typically becomes apparent in children or young adults after a normal early developmental period (49). Signs and symptoms include recurrent vomiting, migraine-like headache, and stroke-like episodes causing cortical blindness, hemiparesis, or hemianopia. MRI of the brain shows regions of hypoperfusion that do not correspond to a vascular distribution, and it has been suggested that the underlying defect is one of endothelial function due to a functional deficiency in nitric oxide (50). Later features include hearing loss, short stature, diabetes, retinopathy, muscle fatigue, and lactic acidosis. The aberrant molecular mechanisms of the mutant tRNA underlying the disorder are varied and somewhat controversial, including a reduction in the aminoacylation of the tRNA and a lack of wobble-base hypermodification. This posttranscriptional taurine modification at the anticodon wobble position is needed to restrict decoding to leucine$^{UUR}$ codons, and loss of this modification leads to a combination of a decoding defect of UUG and UUA codons and amino acid misincorporation into proteins (51). Additionally, the 3243A>G mutation has been shown to diminish 16S rRNA transcription termination and alter processing of the primary transcript (52). It is worth noting that the 3243A>G mutation is a common, recurrent mutation that appears to arise on a variety of haplogroup backgrounds and thus does not represent a founder mutation. While the most common mtDNA mutation causing MELAS syndrome is 3243A>G, and it is always found in the heteroplasmic state, a number of other mutations have also been associated with MELAS syndrome, including a missense mutation in the ND5 gene that encodes subunit 5 of NADH dehydrogenase (53) and an intriguing mutation that abolishes the binding site of the transcription termination factor MTERF1 to the tRNA$^{Leu(UUR)}$ gene (54).

A second common site for tRNA mutation is that of tRNA$^{Lys}$ (*MT-TK*). The most common mutation is 8344A>G, which is associated with MERRF (myoclonic epilepsy with ragged-red fibers) syndrome, and this mutation accounts for 80% of affected individuals. The disorder is characterized by myoclonus, generalized seizures, mitochondrial myopathy, and cerebellar ataxia. Other clinical signs include short stature, dementia, hearing loss, a peripheral neuropathy, and cardiomyopathy with Wolff–Parkinson–White syndrome, a cardiac conduction defect. Occasionally, pigmentary retinopathy and lipomatosis are present. Similar to 3243A>G, it has been reported to affect both aminoacylation and taurine modification of the wobble-base U, the latter disrupting codon-anticodon pairing on the mitochondrial ribosome for both of the tRNA$^{Lys}$ codons (55). Two additional mutations in the tRNA$^{Lys}$ gene have been associated with MERRF syndrome (8356T>C and 8363G>A), and, like the 3243A>G mutation, MERFF mutations exist in the heteroplasmic state.

A third common point mutation leads to a missense substitution in *MT-ATP6*; most commonly 8993T>G or 8993T>C, with the former generally being clinically more severe. The clinical syndromes associated with this mutation are defined by the degree of heteroplasmy: lower mutation burdens cause NARP (neurogenic muscle weakness, ataxia, retinitis pigmentosa) syndrome, which usually affects young adults and causes retinitis pigmentosa, dementia, seizures, ataxia, proximal muscle weakness, and a sensory neuropathy (56). When there is a greater percentage of mutant mtDNA molecules present, maternally inherited Leigh syndrome (MILS) is observed, which is a severe infantile encephalopathy with characteristic symmetrical lesions in the basal ganglia and the brainstem and typically leads to early death (57).

An additional example of a class of mtDNA missense mutations, in this case, mutations that are uniformly homoplasmic, causes Leber's hereditary optic neuropathy (LHON). The disorder is characterized by acute or subacute, painless loss of vision in young adults due to bilateral optic atrophy, with reduced penetrance and a four- to fivefold greater frequency in males due to as yet unidentified nuclear gene modifiers that have been mapped to the X chromosome by linkage analysis (58). Three mtDNA point mutations in complex I subunit genes account for more than 90% of LHON cases. The causative mutations are 11778G>A in ND4, 3460G>A in ND1, and 14484T>C in ND6. Because ETC bioenergetics appears minimally impaired, it has been suggested that excess reactive oxygen species in conjunction with a unique retinal ganglion cell sensitivity accounts for the disease pathogenesis (59). In addition, there is a clear effect of the mtDNA haplogroup on the penetrance of specific mutations (60).

One final example of an mtDNA point mutation that is representative of a class of mutations is the 1555A>G mutation in the 12S rRNA (*MT-RNR1*). Mammalian mitochondrial ribosomes differ notably from cytosolic or bacterial ribosomes and even from ribosomes from other mitochondria. They lack nearly half the rRNA present in bacterial ribosomes and contain a correspondingly higher protein content due to the incorporation of larger proteins and numerous additional proteins, causing a greater molecular mass and size than bacterial ribosomes (61). The 1555A>G mutation is located in the decoding site of the mitochondrial small subunit (SSU) ribosomal RNA and is predicted to cause a change in the secondary rRNA structure to one that more closely resembles the corresponding region of the bacterial 16S rRNA. This alteration impairs protein synthesis and enhances an interaction with aminoglycoside antibiotics, which further exacerbates the translation defect. The mutation alone typically does not lead to disease, but in combination with environmental modifiers such as the aminoglycosides or perhaps genetic modifiers such as mitochondrial haplogroups (62), varying degrees of hearing loss is observed. In addition to mtDNA-encoded modifiers, nuclear modifier genes have been putatively identified, making this class of mutation currently unique. *TFB1M*, encoding a mitochondrial rRNA methyltransferase, has been identified as a possible nuclear modifier of the 1555A>G mutation (63), as has a second RNA modifying enzyme TRMU (64). TRMU was recently identified as a tRNA 5-methylaminomethyl-2-thiouridylate methyltransferase that when deficient causes transient liver failure (65). Presumably, alterations in RNA methylation due to malfunctioning TFB1M or TRMU can diminish the deleterious effect of the 1555A>G mutation on the ribosome conformation, although additional supporting evidence is needed to firmly establish their role.

In summary, a variety of mutations can arise in mtDNA due in large part to the high mutation rate, and these mutations can exhibit a complex and broad spectrum of disease manifestations. The properties of maternal inheritance, heteroplasmy, tissue- and cell-specific threshold effects, and mitotic segregation all interact to make for a unique set of clinical challenges in the diagnosis and management of mitochondrial disorders. Future work in understanding the forces that influence mtDNA segregation holds the promise of potential therapies that could shift the distribution of mutant mtDNA, thus reducing disease severity.

## Acknowledgment

Thanks to Sha Tang, PhD, for creating Fig. 1.

## References

1. Liu, Z., and Butow, R. A. (2006) Mitochondrial retrograde signaling, *Annu Rev Genet 40*, 159–185.
2. Brookes, P. S., Yoon, Y., Robotham, J. L., et al. (2004) Calcium, ATP, and ROS: a mitochondrial love-hate triangle, *Am J Physiol Cell Physiol 287*, C817–833.
3. Acin-Perez, R., Salazar, E., Kamenetsky, M., et al. (2009) Cyclic AMP produced inside mitochondria regulates oxidative phosphorylation, *Cell Metab 9*, 265–276.
4. Acin-Perez, R., Hoyos, B., Gong, J., et al. (2010) Regulation of intermediary metabolism by the PKCdelta signalosome in mitochondria, *FASEB J 24*, 5033–5042.
5. Yao, Z., and Seger, R. (2009) The ERK signaling cascade--views from different subcellular compartments, *Biofactors 35*, 407–416.
6. Shaw, P. E. (2010) Could STAT3 provide a link between respiration and cell cycle progression?, *Cell Cycle 9*, 4294–4296.
7. Mookherjee, P., Quintanilla, R., Roh, M. S., et al. (2007) Mitochondrial-targeted active Akt protects SH-SY5Y neuroblastoma cells from staurosporine-induced apoptotic cell death, *J Cell Biochem 102*, 196–210.
8. Reeve, A. K., Krishnan, K. J., and Turnbull, D. (2008) Mitochondrial DNA mutations in disease, aging, and neurodegeneration, *Ann N Y Acad Sci 1147*, 21–29.
9. Smits, P., Smeitink, J., and van den Heuvel, L. (2010) Mitochondrial translation and beyond: processes implicated in combined oxidative phosphorylation deficiencies, *J Biomed Biotechnol 2010*, 737385.
10. Zeviani, M., and Di Donato, S. (2004) Mitochondrial disorders, *Brain 127*, 2153–2172.
11. Schatz, G. (1963) The Isolation of Possible Mitochondrial Precursor Structures from Aerobically Grown Baker's Yeast, *Biochem Biophys Res Commun 12*, 448–451.
12. Nass, M. M., and Nass, S. (1963) Intramitochondrial Fibers with DNA Characteristics. I. Fixation and Electron Staining Reactions, *J Cell Biol 19*, 593–611.
13. Holt, I. J., Harding, A. E., and Morgan-Hughes, J. A. (1988) Deletions of muscle mitochondrial DNA in patients with mitochondrial myopathies, *Nature 331*, 717–719.
14. Zeviani, M., Moraes, C. T., DiMauro, S., et al. (1988) Deletions of mitochondrial DNA in Kearns-Sayre syndrome, *Neurology 38*, 1339–1346.
15. Lestienne, P., and Ponsot, G. (1988) Kearns-Sayre syndrome with muscle mitochondrial DNA deletion, *Lancet 1*, 885.
16. Wallace, D. C., Singh, G., Lott, M. T., et al. (1988) Mitochondrial DNA mutation associated with Leber's hereditary optic neuropathy, *Science 242*, 1427–1430.
17. Shoffner, J. M., Lott, M. T., Lezza, A. M., et al. (1990) Myoclonic epilepsy and ragged-red fiber disease (MERRF) is associated with a mitochondrial DNA tRNA(Lys) mutation, *Cell 61*, 931–937.
18. Goto, Y., Nonaka, I., and Horai, S. (1990) A mutation in the tRNA(Leu)(UUR) gene associated with the MELAS subgroup of mitochondrial encephalomyopathies, *Nature 348*, 651–653.
19. Prezant, T. R., Agapian, J. V., Bohlman, M. C., et al. (1993) Mitochondrial ribosomal RNA mutation associated with both antibiotic-induced and non-syndromic deafness, *Nat Genet 4*, 289–294.

20. Brown, W. M., Prager, E. M., Wang, A., et al. (1982) Mitochondrial DNA sequences of primates: tempo and mode of evolution, *J Mol Evol 18*, 225–239.

21. Parson, W., and Bandelt, H. J. (2007) Extended guidelines for mtDNA typing of population data in forensic science, *Forensic Sci Int Genet 1*, 13–19.

22. Ghelli, A., Porcelli, A. M., Zanna, C., et al. (2009) The background of mitochondrial DNA haplogroup J increases the sensitivity of Leber's hereditary optic neuropathy cells to 2,5-hexanedione toxicity, *PLoS One 4*, e7922.

23. Gomez-Duran, A., Pacheu-Grau, D., Lopez-Gallardo, E., et al. (2010) Unmasking the causes of multifactorial disorders: OXPHOS differences between mitochondrial haplogroups, *Hum Mol Genet 19*, 3343–3353.

24. Thorburn, D. R. (2004) Mitochondrial disorders: prevalence, myths and advances, *J Inherit Metab Dis 27*, 349–362.

25. Skladal, D., Halliday, J., and Thorburn, D. R. (2003) Minimum birth prevalence of mitochondrial respiratory chain disorders in children, *Brain 126*, 1905–1912.

26. Diogo, L., Grazina, M., Garcia, P., et al. (2009) Pediatric mitochondrial respiratory chain disorders in the Centro region of Portugal, *Pediatr Neurol 40*, 351–356.

27. Chinnery, P. F., Johnson, M. A., Wardell, T. M., et al. (2000) The epidemiology of pathogenic mitochondrial DNA mutations, *Ann Neurol 48*, 188–193.

28. Schaefer, A. M., McFarland, R., Blakely, E. L., et al. (2008) Prevalence of mitochondrial DNA disease in adults, *Ann Neurol 63*, 35–39.

29. Elliott, H. R., Samuels, D. C., Eden, J. A., et al. (2008) Pathogenic mitochondrial DNA mutations are common in the general population, *Am J Hum Genet 83*, 254–260.

30. Schwartz, M., and Vissing, J. (2002) Paternal inheritance of mitochondrial DNA, *N Engl J Med 347*, 576–580.

31. Kraytsberg, Y., Schwartz, M., Brown, T. A., et al. (2004) Recombination of human mitochondrial DNA, *Science 304*, 981.

32. Cree, L. M., Samuels, D. C., de Sousa Lopes, S. C., et al. (2008) A reduction of mitochondrial DNA molecules during embryogenesis explains the rapid segregation of genotypes, *Nat Genet 40*, 249–254.

33. Wai, T., Teoli, D., and Shoubridge, E. A. (2008) The mitochondrial DNA genetic bottleneck results from replication of a subpopulation of genomes, *Nat Genet 40*, 1484–1488.

34. Cao, L., Shitara, H., Sugimoto, M., et al. (2009) New evidence confirms that the mitochondrial bottleneck is generated without reduction of mitochondrial DNA content in early primordial germ cells of mice, *PLoS Genet 5*, e1000756.

35. Battersby, B. J., Loredo-Osti, J. C., and Shoubridge, E. A. (2003) Nuclear genetic control of mitochondrial DNA segregation, *Nat Genet 33*, 183–186.

36. Jenuth, J. P., Peterson, A. C., and Shoubridge, E. A. (1997) Tissue-specific selection for different mtDNA genotypes in heteroplasmic mice, *Nat Genet 16*, 93–95.

37. Jokinen, R., Marttinen, P., Sandell, H. K., et al. (2010) Gimap3 regulates tissue-specific mitochondrial DNA segregation, *PLoS Genet 6*, e1001161.

38. Pearson, H. A., Lobel, J. S., Kocoshis, S. A., et al. (1979) A new syndrome of refractory sideroblastic anemia with vacuolization of marrow precursors and exocrine pancreatic dysfunction, *J Pediatr 95*, 976–984.

39. Hammans, S. R. (1994) Mitochondrial DNA and disease, *Essays Biochem 28*, 99–112.

40. Schroder, R., Vielhaber, S., Wiedemann, F. R., et al. (2000) New insights into the metabolic consequences of large-scale mtDNA deletions: a quantitative analysis of biochemical, morphological, and genetic findings in human skeletal muscle, *J Neuropathol Exp Neurol 59*, 353–360.

41. Wong, L. J. (2001) Recognition of mitochondrial DNA deletion syndrome with non-neuromuscular multisystemic manifestation, *Genet Med 3*, 399–404.

42. Samuels, D. C., Schon, E. A., and Chinnery, P. F. (2004) Two direct repeats cause most human mtDNA deletions, *Trends Genet 20*, 393–398.

43. Degoul, F., Nelson, I., Amselem, S., et al. (1991) Different mechanisms inferred from sequences of human mitochondrial DNA deletions in ocular myopathies, *Nucleic Acids Res 19*, 493–496.

44. Holt, I. J., Lorimer, H. E., and Jacobs, H. T. (2000) Coupled leading- and lagging-strand synthesis of mammalian mitochondrial DNA, *Cell 100*, 515–524.

45. Krishnan, K. J., Reeve, A. K., Samuels, D. C., et al. (2008) What causes mitochondrial DNA deletions in human cells?, *Nat Genet 40*, 275–279.

46. Sadikovic, B., Wang, J., El-Hattab, A., et al. (2010) Sequence homology at the breakpoint and clinical phenotype of mitochondrial DNA deletion syndromes, *PLoS One 5*, e15687.

47. Helm, M. (2006) Post-transcriptional nucleotide modification and alternative folding of RNA, *Nucleic Acids Res 34*, 721–733.

48. Florentz, C., Sohm, B., Tryoen-Toth, P., et al. (2003) Human mitochondrial tRNAs in health and disease, *Cell Mol Life Sci 60*, 1356–1375.

49. Finsterer, J. (2007) Genetic, pathogenetic, and phenotypic implications of the mitochondrial A3243G tRNALeu(UUR) mutation, *Acta Neurol Scand 116*, 1–14.

50. Koga, Y., Akita, Y., Nishioka, J., et al. (2005) L-arginine improves the symptoms of strokelike episodes in MELAS, *Neurology 64*, 710–712.

51. Kirino, Y., Yasukawa, T., Ohta, S., et al. (2004) Codon-specific translational defect caused by a wobble modification deficiency in mutant tRNA from a human mitochondrial disease, *Proc Natl Acad Sci USA 101*, 15070–15075.

52. Hess, J. F., Parisi, M. A., Bennett, J. L., et al. (1991) Impairment of mitochondrial transcription termination by a point mutation associated with the MELAS subgroup of mitochondrial encephalomyopathies, *Nature 351*, 236–239.

53. Shanske, S., Coku, J., Lu, J., et al. (2008) The G13513A mutation in the ND5 gene of mitochondrial DNA as a common cause of MELAS or Leigh syndrome: evidence from 12 cases, *Arch Neurol 65*, 368–372.

54. Chomyn, A., Martinuzzi, A., Yoneda, M., et al. (1992) MELAS mutation in mtDNA binding site for transcription termination factor causes defects in protein synthesis and in respiration but no change in levels of upstream and downstream mature transcripts, *Proc Natl Acad Sci USA 89*, 4221–4225.

55. Enriquez, J. A., Chomyn, A., and Attardi, G. (1995) MtDNA mutation in MERRF syndrome causes defective aminoacylation of tRNA(Lys) and premature translation termination, *Nat Genet 10*, 47–55.

56. Holt, I. J., Harding, A. E., Petty, R. K., et al. (1990) A new mitochondrial disease associated with mitochondrial DNA heteroplasmy, *Am J Hum Genet 46*, 428–433.

57. Makela-Bengs, P., Suomalainen, A., Majander, A., et al. (1995) Correlation between the clinical symptoms and the proportion of mitochondrial DNA carrying the 8993 point mutation in the NARP syndrome, *Pediatr Res 37*, 634–639.

58. Shankar, S. P., Fingert, J. H., Carelli, V., et al. (2008) Evidence for a novel x-linked modifier locus for leber hereditary optic neuropathy, *Ophthalmic Genet 29*, 17–24.

59. Tonska, K., Kodron, A., and Bartnik, E. (2010) Genotype-phenotype correlations in Leber hereditary optic neuropathy, *Biochim Biophys Acta 1797*, 1119–1123.

60. Torroni, A., Petrozzi, M., D'Urbano, L., et al. (1997) Haplotype and phylogenetic analyses suggest that one European-specific mtDNA background plays a role in the expression of Leber hereditary optic neuropathy by increasing the penetrance of the primary mutations 11778 and 14484, *Am J Hum Genet 60*, 1107–1121.

61. O'Brien, T. W. (2003) Properties of human mitochondrial ribosomes, *IUBMB Life 55*, 505–513.

62. Lu, J., Qian, Y., Li, Z., et al. (2010) Mitochondrial haplotypes may modulate the phenotypic manifestation of the deafness-associated 12 S rRNA 1555A > G mutation, *Mitochondrion 10*, 69–81.

63. Bykhovskaya, Y., Mengesha, E., Wang, D., et al. (2004) Human mitochondrial transcription factor B1 as a modifier gene for hearing loss associated with the mitochondrial A1555G mutation, *Mol Genet Metab 82*, 27–32.

64. Guan, M. X., Yan, Q., Li, X., et al. (2006) Mutation in TRMU related to transfer RNA modification modulates the phenotypic expression of the deafness-associated mitochondrial 12 S ribosomal RNA mutations, *Am J Hum Genet 79*, 291–302.

65. Zeharia, A., Shaag, A., Pappo, O., et al. (2009) Acute infantile liver failure due to mutations in the TRMU gene, *Am J Hum Genet 85*, 401–407.

# Chapter 2

# Nuclear Gene Defects in Mitochondrial Disorders

## Fernando Scaglia

## Abstract

Most mitochondrial cytopathies in infants are caused by mutations in nuclear genes encoding proteins targeted to the mitochondria rather than by primary mutations in the mitochondrial DNA. Over the past few years, the awareness of the number of disease-causing mutations in different nuclear genes has grown exponentially. These genes encode the various subunits of each respiratory chain complex, the ancillary proteins involved in the assembly of these subunits, proteins involved in mitochondrial DNA replication and maintenance, proteins involved in mitochondrial protein synthesis, and proteins involved in mitochondrial dynamics. This increased awareness has added a challenging dimension to the current diagnostic workup of mitochondrial cytopathies. The advent of new technologies such as next-generation sequencing should facilitate the resolution of this dilemma.

**Key words:** Mitochondria, Respiratory chain defects, Mitochondrial cytopathies, Nuclear genes

## 1. Introduction

Mitochondria participate in an important number of cellular functions. The primary function of mitochondria is to provide most of the cellular ATP requirements through a process called oxidative phosphorylation. This process involves five protein complexes located in the inner mitochondrial membrane. Mitochondrial DNA (mtDNA) encodes 13 of the core structural polypeptides that form the multimeric subunits of the respiratory chain complexes, 2 ribosomal RNAs (rRNAs), and 22 transfer RNAs (tRNAs) required for protein synthesis. Most of the remaining proteins are encoded by nuclear DNA (nDNA). Thus, although human mtDNA encodes the basic machinery for protein synthesis, it depends on nDNA for the provision of enzymes involved in mtDNA replication, repair, transcription, and translation. This complex interaction and dependency would explain newly recognized

Lee-Jun C. Wong (ed.), *Mitochondrial Disorders: Biochemical and Molecular Analysis*, Methods in Molecular Biology, vol. 837, DOI 10.1007/978-1-61779-504-6_2, © Springer Science+Business Media, LLC 2012

syndromes that are characterized by secondary abnormalities of mtDNA. Thus, mitochondrial disorders of oxidative phosphorylation can be classified into two categories: disorders caused by mutations in mtDNA that are regulated by rules of mitochondrial inheritance and disorders caused by mutations in nDNA which follow the rules of Mendelian genetics. Moreover, deficient cross talk between the two genomes may affect the integrity of the mitochondrial genome. The number of disease-causing molecular alterations in nuclear genes is growing exponentially, and mutations in these genes underlie the vast majority of respiratory chain (RC) defects in children. In one retrospective study done on children with definite diagnosis of RC defects, the diagnostic yield of a panel of common mtDNA point mutations, and single deletions was close to 12% (1). Furthermore, it should be considered that mtDNA point mutations and single deletions probably do not account for more than 20–25% of pediatric cases of RC deficiencies when a more comprehensive diagnostic approach is used (2). Therefore, nuclear gene defects are responsible for the majority of pediatric RC defects. Adequate function of the RC requires the presence of various subunits for each complex, ancillary proteins such as chaperones, and proteins involved in the assembly of complex subunits. In addition, it also requires enzymes involved in maintaining mtDNA homeostasis and mitochondrial protein synthesis, proteins that regulate mito-chondrial dynamics, proteins involved in the stability of the phos-pholipid component of the mitochondrial inner membrane, and proteins that intervene in the import system of components of the mitochondrial respiratory chain. Due to the ever increasing aware-ness of the large numbers of nuclear genes involved in pediatric RC defects, it will not be the purpose of this chapter to exhaustively cover all existing nuclear gene defects. In that note, the aim will be to enlighten the reader on the wide spectrum of nuclear genes involved, their associated clinical phenotypes, and possible pathogenic mecha-nisms responsible for infantile disorders of oxidative phosphorylation.

## 2. Defects in Structural Respiratory Chain Genes

The vast majority of mitochondrial respiratory chain proteins are encoded by nuclear genes. Mutations in some of the nuclear genes encoding respiratory chain subunits have been found in patients with mitochondrial cytopathies (3). The overwhelming majority of these mutations occur in nuclear genes encoding complex I subunits (4). However, not many mutations have been found in other genes encoding other complex subunits. These nuclear gene mutations will lead to an isolated deficiency of a specific mitochon-drial RC complex.

The first report of a mutation encoding a mitochondrial respi-ratory chain subunit originated from two siblings affected with Leigh syndrome and complex II deficiency who carried a homozygous mutation in the *SDHA* gene (5). Mutations in the same gene were reported in an additional patient with Leigh syndrome (6). Mutations in *SDHB*, *SDHC*, and *SDHD* encoding subunits B, C, and D of complex II have been reported in patients presenting with pheochromocytoma and paraganglioma. Mutations in these genes would cause an accumulation of succinate and reactive oxygen species that could then result in the overexpression of hypoxia-inducible factor 1 with ensuing formation of these tumors (7).

Complex I is an RC complex that is composed by 7 mtDNA-encoded subunits and at least 35 nuclear encoded subunits. Complex I deficiency is one of the most common causes of mito-chondrial diseases (8). Complex I defects caused by mutations in nuclear genes are associated with early onset of severe multiorgan disorders (9). Furthermore, the majority of children present with Leigh disease. Accompanying clinical signs and symptoms include muscular hypotonia, dystonia, developmental delay, abnormal eye movements, epilepsy, respiratory difficulties, lactic acidosis, and failure to thrive (4). Approximately 40% of complex I deficiencies are caused by mutations in nuclear genes (10).

Mutations in two nuclear genes (*UQCRB* and *UQCRQ*) encoding complex III subunits have been characterized, and the respective phenotypes have been consistent with hypoglycemia and lactic acidosis in the first case and severe psychomotor retardation, extrapyramidal signs, dystonia, athetosis, ataxia, mild axial hypotonia, and defects in verbal and expressive communication skills in the second case (11, 12).

Although initially elusive, mutations in nuclear genes encoding structural subunits for complex IV have finally been identified. Mutations in *COX4I2* have been found associated with a phenotype of exocrine pancreatic insufficiency, dyserythropoietic anemia, and calvarial hyperostosis (13), and mutations in *COX6B1* have been reported in patients with severe infantile encephalomyopathy (14).

Primary coenzyme $Q_{10}$ deficiency is associated with nuclear gene defects, and making a diagnosis is relevant as most patients with this condition respond to high dose of coenzyme $Q_{10}$ supple-mentation. On mitochondrial enzyme assays, this condition should be suspected if complex I + III and complex II + III activities are reduced with normal activity in the remainder of the respiratory chain complexes. Five major clinical phenotypes can be associated with this deficiency: (1) predominantly myopathic disorder with myoglobinuria and involvement of the central nervous system, (2) predominantly encephalopathic disorder with ataxia and cere-bellar atrophy, (3) an isolated myopathy with ragged red fibers and lipid myopathy, (4) a generalized early-onset mitochondrial encephalomyopathy, and (5) nephrotic syndrome associated with encephalopathy (15). This deficiency can be caused by mutations in biosynthetic genes such as *PDSS1* and *PDSS2* (16, 17), *CoQ2* (18), *CABC1/ADCK3* (19, 20), *CoQ6* (21), and *CoQ9* (22).

## 3. Defects in Genes Involved in the Assembly of Mitochondrial Respiratory Chain Complexes

The normal function of a mitochondrial respiratory chain complex involves regulation of its subunits and the maintenance of the structural integrity of the complex. The disturbance of the mechanisms regulating the integrity of a particular complex may lead to instability of the assembly of the different subunits. The dysfunction of different nuclear genes could compromise the incorporation of iron, copper, or heme to a particular complex, the assembly of complex subunits, or the translation of specific subunits. From a diagnostic point of view and with rare exceptions, one would expect to find an isolated and selective deficiency of a mitochondrial respiratory chain enzyme and an accompanying recognizable clinical phenotype that may guide a clinician to specifically test for a nuclear gene defect.

Isolated complex I deficiency is the most frequently encountered respiratory chain defect. Disease-causing mutations have been identified in genes encoding accessory subunits and assembly factors for complex I such as *NDUFAF2*, *NDUFAF1*, and *C6orf66* (23–25).

Regarding complex II, molecular defects in two genes involved in its assembly have been reported in humans. Mutations in the *SDHAF1* gene have been identified in patients with infantile leukoencephalopathy and isolated complex II deficiency (26). An additional gene, *SDH5*, encodes a mitochondrial protein required for the flavination of the SDH1 subunit. Mutations in this gene have been found in paraganglioma (27).

Only one gene involved in complex III assembly is known in humans, *BCS1L*. *BCS1L* mutations have been reported in different phenotypes including: hepatic failure with tubulopathy (28); growth retardation, aminoaciduria, cholestasis, iron overload, lactic acidosis, and early death (GRACILE) syndrome (29); and Björnstad syndrome which is characterized by pili torti and sensorineural hearing loss (30).

Several nuclear genes encoding assembly factors for complex IV have been identified as disease-causing. Mutations in these genes have been associated with complex IV deficiency. *SURF1* represents a major gene associated with Leigh syndrome and COX deficiency, and up to 75% of patients with Leigh syndrome and COX deficiency may harbor mutations in *SURF1* (31). *SCO1* and *SCO2* genes are involved in mitochondrial copper maturation and synthesis of subunit II of COX (32). Mutations in *SCO1* lead to hepatopathy and ketoacidotic coma (33), whereas mutations in *SCO2* gene lead to lethal infantile encephalocardiomyopathy (34). *COX10* encodes a heme A: farnesyltransferase that facilitates the first step in the conversion of protoheme to heme A prosthetic group required for the activity of COX. Mutations in this gene are associated with renal tubulopathy and leukodystrophy (35). *COX15* intervenes in the hydroxylation of heme O to form heme A. Patients with mutations in this gene may present with cardiomyopathy or Leigh syndrome (36, 37). *LRPPRC* encodes an ancillary protein thought to be involved in the stability of subunits I and III of COX (38). Mutations in this gene have been demonstrated in the French-Canadian type of Leigh syndrome (39). Another nuclear gene, *TACO1* which encodes a mitochondrial translational activator and is essential for the efficient synthesis of full-length COX subunit I, is also critical for the assembly of COX. A homozygous single base insertion in this gene was found in five children of Kurdish ancestry, manifesting a phenotype suggestive of late-onset Leigh syndrome characterized by short stature, intellectual disability, autistic-like features, progressive motor symptoms associated with basal ganglia involvement, and isolated cytochrome c oxidase deficiency (40). The mutation seems to result in premature protein truncation and compromise of the assembly of COX.

Mutations in two nuclear genes involved in complex V assembly have been reported. The *ATP12* gene encodes a protein required for the assembly of α and β subunits. Mutations in this gene resulted in dysmorphic features, neurological features, and 3-methylglutaconic aciduria (41). In addition, mutations in *TMEM70*, a gene encoding a transmembrane mitochondrial protein of unknown function involved in the assembly of complex V, have been reported in patients of mostly Roma ethnic ancestry associated with complex V deficiency, neonatal encephalocardiomyopathy, and transient hyperammonemia (42, 43).

A particular exception to these cases of isolated respiratory chain defects is exemplified by the deficiency of the assembly of iron–sulfur clusters that may result in dysfunction of complexes I, II, and III of the mitochondrial respiratory chain. These complexes I, II, and III contain iron–sulfur proteins, and deficiencies in the assembly of the iron–sulfur cluster can lead to multiple respiratory chain deficiencies. Mutations in the *ISCU* gene encoding the iron–sulfur cluster scaffold protein that interacts with frataxin in iron–sulfur cluster biosynthesis are the cause of a mitochondrial syndrome characterized by myopathy, exercise intolerance, and myoglobinuria (44, 45).

## 4. Defects in Genes Involved in mtDNA Stability

The mitochondrial and nuclear genomes have a dual genetic control on oxidative phosphorylation. Defects in any of the proteins involved in the mitochondrial replisome can affect the mtDNA copy number. Furthermore, such replication also depends on the supply of mitochondrial deoxyribonucleotide triphosphate (dNTP).

Thus, molecular defects in several genes involved in mitochondrial dNTP synthesis compromise the stability of mtDNA (46), leading to mtDNA depletion. This group of disorders is characterized by early-onset autosomal recessive conditions associated with a myopathic, encephalomyopathic, or hepatocerebral phenotype (47). In other cases, mutations in nuclear genes that control mtDNA stability may cause qualitative alterations, leading to multiple mtDNA deletions and autosomal dominant and recessive phenotypes of progressive external ophthalmoplegia (3).

POLG1 gene is the most frequently mutated nuclear gene causing mitochondrial disorders with approximately 150 described mutations (http://tools.niehs.nih.gov/polg/index.cfm?do=main.view). It encodes the enzyme pol γ (the only mtDNA polymerase). This enzyme plays an essential role in mtDNA replication and repair, thus pol γ-related syndromes are associated with mtDNA deletions and mtDNA depletion (48). Disorders associated with POLG1 mutations present with a heterogeneous spectrum of clinical presentations (49). Many cases of autosomal dominant progressive external ophthalmoplegia (adPEO) with multiple mtDNA deletions are associated with mutations in POLG1 (50). Recessive mutations in this gene are also responsible for Alpers-Huttenlocher syndrome, an early-onset mtDNA depletion syndrome with features consisting of psychomotor retardation, epilepsia partialis continua, and liver failure in infants (51). Patients present with RC defects in multiple complexes and mtDNA depletion in liver. The clinical picture overlaps with hepatocerebral syndrome, which is probably among the most severe mitochondrial diseases in infancy (49).

C10ORF2 encodes the mitochondrial protein Twinkle, an mtDNA replicative helicase bound to mtDNA in mitochondrial nucleoids (52). Mutations in this gene also cause adPEO associated with multiple mtDNA deletions (52). In addition, recessive mutations cause severe neonatal- or infantile-onset hepatoencephalopathy or infantile-onset spinocerebellar ataxia associated with mtDNA depletion in brain and liver, but not in skeletal muscle (53). Spinocerebellar ataxia of infancy is a severe recessively inherited neurodegenerative disorder that manifests after 9–18 months of age. Patients usually survive until their adult age. The observed phenotype in this condition suggests that the Twinkle protein may be involved in a crucial role in the maintenance of specific affected neuronal subpopulations (54). The hepatocerebral form of mtDNA depletion syndrome can also be caused by different recessive mutations in C10ORF2 (55).

DGUOK gene encodes deoxyguanosinase kinase which is an enzyme that catalyzes the first step of the mitochondrial deoxypurine salvage pathway (56). The typical phenotype associated with DGUOK mutations is characterized by neonatal onset of liver failure associated with neurological dysfunction (hypotonia, nystagmus, and motor retardation). Peripheral neuropathy and renal tubulopathy have been reported (57, 58). Many patients may have elevated levels of tyrosine and phenylalanine in their plasma as marker of liver dysfunction, and some of them may have elevated tyrosine levels on newborn screening (58). mtDNA depletion and combined RC complex deficiencies have been observed in the liver. Liver biopsy results may reveal microvesicular steatosis, cholestasis, fibrosis, and cirrhosis. In the majority of cases, there is a fast progressive course, leading to death by the age of 12 months (59). Although liver transplantation has been considered for cases with isolated and relatively stable liver disease, the presence of neurological features such as nystagmus, severe hypotonia, and psychomotor retardation would preclude the use of liver transplantation (59).

MPV17 encodes a mitochondrial inner membrane protein of unknown function recently associated with mtDNA depletion. The clinical phenotype of this syndrome consists of an early presentation in the first year of life associated with severe liver disease, hypoglycemia, growth retardation, neurological symptoms, and multiple brain lesions (60). There is a remarkable depletion of mtDNA in the liver in association with deficiencies of multiple RC complexes (61). MPV17 mutations are responsible for Navajo neurohepatopathy which is an autosomal recessive condition found in the southwestern USA (62). The clinical spectrum of MPV17 mutations has been associated with different phenotypes: an infantile-onset and childhood-onset forms associated with hypoglycemia and progressive liver disease and a classic form with more moderate liver disease and axonal neuropathy (63). In cases of isolated and progressive liver disease, orthotopic liver transplantation has been attempted. In few cases, the condition has been associated with the onset of hepatocellular carcinoma (63). The same pathogenic mutation (p.R50Q) detected in patients affected with Navajo neurohepatopathy has been detected in Italian patients which could be explained by the possibility of a mutation hot spot or an ancient founder effect (54).

TK2 gene encodes thymidine kinase 2 which is an intramitochondrial nucleoside kinase that phosphorylates deoxythymidine, deoxycytidine, and deoxyuridine. Mutations in TK2 produce primarily a myopathic mtDNA depletion syndrome. Classical phenotype is that of a severe infantile-onset progressive myopathy (64, 65). The course is complicated by motor regression and early death from respiratory insufficiency (66). A clinical phenotype associated with spinal muscular atrophy has been found (65). Furthermore, milder clinical phenotypes associated with slower progression and longer survival have been described (53). Existing clinical phenotypes may be explained by variable range of residual enzymatic activity. Electromyography demonstrates the presence of myopathy, and creatine kinase is elevated (usually above 1,000 U/L) with mild elevation of lactate levels. On muscle biopsy, there is a predominance of cytochrome c oxidase negative ragged red fibers.

*RRM2B* encodes the small subunit of a p53-inducible ribonucleotide reductase, a heterotetrameric enzyme responsible for de novo conversion of ribonucleoside diphosphates into deoxyribonucleoside diphosphates that are relevant for DNA synthesis (67). This enzyme is the key regulator of the cytoplasmic nucleotide pools, and its small subunit has a key function in maintaining the mitochondrial deoxynucleotide pool for mtDNA synthesis. The associated clinical phenotype presents with hypotonia, failure to thrive, renal tubulopathy, and lactic acidosis in the first months of life, with profound mtDNA copy number reduction in skeletal muscle (67). Furthermore, a MNGIE-like phenotype has also been reported to be associated with mutations in *RRM2B* (68).

Succinyl CoA synthase is a mitochondrial matrix enzyme that mediates the synthesis of succinate and adenosine triphosphate (ATP) or guanosine triphosphate (GTP) from succinyl CoA and adenosine diphosphate (ADP) in the Krebs cycle. Succinyl CoA enzyme is formed by two subunits, α and β, encoded by *SUCLG1* and *SUCLA2* respectively. *SUCLA2* and *SUCLG1* mutations disrupt an association between succinyl CoA synthase and mitochondrial diphosphate kinase, leading to an unbalanced mitochondrial dNTP pool and mild mtDNA depletion. Mutations in these two genes are associated with a hepatoencephalomyopathic form of infantile mtDNA depletion syndrome (69). Secondary to the metabolic block, a mild elevation of urinary methylmalonic acid and Krebs cycle intermediates have been observed in the majority of cases. In few cases, an elevation of propionylcarnitine ascertained by newborn screening has been determined (70). The majority of patients exhibit mutations in *SUCLA2*. The clinical phenotype of patients with *SUCLA2* mutations includes early childhood hypotonia, developmental delay, dystonia, and sensorineural hearing loss (71). A founder mutation has been found in the Faroe Islands (71). *SUCLG1* mutations have been reported in fewer families associated with neonatal metabolic crises and early death; however, the clinical severity may correlate with the residual activity of the protein (72).

*TP* gene encodes thymidine phosphorylase (73). Deficiency of thymidine phosphorylase causes mitochondrial neurogastrointestinal encephalomyopathy (MNGIE). MNGIE syndrome is a multisystemic disorder with onset between the second and fifth decades of life and whose clinical features are characterized by severe gastrointestinal dysmotility, leukoencephalopathy, ptosis, PEO, peripheral neuropathy, and myopathy (73). In this syndrome, the altered balance of intramitochondrial dNTP pool leads to multiple mtDNA deletions and mtDNA depletion (54). Elevated plasma thymidine and deoxyuridine values are a useful first screening tool if the diagnosis is suspected (74). Although the muscle histology may reveal ragged red fibers and COX deficient fibers, normal muscle histology should not preclude the consideration of this diagnosis (75).

## 5. Defects in Genes Involved in Mitochondrial Protein Synthesis

Another category of mitochondrial cytopathies may be caused by nuclear gene mutations deranging mitochondrial protein synthesis disorders without loss of mtDNA integrity. These defects involve nuclear genes encoding proteins that mediate mitochondrial protein synthesis (transfer RNA modification, initiation, elongation, and termination factors, ribosomal proteins, and aminoacyl-transfer RNA synthetases) (76). This group is genetically heterogeneous and clinically diverse, and most patients will exhibit neurological features with associated combined respiratory chain defects.

### 5.1. Molecular Defects in Genes Encoding Mitochondrial tRNA Modifying Enzymes

The *PUS1* gene encodes an enzyme that converts uridine into pseudouridine at several cytoplasmic and mitochondrial tRNA positions, improving the translation efficiency in the cytosol and in the mitochondrion. PUS1 protein is required for posttranslational modification of tRNA. Pseudouridylation stabilizes both base pairing in stems and base stacking in the anticodon loop. PUS1 may also be required for the interaction of the tRNA with its cognate aminoacyl tRNA synthetase (77). Mutations in this gene lead to decreased pseudouridylation of cytoplasmic and mitochondrial tRNA, resulting in impairment of mitochondrial protein translation (78). *PUS1* mutations are responsible for the rare MLASA (myopathy, lactic acidosis, and sideroblastic anemia) syndrome (78). There is variable clinical severity, and the clinical spectrum may also include intellectual disability.

*TRMU* gene encodes an evolutionarily conserved protein involved in mitochondrial tRNA modification, tRNA 5-methylaminomethyl-2-thiouridylate methyltransferase (TRMU) which is important for mitochondrial translation. This mitochondrial specific enzyme is required for the 2-thiolation on the wobble position of the tRNA anticodon, leading to reduced steady state levels of tRNA$^{Lys}$, tRNA$^{Gln}$, and tRNA$^{Glu}$, leading to impaired mitochondrial protein synthesis (79). Patients with *TRMU* mutations exhibit combined respiratory chain defects and deficient mitochondrial translation, leading to acute liver failure in infancy (79). Moreover, a mutation in *TRMU* may have a modifier effect on the phenotypic expression of the deafness-associated mitochondrial 12S ribosomal RNA mutations (80).

### 5.2. Molecular Defects in Genes Encoding Mitochondrial Elongation Factors

Mutations in genes encoding components of the mitochondrial translation elongation machinery have been found in the *TUFM*, *TSFM*, and *GFM1* genes encoding elongation factor Tu (EF-Tu), elongation factor Ts (EF-Ts), and elongation factor G1 (EFG1) respectively. In general, these patients exhibit a severe phenotype with a lethal outcome and present combined deficiencies of the mitochondrial respiratory chain complexes. A patient with mutations in *TUFM* exhibited severe infantile macrocystic leukodystrophy with micropolygyria (81), and a neonate carrying mutations in the *GFM1* gene presented with neonatal lactic acidosis and Leigh-like encephalopathy. Mutations in *TSFM* have been associated with a phenotype of encephalomyopathy in one neonate and with the presence of hypertrophic cardiomyopathy in another infant (82).

**5.3. Molecular Defects in MRPS16 and MRPS22**

Mutations have been found in two genes encoding mitochondrial ribosomal proteins, *MRPS16* and *MRPS22* (83, 84). These mutations led to a decrease in the 12s rRNA level caused by impaired assembly of the mitoribosomal small subunit, compromising the stability of 12s rRNA. This effect in turn will lead to the degradation of the components of the mitoribosome. The observed clinical phenotype is severe and consistent with agenesis of the corpus callosum, dysmorphic features, hypertrophic cardiomyopathy, and neonatal lactic acidosis.

**5.4. Molecular Defects in Genes Encoding Mitochondrial Aminoacyl-Transfer RNA Synthetases (RARS2, DARS2, and YARS2)**

Mutations in the *RARS2* gene encoding the mitochondrial arginyl-transfer RNA are associated with severe encephalopathy and pontocerebellar hypoplasia (85), whereas mutations in the *DARS2* gene encoding the mitochondrial aspartyl-transfer RNA synthetases have been described in subjects with leukoencephalopathy with brain stem and spinal cord involvement and lactate elevation (86). The specific neuroradiological findings may aid in the diagnosis. The majority of these patients exhibit onset of disease in childhood, and the features consist of delayed motor development, epilepsy, tremor, dysarthria, and spasticity. In addition, these patients may present with axonal peripheral neuropathy with distal weakness and decreased proprioception. Furthermore, mutations in the *YARS2* gene encoding the mitochondrial tyrosyl transfer RNA synthetase have been associated with the MLASA phenotype (87).

**6. Molecular Defects in Genes Encoding Proteins Involved in Mitochondrial Motility, Fusion, and Fission**

Mutations in genes encoding proteins involved in mitochondrial dynamics have been linked to neurodegenerative diseases (3). A molecular defect in the gene *KIF5A* encoding one of the mitochondrial kinesins was discovered to be associated with autosomal dominant hereditary spastic paraplegia. It was found that this mutation would affect mitochondrial motility (88). *OPA1* mutations cause autosomal dominant optic atrophy (89), whereas mutations in *MFN2* which encodes mitofusin 2 cause autosomal dominant axonal variant of Charcot-Marie-Tooth (90). Moreover, mutations in *GDAP1* encoding ganglioside-induced differentiation protein 1, a protein that regulates the mitochondrial network, cause an autosomal recessive, early-onset type of either demyelinating or axonal neuropathy (91).

## 7. Molecular Defects in Genes Associated with Secondary Respiratory Chain Defects

In this particular group, the molecular defects could impact the function of the respiratory chain by distorting the lipid structure of the mitochondrial inner membrane, where the respiratory chain subunits are embedded, or by affecting the importation of one or more subunits of the respiratory chain.

Mutations in the *G4.5* gene cause Barth syndrome, an X-linked disorder with features consistent with mitochondrial myopathy, neutropenia, 3-methylglutaconic aciduria, cardiac involvement including left ventricular noncompaction or dilated cardiomyopathy, and failure to thrive (92). This gene encodes the protein taffazin which is involved in the synthesis of phospholipids. In Barth syndrome, there are altered amounts of cardiolipin, which is a key phospholipid component of the mitochondrial inner membrane (93).

Defects in importation of components of the respiratory chain could be exemplified by mutations in *TIMM8A*, which encodes the deafness-dystonia peptide (DDP1), a component of the mitochondrial protein import machinery located in the intermembrane space (94). These molecular defects could lead to the X-linked Mohr-Tranebjaerg syndrome which is characterized by progressive sensorineural deafness, dystonia, and psychiatric features (94).

## 8. Other Nuclear Genes Involved in Mitochondrial Function

*SLC25A19* encodes a mitochondria inner membrane transporter for both deoxynucleotides and thiamine pyrophosphate (TPP) (95). Mutations in this gene have been associated with Amish microcephaly, a metabolic disorder previously characterized by severe infantile lethal congenital microcephaly and alpha-ketoglutaric aciduria. All reported patients have been from the Pennsylvania Amish community and homozygous for a p.Gly177Ala mutation in *SLC25A19*. The biochemical phenotype may be attributable to decreased activity of the three mitochondrial enzymes that require TPP as a cofactor: pyruvate dehydrogenase, alpha-ketoglutarate dehydrogenase, and branched-chain amino acid dehydrogenase (96). A different phenotype associated with another missense mutation in the same gene consists of neuropathy and bilateral striatal necrosis (97).

*SLC25A3* encodes a mitochondrial phosphate carrier that transports inorganic phosphate into the mitochondrial matrix, which is essential for the aerobic synthesis of adenosine triphosphate (ATP). A homozygous mutation c.215G→A (p.Gly72Glu) was reported in the alternatively spliced exon 3A of this gene in two siblings presenting with lactic acidosis, hypertrophic cardiomyopathy, and muscular hypotonia who died within the first year of life. Functional investigation of intact mitochondria demonstrated a deficiency of ATP synthesis in muscle, which correlated with the tissue-specific expression of exon 3A in muscle (98).

## 9. Conclusions

The increasing number of Mendelian mitochondrial syndromes and nuclear genes leading to mitochondrial respiratory chain defects continues to shed light on the genetic and clinical phenotypic heterogeneity of mitochondrial cytopathies. Identifying nuclear genes is not only important for proper genetic counseling and prenatal diagnosis but also to establish a good understanding of the pathophysiology of these disorders that could eventually help to develop adequate therapeutic strategies. More importantly, the increasing awareness of the large number of nuclear genes involved in mitochondrial syndromes will require a better strategy for detection of molecular alterations that should include the development of large-scale, high-throughput technologies that can increase our insight into these complicated pathologies.

## References

1. Scaglia, F., Towbin, J. A., Craigen, W. J., Belmont, J. W., Smith, E. O., Neish, S. R., Ware, S. M., Hunter, J. V., Fernbach, S. D., Vladutiu, G. D., Wong, L. J., and Vogel, H. (2004) Clinical spectrum, morbidity, and mortality in 113 pediatric patients with mitochondrial disease, *Pediatrics 114*, 925–931.

2. Thorburn, D. R. (2004) Mitochondrial disorders: prevalence, myths and advances, *J Inherit Metab Dis 27*, 349–362.

3. DiMauro, S., and Hirano, M. (2009) Pathogenesis and treatment of mitochondrial disorders, *Adv Exp Med Biol 652*, 139–170.

4. Distelmaier, F., Koopman, W. J., van den Heuvel, L. P., Rodenburg, R. J., Mayatepek, E., Willems, P. H., and Smeitink, J. A. (2009) Mitochondrial complex I deficiency: from organelle dysfunction to clinical disease, *Brain 132*, 833–842.

5. Bourgeron, T., Rustin, P., Chretien, D., Birch-Machin, M., Bourgeois, M., Viegas-Pequignot, E., Munnich, A., and Rotig, A. (1995) Mutation of a nuclear succinate dehydrogenase gene results in mitochondrial respiratory chain deficiency, *Nat Genet 11*, 144–149.

6. Horvath, R., Abicht, A., Holinski-Feder, E., Laner, A., Gempel, K., Prokisch, H., Lochmuller, H., Klopstock, T., and Jaksch, M. (2006) Leigh syndrome caused by mutations in the flavoprotein (Fp) subunit of succinate dehydrogenase (SDHA), *J Neurol Neurosurg Psychiatry 77*, 74–76.

7. Kantorovich, V., King, K. S., and Pacak, K. SDH-related pheochromocytoma and paraganglioma, *Best Pract Res Clin Endocrinol Metab 24*, 415–424.

8. Smeitink, J., van den Heuvel, L., and DiMauro, S. (2001) The genetics and pathology of oxidative

phosphorylation, *Nat Rev Genet 2*, 342–352.

9. Smeitink, J. A., van den Heuvel, L. W., Koopman, W. J., Nijtmans, L. G., Ugalde, C., and Willems, P. H. (2004) Cell biological consequences of mitochondrial NADH: ubiquinone oxidoreductase deficiency, *Curr Neurovasc Res 1*, 29–40.

10. Rotig, A. Genetic bases of mitochondrial respiratory chain disorders, *Diabetes Metab 36*, 97–107.

11. Haut, S., Brivet, M., Touati, G., Rustin, P., Lebon, S., Garcia-Cazorla, A., Saudubray, J. M., Boutron, A., Legrand, A., and Slama, A. (2003) A deletion in the human QP-C gene causes a complex III deficiency resulting in hypoglycaemia and lactic acidosis, *Hum Genet 113*, 118–122.

12. Barel, O., Shorer, Z., Flusser, H., Ofir, R., Narkis, G., Finer, G., Shalev, H., Nasasra, A., Saada, A., and Birk, O. S. (2008) Mitochondrial complex III deficiency associated with a homozygous mutation in UQCRQ, *Am J Hum Genet 82*, 1211–1216.

13. Shteyer, E., Saada, A., Shaag, A., Al-Hijawi, F. A., Kidess, R., Revel-Vilk, S., and Elpeleg, O. (2009) Exocrine pancreatic insufficiency, dyserythropoeitic anemia, and calvarial hyperostosis are caused by a mutation in the COX4I2 gene, *Am J Hum Genet 84*, 412–417.

14. Massa, V., Fernandez-Vizarra, E., Alshahwan, S., Bakhsh, E., Goffrini, P., Ferrero, I., Mereghetti, P., D'Adamo, P., Gasparini, P., and Zeviani, M. (2008) Severe infantile encephalomyopathy caused by a mutation in COX6B1, a nucleus-encoded subunit of cytochrome c oxidase, *Am J Hum Genet 82*, 1281–1289.

15. Quinzii, C. M., and Hirano, M. Coenzyme Q and mitochondrial disease, *Dev Disabil Res Rev 16*, 183–188.

16. Lopez, L. C., Schuelke, M., Quinzii, C. M., Kanki, T., Rodenburg, R. J., Naini, A., Dimauro, S., and Hirano, M. (2006) Leigh syndrome with nephropathy and CoQ10 deficiency due to decaprenyl diphosphate synthase subunit 2 (PDSS2) mutations, *Am J Hum Genet 79*, 1125–1129.

17. Mollet, J., Giurgea, I., Schlemmer, D., Dallner, G., Chretien, D., Delahodde, A., Bacq, D., de Lonlay, P., Munnich, A., and Rotig, A. (2007) Prenyldiphosphate synthase, subunit 1 (PDSS1) and OH-benzoate polyprenyltransferase (COQ2) mutations in ubiquinone deficiency and oxidative phosphorylation disorders, *J Clin Invest 117*, 765–772.

18. Quinzii, C., Naini, A., Salviati, L., Trevisson, E., Navas, P., Dimauro, S., and Hirano, M. (2006) A mutation in para-hydroxybenzoate-polyprenyl transferase (COQ2) causes primary coenzyme Q10 deficiency, *Am J Hum Genet 78*, 345–349.

19. Lagier-Tourenne, C., Tazir, M., Lopez, L. C., Quinzii, C. M., Assoum, M., Drouot, N., Busso, C., Makri, S., Ali-Pacha, L., Benhassine, T., Anheim, M., Lynch, D. R., Thibault, C., Plewniak, F., Bianchetti, L., Tranchant, C., Poch, O., DiMauro, S., Mandel, J. L., Barros, M. H., Hirano, M., and Koenig, M. (2008) ADCK3, an ancestral kinase, is mutated in a form of recessive ataxia associated with coenzyme Q10 deficiency, *Am J Hum Genet 82*, 661–672.

20. Mollet, J., Delahodde, A., Serre, V., Chretien, D., Schlemmer, D., Lombes, A., Boddaert, N., Desguerre, I., de Lonlay, P., de Baulny, H. O., Munnich, A., and Rotig, A. (2008) CABC1 gene mutations cause ubiquinone deficiency with cerebellar ataxia and seizures, *Am J Hum Genet 82*, 623–630.

21. Heeringa, S. F., Chernin, G., Chaki, M., Zhou, W., Sloan, A. J., Ji, Z., Xie, L. X., Salviati, L., Hurd, T. W., Vega-Warner, V., Killen, P. D., Raphael, Y., Ashraf, S., Ovunc, B., Schoeb, D. S., McLaughlin, H. M., Airik, R., Vlangos, C. N., Gbadegesin, R., Hinkes, B., Saisawat, P., Trevisson, E., Doimo, M., Casarin, A., Pertegato, V., Giorgi, G., Prokisch, H., Rotig, A., Nurnberg, G., Becker, C., Wang, S., Ozaltin, F., Topaloglu, R., Bakkaloglu, A., Bakkaloglu, S. A., Muller, D., Beissert, A., Mir, S., Berdeli, A., Varpizen, S., Zenker, M., Matejas, V., Santos-Ocana, C., Navas, P., Kusakabe, T., Kispert, A., Akman, S., Soliman, N. A., Krick, S., Mundel, P., Reiser, J., Nurnberg, P., Clarke, C. F., Wiggins, R. C., Faul, C., and Hildebrandt, F. COQ6 mutations in human patients produce nephrotic syndrome with sensorineural deafness, *J Clin Invest 121*, 2013–2024.

22. Duncan, A. J., Bitner-Glindzicz, M., Meunier, B., Costello, H., Hargreaves, I. P., Lopez, L. C., Hirano, M., Quinzii, C. M., Sadowski, M. I., Hardy, J., Singleton, A., Clayton, P. T., and Rahman, S. (2009) A nonsense mutation in COQ9 causes autosomal-recessive neonatal-onset primary coenzyme Q10 deficiency: a potentially treatable form of mitochondrial disease, *Am J Hum Genet 84*, 558–566.

23. Ogilvie, I., Kennaway, N. G., and Shoubridge, E. A. (2005) A molecular chaperone for mitochondrial complex I assembly is mutated in a progressive encephalopathy, *J Clin Invest 115*, 2784–2792.

24. Dunning, C. J., McKenzie, M., Sugiana, C., Lazarou, M., Silke, J., Connelly, A., Fletcher, J. M., Kirby, D. M., Thorburn, D. R., and

Ryan, M. T. (2007) Human CIA30 is involved in the early assembly of mitochondrial complex I and mutations in its gene cause disease, *EMBO J 26*, 3227–3237.

25. Saada, A., Edvardson, S., Rapoport, M., Shaag, A., Amry, K., Miller, C., Lorberboum-Galski, H., and Elpeleg, O. (2008) C6ORF66 is an assembly factor of mitochondrial complex I, *Am J Hum Genet 82*, 32–38.

26. Ghezzi, D., Goffrini, P., Uziel, G., Horvath, R., Klopstock, T., Lochmuller, H., D'Adamo, P., Gasparini, P., Strom, T. M., Prokisch, H., Invernizzi, F., Ferrero, I., and Zeviani, M. (2009) SDHAF1, encoding a LYR complex-II specific assembly factor, is mutated in SDH-defective infantile leukoencephalopathy, *Nat Genet 41*, 654–656.

27. Hao, H. X., Khalimonchuk, O., Schraders, M., Dephoure, N., Bayley, J. P., Kunst, H., Devilee, P., Cremers, C. W., Schiffman, J. D., Bentz, B. G., Gygi, S. P., Winge, D. R., Kremer, H., and Rutter, J. (2009) SDH5, a gene required for flavination of succinate dehydrogenase, is mutated in paraganglioma, *Science 325*, 1139–1142.

28. de Lonlay, P., Valnot, I., Barrientos, A., Gorbatyuk, M., Tzagoloff, A., Taanman, J. W., Benayoun, E., Chretien, D., Kadhom, N., Lombes, A., de Baulny, H. O., Niaudet, P., Munnich, A., Rustin, P., and Rotig, A. (2001) A mutant mitochondrial respiratory chain assembly protein causes complex III deficiency in patients with tubulopathy, encephalopathy and liver failure, *Nat Genet 29*, 57–60.

29. Visapaa, I., Fellman, V., Vesa, J., Dasvarma, A., Hutton, J. L., Kumar, V., Payne, G. S., Makarow, M., Van Coster, R., Taylor, R. W., Turnbull, D. M., Suomalainen, A., and Peltonen, L. (2002) GRACILE syndrome, a lethal metabolic disorder with iron overload, is caused by a point mutation in BCS1L, *Am J Hum Genet 71*, 863–876.

30. Hinson, J. T., Fantin, V. R., Schonberger, J., Breivik, N., Siem, G., McDonough, B., Sharma, P., Keogh, I., Godinho, R., Santos, F., Esparza, A., Nicolau, Y., Selvaag, E., Cohen, B. H., Hoppel, C. L., Tranebjaerg, L., Eavey, R. D., Seidman, J. G., and Seidman, C. E. (2007) Missense mutations in the BCS1L gene as a cause of the Bjornstad syndrome, *N Engl J Med 356*, 809–819.

31. Sue, C. M., Karadimas, C., Checcarelli, N., Tanji, K., Papadopoulou, L. C., Pallotti, F., Guo, F. L., Shanske, S., Hirano, M., De Vivo, D. C., Van Coster, R., Kaplan, P., Bonilla, E., and DiMauro, S. (2000) Differential features of patients with mutations in two COX assembly genes, SURF-1 and SCO2, *Ann Neurol 47*, 589–595.

32. Leary, S. C., Sasarman, F., Nishimura, T., and Shoubridge, E. A. (2009) Human SCO2 is required for the synthesis of CO II and as a thiol-disulphide oxidoreductase for SCO1, *Hum Mol Genet 18*, 2230–2240.

33. Valnot, I., Osmond, S., Gigarel, N., Mehaye, B., Amiel, J., Cormier-Daire, V., Munnich, A., Bonnefont, J. P., Rustin, P., and Rotig, A. (2000) Mutations of the SCO1 gene in mitochondrial cytochrome c oxidase deficiency with neonatal-onset hepatic failure and encephalopathy, *Am J Hum Genet 67*, 1104–1109.

34. Papadopoulou, L. C., Sue, C. M., Davidson, M. M., Tanji, K., Nishino, I., Sadlock, J. E., Krishna, S., Walker, W., Selby, J., Glerum, D. M., Coster, R. V., Lyon, G., Scalais, E., Lebel, R., Kaplan, P., Shanske, S., De Vivo, D. C., Bonilla, E., Hirano, M., DiMauro, S., and Schon, E. A. (1999) Fatal infantile cardioencephalomyopathy with COX deficiency and mutations in SCO2, a COX assembly gene, *Nat Genet 23*, 333–337.

35. Valnot, I., von Kleist-Retzow, J. C., Barrientos, A., Gorbatyuk, M., Taanman, J. W., Mehaye, B., Rustin, P., Tzagoloff, A., Munnich, A., and Rotig, A. (2000) A mutation in the human heme A:farnesyltransferase gene (COX10) causes cytochrome c oxidase deficiency, *Hum Mol Genet 9*, 1245–1249.

36. Antonicka, H., Mattman, A., Carlson, C. G., Glerum, D. M., Hoffbuhr, K. C., Leary, S. C., Kennaway, N. G., and Shoubridge, E. A. (2003) Mutations in COX15 produce a defect in the mitochondrial heme biosynthetic pathway, causing early-onset fatal hypertrophic cardiomyopathy, *Am J Hum Genet 72*, 101–114.

37. Oquendo, C. E., Antonicka, H., Shoubridge, E. A., Reardon, W., and Brown, G. K. (2004) Functional and genetic studies demonstrate that mutation in the COX15 gene can cause Leigh syndrome, *J Med Genet 41*, 540–544.

38. Xu, F., Morin, C., Mitchell, G., Ackerley, C., and Robinson, B. H. (2004) The role of the LRPPRC (leucine-rich pentatricopeptide repeat cassette) gene in cytochrome oxidase assembly: mutation causes lowered levels of COX (cytochrome c oxidase) I and COX III mRNA, *Biochem J 382*, 331–336.

39. Mootha, V. K., Lepage, P., Miller, K., Bunkenborg, J., Reich, M., Hjerrild, M., Delmonte, T., Villeneuve, A., Sladek, R., Xu, F., Mitchell, G. A., Morin, C., Mann, M., Hudson, T. J., Robinson, B., Rioux, J. D., and Lander, E. S. (2003) Identification of a gene causing human cytochrome c oxidase deficiency

by integrative genomics, *Proc Natl Acad Sci USA 100*, 605–610.

40. Seeger, J., Schrank, B., Pyle, A., Stucka, R., Lorcher, U., Muller-Ziermann, S., Abicht, A., Czermin, B., Holinski-Feder, E., Lochmuller, H., and Horvath, R. Clinical and neuropathological findings in patients with TACO1 mutations, *Neuromuscul Disord 20*, 720–724.

41. De Meirleir, L., Seneca, S., Lissens, W., De Clercq, I., Eyskens, F., Gerlo, E., Smet, J., and Van Coster, R. (2004) Respiratory chain complex V deficiency due to a mutation in the assembly gene ATP12, *J Med Genet 41*, 120–124.

42. Cizkova, A., Stranecky, V., Mayr, J. A., Tesarova, M., Havlickova, V., Paul, J., Ivanek, R., Kuss, A. W., Hansikova, H., Kaplanova, V., Vrbacky, M., Hartmannova, H., Noskova, L., Honzik, T., Drahota, Z., Magner, M., Hejzlarova, K., Sperl, W., Zeman, J., Houstek, J., and Kmoch, S. (2008) TMEM70 mutations cause isolated ATP synthase deficiency and neonatal mitochondrial encephalocardiomyopathy, *Nat Genet 40*, 1288–1290.

43. Shchelochkov, O. A., Li, F. Y., Wang, J., Zhan, H., Towbin, J. A., Jefferies, J. L., Wong, L. J., and Scaglia, F. Milder clinical course of Type IV 3-methylglutaconic aciduria due to a novel mutation in TMEM70, *Mol Genet Metab 101*, 282–285.

44. Mochel, F., Knight, M. A., Tong, W. H., Hernandez, D., Ayyad, K., Taivassalo, T., Andersen, P. M., Singleton, A., Rouault, T. A., Fischbeck, K. H., and Haller, R. G. (2008) Splice mutation in the iron-sulfur cluster scaffold protein ISCU causes myopathy with exercise intolerance, *Am J Hum Genet 82*, 652–660.

45. Olsson, A., Lind, L., Thornell, L. E., and Holmberg, M. (2008) Myopathy with lactic acidosis is linked to chromosome 12q23.3-24.11 and caused by an intron mutation in the ISCU gene resulting in a splicing defect, *Hum Mol Genet 17*, 1666–1672.

46. DiMauro, S., and Mancuso, M. (2007) Mitochondrial diseases: therapeutic approaches, *Biosci Rep 27*, 125–137.

47. Spinazzola, A., and Zeviani, M. (2005) Disorders of nuclear-mitochondrial intergenomic signaling, *Gene 354*, 162–168.

48. Chan, S. S., and Copeland, W. C. (2009) DNA polymerase gamma and mitochondrial disease: understanding the consequence of POLG mutations, *Biochim Biophys Acta 1787*, 312–319.

49. Rotig, A., and Poulton, J. (2009) Genetic causes of mitochondrial DNA depletion in humans, *Biochim Biophys Acta 1792*, 1103–1108.

50. Van Goethem, G., Dermaut, B., Lofgren, A., Martin, J. J., and Van Broeckhoven, C. (2001) Mutation of POLG is associated with progressive external ophthalmoplegia characterized by mtDNA deletions, *Nat Genet 28*, 211–212.

51. Naviaux, R. K., Nyhan, W. L., Barshop, B. A., Poulton, J., Markusic, D., Karpinski, N. C., and Haas, R. H. (1999) Mitochondrial DNA polymerase gamma deficiency and mtDNA depletion in a child with Alpers' syndrome, *Ann Neurol 45*, 54–58.

52. Spelbrink, J. N., Li, F. Y., Tiranti, V., Nikali, K., Yuan, Q. P., Tariq, M., Wanrooij, S., Garrido, N., Comi, G., Morandi, L., Santoro, L., Toscano, A., Fabrizi, G. M., Somer, H., Croxen, R., Beeson, D., Poulton, J., Suomalainen, A., Jacobs, H. T., Zeviani, M., and Larsson, C. (2001) Human mitochondrial DNA deletions associated with mutations in the gene encoding Twinkle, a phage T7 gene 4-like protein localized in mitochondria, *Nat Genet 28*, 223–231.

53. Suomalainen, A., and Isohanni, P. Mitochondrial DNA depletion syndromes--many genes, common mechanisms, *Neuromuscul Disord 20*, 429–437.

54. Spinazzola, A., Invernizzi, F., Carrara, F., Lamantea, E., Donati, A., Dirocco, M., Giordano, I., Meznaric-Petrusa, M., Baruffini, E., Ferrero, I., and Zeviani, M. (2009) Clinical and molecular features of mitochondrial DNA depletion syndromes, *J Inherit Metab Dis 32*, 143–158.

55. Hakonen, A. H., Goffart, S., Marjavaara, S., Paetau, A., Cooper, H., Mattila, K., Lampinen, M., Sajantila, A., Lonnqvist, T., Spelbrink, J. N., and Suomalainen, A. (2008) Infantile-onset spinocerebellar ataxia and mitochondrial recessive ataxia syndrome are associated with neuronal complex I defect and mtDNA depletion, *Hum Mol Genet 17*, 3822–3835.

56. Spinazzola, A., and Zeviani, M. (2009) Disorders from perturbations of nuclear-mitochondrial intergenomic cross-talk, *J Intern Med 265*, 174–192.

57. Scaglia, F., Dimmock, D., and Wong, L. J. (1993).

58. Dimmock, D. P., Zhang, Q., Dionisi-Vici, C., Carrozzo, R., Shieh, J., Tang, L. Y., Truong, C., Schmitt, E., Sifry-Platt, M., Lucioli, S., Santorelli, F. M., Ficicioglu, C. H., Rodriguez, M., Wierenga, K., Enns, G. M., Longo, N., Lipson, M. H., Vallance, H., Craigen, W. J., Scaglia, F., and Wong, L. J. (2008) Clinical and

molecular features of mitochondrial DNA depletion due to mutations in deoxyguanosine kinase, *Hum Mutat 29*, 330–331.

59. Dimmock, D. P., Dunn, J. K., Feigenbaum, A., Rupar, A., Horvath, R., Freisinger, P., Mousson de Camaret, B., Wong, L. J., and Scaglia, F. (2008) Abnormal neurological features predict poor survival and should preclude liver transplantation in patients with deoxyguanosine kinase deficiency, *Liver Transpl 14*, 1480–1485.

60. Wong, L. J., Brunetti-Pierri, N., Zhang, Q., Yazigi, N., Bove, K. E., Dahms, B. B., Puchowicz, M. A., Gonzalez-Gomez, I., Schmitt, E. S., Truong, C. K., Hoppel, C. L., Chou, P. C., Wang, J., Baldwin, E. E., Adams, D., Leslie, N., Boles, R. G., Kerr, D. S., and Craigen, W. J. (2007) Mutations in the MPV17 gene are responsible for rapidly progressive liver failure in infancy, *Hepatology 46*, 1218–1227.

61. Alberio, S., Mineri, R., Tiranti, V., and Zeviani, M. (2007) Depletion of mtDNA: syndromes and genes, *Mitochondrion 7*, 6–12.

62. Copeland, W. C. (2008) Inherited mitochondrial diseases of DNA replication, *Annu Rev Med 59*, 131–146.

63. El-Hattab, A. W., Li, F. Y., Schmitt, E., Zhang, S., Craigen, W. J., and Wong, L. J. MPV17-associated hepatocerebral mitochondrial DNA depletion syndrome: new patients and novel mutations, *Mol Genet Metab 99*, 300–308.

64. Saada, A., Shaag, A., Mandel, H., Nevo, Y., Eriksson, S., and Elpeleg, O. (2001) Mutant mitochondrial thymidine kinase in mitochondrial DNA depletion myopathy, *Nat Genet 29*, 342–344.

65. Mancuso, M., Salviati, L., Sacconi, S., Otaegui, D., Camano, P., Marina, A., Bacman, S., Moraes, C. T., Carlo, J. R., Garcia, M., Garcia-Alvarez, M., Monzon, L., Naini, A. B., Hirano, M., Bonilla, E., Taratuto, A. L., DiMauro, S., and Vu, T. H. (2002) Mitochondrial DNA depletion: mutations in thymidine kinase gene with myopathy and SMA, *Neurology 59*, 1197–1202.

66. Oskoui, M., Davidzon, G., Pascual, J., Erazo, R., Gurgel-Giannetti, J., Krishna, S., Bonilla, E., De Vivo, D. C., Shanske, S., and DiMauro, S. (2006) Clinical spectrum of mitochondrial DNA depletion due to mutations in the thymidine kinase 2 gene, *Arch Neurol 63*, 1122–1126.

67. Bourdon, A., Minai, L., Serre, V., Jais, J. P., Sarzi, E., Aubert, S., Chretien, D., de Lonlay, P., Paquis-Flucklinger, V., Arakawa, H., Nakamura, Y., Munnich, A., and Rotig, A. (2007) Mutation of RRM2B, encoding p53-controlled ribonucleotide reductase (p53R2), causes severe mitochondrial DNA depletion, *Nat Genet 39*, 776–780.

68. Shaibani, A., Shchelochkov, O. A., Zhang, S., Katsonis, P., Lichtarge, O., Wong, L. J., and Shinawi, M. (2009) Mitochondrial neurogastrointestinal encephalopathy due to mutations in RRM2B, *Arch Neurol 66*, 1028–1032.

69. Van Hove, J. L., Saenz, M. S., Thomas, J. A., Gallagher, R. C., Lovell, M. A., Fenton, L. Z., Shanske, S., Myers, S. M., Wanders, R. J., Ruiter, J., Turkenburg, M., and Waterham, H. R. Succinyl-CoA ligase deficiency: a mitochondrial hepatoencephalomyopathy, *Pediatr Res 68*, 159–164.

70. Randolph, L. M., Jackson, H. A., Wang, J., Shimada, H., Sanchez-Lara, P. A., Wong, D. A., Wong, L. J., and Boles, R. G. Fatal infantile lactic acidosis and a novel homozygous mutation in the SUCLG1 gene: a mitochondrial DNA depletion disorder, *Mol Genet Metab 102*, 149–152.

71. Morava, E., Steuerwald, U., Carrozzo, R., Kluijtmans, L. A., Joensen, F., Santer, R., Dionisi-Vici, C., and Wevers, R. A. (2009) Dystonia and deafness due to SUCLA2 defect; Clinical course and biochemical markers in 16 children, *Mitochondrion 9*, 438–442.

72. Rouzier, C., Le Guedard-Mereuze, S., Fragaki, K., Serre, V., Miro, J., Tuffery-Giraud, S., Chaussenot, A., Bannwarth, S., Caruba, C., Ostergaard, E., Pellissier, J. F., Richelme, C., Espil, C., Chabrol, B., and Paquis-Flucklinger, V. The severity of phenotype linked to SUCLG1 mutations could be correlated with residual amount of SUCLG1 protein, *J Med Genet 47*, 670–676.

73. Nishino, I., Spinazzola, A., and Hirano, M. (1999) Thymidine phosphorylase gene mutations in MNGIE, a human mitochondrial disorder, *Science 283*, 689–692.

74. Marti, R., Nishigaki, Y., and Hirano, M. (2003) Elevated plasma deoxyuridine in patients with thymidine phosphorylase deficiency, *Biochem Biophys Res Commun 303*, 14–18.

75. Szigeti, K., Wong, L. J., Perng, C. L., Saifi, G. M., Eldin, K., Adesina, A. M., Cass, D. L., Hirano, M., Lupski, J. R., and Scaglia, F. (2004) MNGIE with lack of skeletal muscle involvement and a novel TP splice site mutation, *J Med Genet 41*, 125–129.

76. Nogueira, C., Carrozzo, R., Vilarinho, L., and Santorelli, F. M. Infantile-Onset Disorders of Mitochondrial Replication and Protein Synthesis, *J Child Neurol*.

77. Bykhovskaya, Y., Mengesha, E., and Fischel-Ghodsian, N. (2007) Pleiotropic effects and compensation mechanisms determine tissue specificity in mitochondrial myopathy and

sideroblastic anemia (MLASA), *Mol Genet Metab 91*, 148–156.

78. Bykhovskaya, Y., Casas, K., Mengesha, E., Inbal, A., and Fischel-Ghodsian, N. (2004) Missense mutation in pseudouridine synthase 1 (PUS1) causes mitochondrial myopathy and sideroblastic anemia (MLASA), *Am J Hum Genet 74*, 1303–1308.

79. Zeharia, A., Shaag, A., Pappo, O., Mager-Heckel, A. M., Saada, A., Beinat, M., Karicheva, O., Mandel, H., Ofek, N., Segel, R., Marom, D., Rotig, A., Tarassov, I., and Elpeleg, O. (2009) Acute infantile liver failure due to mutations in the TRMU gene, *Am J Hum Genet 85*, 401–407.

80. Guan, M. X., Yan, Q., Li, X., Bykhovskaya, Y., Gallo-Teran, J., Hajek, P., Umeda, N., Zhao, H., Garrido, G., Mengesha, E., Suzuki, T., del Castillo, I., Peters, J. L., Li, R., Qian, Y., Wang, X., Ballana, E., Shohat, M., Lu, J., Estivill, X., Watanabe, K., and Fischel-Ghodsian, N. (2006) Mutation in TRMU related to transfer RNA modification modulates the phenotypic expression of the deafness-associated mitochondrial 12 S ribosomal RNA mutations, *Am J Hum Genet 79*, 291–302.

81. Valente, L., Tiranti, V., Marsano, R. M., Malfatti, E., Fernandez-Vizarra, E., Donnini, C., Mereghetti, P., De Gioia, L., Burlina, A., Castellan, C., Comi, G. P., Savasta, S., Ferrero, I., and Zeviani, M. (2007) Infantile encephalopathy and defective mitochondrial DNA translation in patients with mutations of mitochondrial elongation factors EFG1 and EFTu, *Am J Hum Genet 80*, 44–58.

82. Smeitink, J. A., Elpeleg, O., Antonicka, H., Diepstra, H., Saada, A., Smits, P., Sasarman, F., Vriend, G., Jacob-Hirsch, J., Shaag, A., Rechavi, G., Welling, B., Horst, J., Rodenburg, R. J., van den Heuvel, B., and Shoubridge, E. A. (2006) Distinct clinical phenotypes associated with a mutation in the mitochondrial translation elongation factor EFTs, *Am J Hum Genet 79*, 869–877.

83. Miller, C., Saada, A., Shaul, N., Shabtai, N., Ben-Shalom, E., Shaag, A., Hershkovitz, E., and Elpeleg, O. (2004) Defective mitochondrial translation caused by a ribosomal protein (MRPS16) mutation, *Ann Neurol 56*, 734–738.

84. Saada, A., Shaag, A., Arnon, S., Dolfin, T., Miller, C., Fuchs-Telem, D., Lombes, A., and Elpeleg, O. (2007) Antenatal mitochondrial disease caused by mitochondrial ribosomal protein (MRPS22) mutation, *J Med Genet 44*, 784–786.

85. Edvardson, S., Shaag, A., Kolesnikova, O., Gomori, J. M., Tarassov, I., Einbinder, T.,

Saada, A., and Elpeleg, O. (2007) Deleterious mutation in the mitochondrial arginyl-transfer RNA synthetase gene is associated with pontocerebellar hypoplasia, *Am J Hum Genet 81*, 857–862.

86. Lin, J., Chiconelli Faria, E., Da Rocha, A. J., Rodrigues Masruha, M., Pereira Vilanova, L. C., Scheper, G. C., and Van der Knaap, M. S. Leukoencephalopathy with brainstem and spinal cord involvement and normal lactate: a new mutation in the DARS2 gene, *J Child Neurol 25*, 1425–1428.

87. Riley, L. G., Cooper, S., Hickey, P., Rudinger-Thirion, J., McKenzie, M., Compton, A., Lim, S. C., Thorburn, D., Ryan, M. T., Giege, R., Bahlo, M., and Christodoulou, J. Mutation of the mitochondrial tyrosyl-tRNA synthetase gene, YARS2, causes myopathy, lactic acidosis, and sideroblastic anemia--MLASA syndrome, *Am J Hum Genet 87*, 52–59.

88. Fichera, M., Lo Giudice, M., Falco, M., Sturnio, M., Amata, S., Calabrese, O., Bigoni, S., Calzolari, E., and Neri, M. (2004) Evidence of kinesin heavy chain (KIF5A) involvement in pure hereditary spastic paraplegia, *Neurology 63*, 1108–1110.

89. Amati-Bonneau, P., Milea, D., Bonneau, D., Chevrollier, A., Ferre, M., Guillet, V., Gueguen, N., Loiseau, D., de Crescenzo, M. A., Verny, C., Procaccio, V., Lenaers, G., and Reynier, P. (2009) OPA1-associated disorders: phenotypes and pathophysiology, *Int J Biochem Cell Biol 41*, 1855–1865.

90. Zuchner, S., Mersiyanova, I. V., Muglia, M., Bissar-Tadmouri, N., Rochelle, J., Dadali, E. L., Zappia, M., Nelis, E., Patitucci, A., Senderek, J., Parman, Y., Evgrafov, O., Jonghe, P. D., Takahashi, Y., Tsuji, S., Pericak-Vance, M. A., Quattrone, A., Battaloglu, E., Polyakov, A. V., Timmerman, V., Schroder, J. M., and Vance, J. M. (2004) Mutations in the mitochondrial GTPase mitofusin 2 cause Charcot-Marie-Tooth neuropathy type 2A, *Nat Genet 36*, 449–451.

91. Niemann, A., Ruegg, M., La Padula, V., Schenone, A., and Suter, U. (2005) Ganglioside-induced differentiation associated protein 1 is a regulator of the mitochondrial network: new implications for Charcot-Marie-Tooth disease, *J Cell Biol 170*, 1067–1078.

92. Schlame, M., and Ren, M. (2006) Barth syndrome, a human disorder of cardiolipin metabolism, *FEBS Lett 580*, 5450–5455.

93. Houtkooper, R. H., Turkenburg, M., Poll-The, B. T., Karall, D., Perez-Cerda, C., Morrone, A., Malvagia, S., Wanders, R. J.,

Kulik, W., and Vaz, F. M. (2009) The enigmatic role of tafazzin in cardiolipin metabolism, *Biochim Biophys Acta 1788*, 2003–2014.

94. Roesch, K., Curran, S. P., Tranebjaerg, L., and Koehler, C. M. (2002) Human deafness dystonia syndrome is caused by a defect in assembly of the DDP1/TIMM8a-TIMM13 complex, *Hum Mol Genet 11*, 477–486.

95. Rosenberg, M. J., Agarwala, R., Bouffard, G., Davis, J., Fiermonte, G., Hilliard, M. S., Koch, T., Kalikin, L. M., Makalowska, I., Morton, D. H., Petty, E. M., Weber, J. L., Palmieri, F., Kelley, R. I., Schaffer, A. A., and Biesecker, L. G. (2002) Mutant deoxynucleotide carrier is associated with congenital microcephaly, *Nat Genet 32*, 175–179.

96. Siu, V. M., Ratko, S., Prasad, A. N., Prasad, C., and Rupar, C. A. Amish microcephaly: Long-term survival and biochemical characterization, *Am J Med Genet A 152A*, 1747–1751.

97. Spiegel, R., Shaag, A., Edvardson, S., Mandel, H., Stepensky, P., Shalev, S. A., Horovitz, Y., Pines, O., and Elpeleg, O. (2009) SLC25A19 mutation as a cause of neuropathy and bilateral striatal necrosis, *Ann Neurol 66*, 419–424.

98. Mayr, J. A., Merkel, O., Kohlwein, S. D., Gebhardt, B. R., Bohles, H., Fotschl, U., Koch, J., Jaksch, M., Lochmuller, H., Horvath, R., Freisinger, P., and Sperl, W. (2007) Mitochondrial phosphate-carrier deficiency: a novel disorder of oxidative phosphorylation, *Am J Hum Genet 80*, 478–484.

# Chapter 3

# Diagnostic Challenges of Mitochondrial Disorders: Complexities of Two Genomes

## Brett H. Graham

## Abstract

Mitochondrial disorders causing respiratory chain dysfunction comprise a group of genetically and clinically heterogeneous diseases. This heterogeneity reflects both the biochemical complexity of oxidative phosphorylation and the genetic contribution of both the nuclear and mitochondrial genomes to the respiratory chain. Current approaches to diagnose and classify mitochondrial disorders incorporate clinical, biochemical, and histological criteria, as well as DNA-based molecular diagnostic testing. While the identification of pathogenic mutations is generally accepted as definitive, the large number of candidate nuclear genes, the involvement of two genomes, and potential heteroplasmy of pathogenic mitochondrial DNA (mtDNA) frequently complicate successful molecular diagnostic confirmation. The strategy for pursuing a diagnosis derives from the integration of family history, clinical findings, biochemical evaluations, histopathological analyses, neuroradiological results, and the availability of different tissues for analyses. Screening for common point mutations and large deletions in mtDNA is usually the first step. Specific subsets of known nuclear disease genes can be screened by direct sequencing for cases of recognizable patterns of respiratory chain deficiencies or clinically identifiable syndromic presentations. Measurement of mtDNA content in affected tissues such as muscle and liver allows screening for mtDNA depletion syndromes. The growing list of known disease-causing genes and the promise of next generation sequencing technologies will undoubtedly improve diagnostic accuracy and genetic counseling for this challenging group of disorders.

**Key words:** Mitochondria, Respiratory chain disorders, Heteroplasmy, mtDNA, Gene testing

## 1. Introduction

The diagnosis of primary mitochondrial respiratory chain diseases represents a significant clinical challenge for physicians. Primary disorders of oxidative phosphorylation frequently present with dysfunction of multiple organ systems, and this observation may be the initial indication of the underlying diagnosis. However, the marked variable expressivity of clinical presentations, the necessity of invasive tissue biopsies for biochemical and histopathological

Lee-Jun C. Wong (ed.), *Mitochondrial Disorders: Biochemical and Molecular Analysis*, Methods in Molecular Biology, vol. 837, DOI 10.1007/978-1-61779-504-6_3, © Springer Science+Business Media, LLC 2012

analyses, and considerable genetic heterogeneity all contribute to diagnostic uncertainty in many cases. Because of the clinical heterogeneity, any of a number of medical specialists may be the first to encounter these patients, including cardiologists, gastroenterologists, neurologists, and ophthalmologists, and patients frequently experience prolonged delays before the correct diagnosis is reached (1, 2). There are several clinical syndromes with stereotypic features such as Leigh syndrome (subacute necrotizing encephalomyelopathy), MELAS (mitochondrial myopathy, encephalopathy, lactic acidosis, and stroke-like episodes), LHON (Leber Hereditary Optic Neuropathy), and Alpers disease (epilepsy and liver failure). However, patients with mitochondrial disease can often present with only nonspecific features such as developmental delay or regression, further delaying correct diagnoses. Attempts to improve the reliable diagnosis of mitochondrial disorders include the proposal of diagnostic criteria that integrate clinical manifestations, enzymatic and physiologic analyses, tissue histochemical results, levels of biochemical analytes, and DNA analysis (3, 4). Obtaining a definitive and specific molecular diagnosis for a patient with clinically suspected mitochondrial disease is important for multiple reasons: (1) to allow more specific medical management and health surveillance of the patient, (2) to allow diagnostic testing of any at risk relatives, and (3) to allow more specific genetic counseling and recurrence risk estimation for the family. While the diagnosis of mitochondrial disease requires an integrative approach, the identification of a molecular pathogenic defect, when possible, can overcome the ambiguities often seen with biochemical analytes, histological, and enzymological evaluations.

Mitochondrial disease can be caused by mutations in either the mitochondrial or nuclear genome. As a manifestation of the complexity deriving from the involvement of two genomes, the mitochondrial genome is maintained by proteins encoded in the nucleus, giving rise to either primary mitochondrial DNA (mtDNA) mutations, which may exhibit a matrilineal pattern of inheritance, or Mendelian traits that present with multiple somatic mtDNA alterations. Mutations of mtDNA are complicated by the fact that each cell contains hundreds to thousands of copies, and different tissues in the same individual can manifest different proportions or heteroplasmy of mutant mtDNA (2). Furthermore, the penetrance or severity of a primary mtDNA mutation is potentially modified by the nuclear genome background in which it coexists or by environmental factors. In the case of Mendelian disorders, autosomal recessive, autosomal dominant, and X-linked patterns have all been observed.

Mitochondrial disorders are much more common than has been historically believed, but attempts to measure the true incidence of mitochondrial diseases have been limited by disease heterogeneity and the lack of definitive biomarkers. Studies in adults and/or children suggest a minimum prevalence of 1 in 7,600 (5). However, a subsequent meta-analysis suggests that the prevalence of combined mitochondrial diseases is at least 1 in 5,000 (6). Strikingly, a recent study of the frequency of more common mitochondrial DNA mutations in newborns and their mothers suggests the possibility of a much higher number, in part due to an apparently high rate of de novo mutations (7). Importantly, it is currently unclear to what degree these inherited and de novo mtDNA mutations detected in otherwise healthy newborns will lead to disease over the course of a lifetime. While adults that develop signs and symptoms of mitochondrial disease are often found to harbor primary mtDNA mutations, the vast majority of pediatric patients likely exhibit Mendelian disease (1). Since current estimates of the number of nuclear genes that contribute to the biogenesis and appropriate function of mitochondria approaches 1,300 (8), there is a significant need for definitive molecular testing, given that currently clinical analysis is available for less than 60 nuclear genes known to cause mitochondrial disease. A comprehensive discussion of the varied clinical presentations of mitochondrial disease, the overall clinical and biochemical evaluation of patients with suspected mitochondrial disease, and the nuclear genes known to cause mitochondrial disease are beyond the scope of this review but have been recently reviewed elsewhere (1, 2, 9) and in the first two chapters of this volume. This review provides an approach for the use of currently available DNA-based testing in order to aid in establishing a definitive diagnosis of a mitochondrial respiratory chain disorder.

## 2. Clinical Presentation Informs Approach to Molecular Testing

Given the biological complexity of mitochondria, a molecular diagnostic plan for individual patients must be informed by the clinical presentation, family history, and biochemical testing (Figs. 1–3). Based on the author's personal experiences and the relevant literature, 80–95% of patients with clinically suspected primary mitochondrial disease do not have a detectable pathogenic mtDNA mutation. These cases are therefore presumed to harbor mutation(s) in a nuclear-encoded mitochondrial-targeted gene (10–15). Currently, a specific molecular diagnosis is obtained for only a small minority of these patients. However, an increased proportion of successful molecular diagnoses has been observed for certain clinical subsets of patients with suspected mitochondrial disease. For example, mutations in *POLG* are responsible for a heterogeneous group of at least six major clinical phenotypes: (1) childhood myocerebrohepatopathy spectrum disorders (MCHS), (2) Alpers syndrome, (3) ataxia neuropathy spectrum (ANS) disorders including

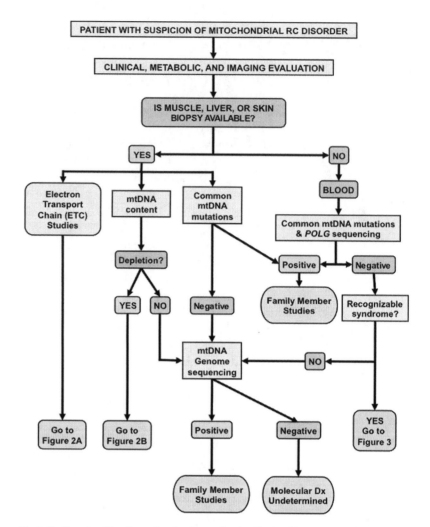

Fig. 1. Testing algorithm for molecular diagnosis of patients with suspected mitochondrial disease. A generalized algorithm for molecular testing based on clinical and biochemical information is presented (Figs. 1–3). This algorithm is designed to be a general guide and is not intended to encompass every potential clinical scenario nor all possible genetic etiologies. If a common mtDNA mutation is not detected in the initial screen, then a tissue biopsy for additional molecular and biochemical studies is required (Fig. 2). If a tissue biopsy is not available, then screening appropriate nuclear genes based on a recognized clinical pattern can be pursued (Fig. 3).

spinocerebellar ataxia with epilepsy (SCAE) and mitochondrial recessive ataxia syndrome without ophthalmoplegia (MIRAS), (4) myoclonus, epilepsy, myopathy, and sensory ataxia (MEMSA), (5) autosomal recessive progressive external ophthalmoplegia (arPEO) including sensory ataxic neuropathy, dysarthria and ophthalmoparesis (SANDO), and (6) autosomal dominant progressive external ophthalmoplegia (adPEO) (16). In one study, 15 patients with PEO were screened for mutations in *POLG*, *TWINKLE*, and *ANT1*. *POLG* mutations were identified in six patients (17).

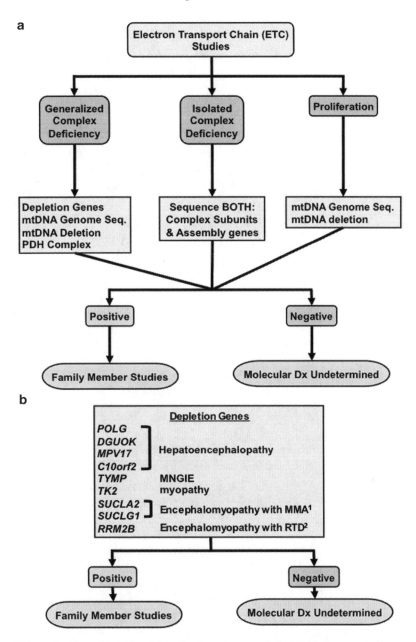

Fig. 2. Algorithms for molecular testing based on respiratory chain enzyme testing or mtDNA quantitation results. (**a**) Gene testing algorithm based on results of electron transport chain testing. *PDH* pyruvate dehygrogenase. (**b**) Gene testing algorithm based on presence of mtDNA depletion. For the genes listed, HUGO gene nomenclature is used. *MNGIE* mitochondrial neurogastrointestinal encephalopathy syndrome; [1]*MMA* methylmalonic aciduria; [2]*RTD* renal tubular dysfunction.

In another series of patients with the above spectrum of clinical features, 61 of 350 individuals had identifiable mutation(s) in *POLG*, making it one of the most frequent nuclear-encoded mitochondrial disease genes to date (16). Given the wide phenotypic

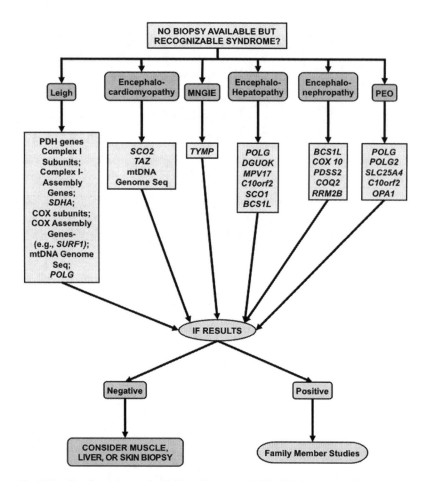

Fig. 3. Algorithm for molecular testing based on recognizable clinical syndrome. For the genes listed, HUGO gene nomenclature is used. *MNGIE* mitochondrial neurogastrointestinal encephalopathy syndrome; *PEO* progressive external ophthalmoplegia syndrome; *PDH* pyruvate dehygrogenase.

spectrum and relatively high mutation detection rate, sequencing *POLG* in patients who present with encephalopathy, seizures, hepatopathy, or PEO should be viewed as a screening test, particularly if mtDNA screening is negative and if there are other diagnostic clues such as multiple deletions or depletion of mtDNA on skeletal muscle and the presence of ragged red fibers in adults (Figs. 2b and 3).

In another example, the detection of *SURF1* mutations in patients with Leigh syndrome and isolated cytochrome c oxidase (COX) deficiency is relatively high (18). Therefore, if the clinical features are consistent with Leigh syndrome with COX deficiency and the family history is consistent with autosomal recessive inheritance, then sequencing of *SURF1* is indicated. In a study of 180 patients with isolated cytochrome c oxidase deficiency, 75 were found to have mutations in either mtDNA or a nuclear gene (19). In addition to cytochrome c oxidase deficiency, abnormalities in complex I, complex II, pyruvate dehydrogenase, or various mtDNA mutations are all possible etiologies for Leigh syndrome. In general, if the patient presents with a family history that suggests matrilineal inheritance and the screening of common mutations is negative, sequence analysis of the whole mitochondrial genome is indicated (Fig. 3).

Screening by quantitative PCR for mtDNA depletion in samples from biopsy of muscle or liver can also be informative (Fig. 2b). In the context of mtDNA depletion with infantile liver dysfunction and/or encephalomyopathy, recent reports have clearly illustrated the success of identifying mutations in nuclear genes required for mtDNA integrity, including *POLG*, *MPV17*, *DGUOK*, *RRM2B*, *C10ORF2*, *SUCLA2*, and *SUCLG1* (16, 20–27). In particular, if the encephalomyopathy is associated with mild methylmalonic aciduria, testing for mutations in the *SUCLA2* and *SUCLG1* genes should be considered (21, 25, 26). If mtDNA depletion in skeletal muscle is associated with a myopathy with elevated creatine kinase that starts in infancy or childhood and follows a slowly progressive clinical course, testing for mutations in the *TK2* gene is warranted (28). In the context of complex I + III and II + III respiratory chain deficiencies with demonstrated skeletal muscle coenzyme $Q_{10}$ deficiency, which can be associated with diverse clinical phenotypes, sequencing of genes involved with coenzyme $Q_{10}$ biosynthesis (*PDSS1*, *PDSS2*, *CoQ2*, *CoQ9*, *CABC1*, or *ADCK3*) should be pursued (29–32). DNA sequencing of *TYMP* is indicated with clinical presentation of mitochondrial neurogastrointestinal encephalomyopathy (MNGIE) and an elevated plasma or urine thymidine level (33). These examples emphasize the importance of obtaining appropriate biochemical and/or clinical information before embarking on molecular testing in order to maximize the chance of successfully obtaining a molecular diagnosis.

Given a 5–20% detection rate of pathogenic mtDNA mutations in pediatric patients with clinically suspected mitochondrial disease (10–14) and a recent study suggesting that the detectable frequency of pathogenic mtDNA mutations in the general population may be as high as 1 in 200 (7), screening for common mtDNA pathogenic mutations in blood is typically the first step because of the ease in obtaining a test sample and wide availability of the test.

If initial mtDNA screening from a blood sample does not establish a molecular diagnosis, then decisions for further molecular diagnostic testing (i.e., whole mtDNA genome sequencing or nuclear gene testing such as *POLG*) should be based on the specific clinical presentation, family history, and results of histological and biochemical testing from appropriate tissues (e.g., skeletal muscle or skin fibroblasts), as described in the next Subheading. There are currently a number of clinical diagnostic laboratories that offer molecular testing of mtDNA and various subsets of nuclear-encoded mitochondrial genes (http://genetests.org).

## 3. Recommendations for Molecular Evaluation of a Patient with Suspected Mitochondrial Disease (Figs. 1–3)

As briefly described above, even though mitochondrial diseases are clinically heterogeneous, specific clinical presentations suggest certain subsets of disease genes for molecular testing. Therefore, in accordance with good medical practice, the evaluation of a patient with suspected mitochondrial disease should begin with a thorough and accurate clinical evaluation. The evaluation should include age of onset, a detailed family history with an emphasis on potential matrilineal inheritance or recessive inheritance, and a comprehensive review of symptoms with particular attention paid to signs and symptoms suggestive of mitochondrial disease (9). In addition, noninvasive evaluations including brain MRI, brain MR spectroscopy, EKG, EEG, brain stem auditory evoked responses (BAERS), and visually evoked responses (VERS) should be performed as clinically appropriate.

The clinical evaluation should also include metabolic screening with measurement of basic blood and urine analytes such as serum transaminases, lactate and pyruvate (if collected appropriately), plasma amino acids, urine organic acids, plasma acylcarnitine profile, and creatine kinase. If available, CSF lactate, pyruvate, and protein concentration should also be measured. It is useful to calculate a lactate-to-pyruvate ratio because an elevated ratio suggests the presence of oxidative phosphorylation dysfunction (2).

Based on the above clinical, metabolic, and imaging studies, if a classical mitochondrial syndrome caused by a mitochondrial DNA (mtDNA) mutation is suspected, screening for common mitochondrial DNA (mtDNA) point mutations and mtDNA deletions should be performed in blood or urine sediment. If maternally inherited sensorineural hearing loss due to aminoglycoside ototoxicity is suspected, then testing for the m.1555A>G mutation in the mitochondrial 12S ribosomal RNA gene may be considered (34).

Since *POLG* appears to be the most frequently mutated nuclear gene in cases of mitochondrial disease (in particular for patients with nonspecific hypotonia, developmental delay, intractable epilepsy, or progressive liver disease triggered by infection in an otherwise developmentally normal child) (16, 22), sequence analysis of *POLG* should be considered in cases of normal mtDNA screening results. In addition, young adults and adults with ataxia, neuropathy, or muscle weakness and normal mtDNA testing may also be candidates for sequence analysis of *POLG* if their clinical features fit within the six major clinical phenotypes mentioned above in Subheading 2.

If screening for mtDNA deletions and common point mutations and *POLG* sequencing are normal, specific nuclear gene or genes can be selected for DNA sequencing based on the results of clinical and metabolic evaluations (Fig. 3). If autosomal recessive mutations in a nuclear disease gene are suspected and sequencing identifies only a single heterozygous pathogenic mutation, then an intragenic deletion of the other allele should be ruled out. Intragenic deletions are effectively detected with the use of clinically available targeted oligonucleotide arrays for comparative genomic hybridization (35).

In most cases, analysis of respiratory chain activities from skeletal muscle biopsy or skin biopsy-derived fibroblasts is important in refining the strategy for molecular testing (Fig. 2a). As an example, if a child would present with Leigh syndrome and a family history suggestive of autosomal recessive inheritance, it would be informative to measure cytochrome c oxidase (COX) in fibroblasts before pursuing DNA testing, since normal COX activity would reliably exclude *SURF1* deficiency as the basis for the disease. Moreover, if nuclear gene analyses in blood do not yield meaningful results, a tissue-specific analysis (biopsy of muscle, liver, or skin) should be considered as the clinically appropriate next step for histological, biochemical, and molecular analyses. The specific clinical presentation and the results from histochemical/electron microscopy structural analysis, respiratory chain enzyme assays, and mtDNA copy number (depletion, over-amplification) analysis should be used to dictate the selection of appropriate molecular tests (Fig. 1). The presence of either generalized respiratory chain deficiency or isolated deficiency of a specific respiratory chain complex can point to an mtDNA depletion disorder, an mtDNA tRNA gene mutation, or a specific complex subunit or assembly factor gene defect (9). As described in Subheading 2, a number of genes are known to cause mtDNA depletion in muscle and/or liver, and quantification of mtDNA copy number by quantitative PCR is a reliable screening test for this group of gene defects (36). A significant increase in mtDNA copy number suggests the presence of mitochondrial proliferation, and whole mitochondrial genome sequence analysis should be considered, in particular, if there is a family history suggestive of matrilineal inheritance (37). When mtDNA mutations are considered, sequencing mtDNA derived from the affected tissue(s), if available, will provide the highest probability of detecting a pathogenic mutation, since, in some cases, pathogenic mtDNA mutations may give low or undetectable heteroplasmy in blood (37, 38).

## 4. Conclusions

Establishing a specific diagnosis in a patient with suspected mitochondrial disease is a challenging endeavor that requires the integration of clinical assessments, family history, biochemical testing, histopathological examination, and directed molecular testing. Close collaboration between primary clinicians, geneticists, pathologists, clinical subspecialists, and diagnostic laboratories with expertise in mitochondrial biochemical and molecular testing is critical to maximize the likelihood of obtaining a correct diagnosis. With the list of nuclear-encoded genes documented to cause mitochondrial disease ever-expanding, the complexity of potential molecular testing increases, but so does the prospect of identifying molecular diagnoses for patients with mitochondrial disease, especially with the promise of incorporating next-generation sequencing technologies into clinical practice.

## References

1. Haas, R. H., Parikh, S., Falk, M. J., Saneto, R. P., Wolf, N. I., Darin, N., and Cohen, B. H. (2007) Mitochondrial disease: a practical approach for primary care physicians, *Pediatrics 120*, 1326–1333.

2. Haas, R. H., Parikh, S., Falk, M. J., Saneto, R. P., Wolf, N. I., Darin, N., Wong, L. J., Cohen, B. H., and Naviaux, R. K. (2008) The in-depth evaluation of suspected mitochondrial disease, *Mol Genet Metab 94*, 16–37.

3. Bernier, F. P., Boneh, A., Dennett, X., Chow, C. W., Cleary, M. A., and Thorburn, D. R. (2002) Diagnostic criteria for respiratory chain disorders in adults and children, *Neurology 59*, 1406–1411.

4. Wolf, N. I., and Smeitink, J. A. (2002) Mitochondrial disorders: a proposal for consensus diagnostic criteria in infants and children, *Neurology 59*, 1402–1405.

5. Skladal, D., Halliday, J., and Thorburn, D. R. (2003) Minimum birth prevalence of mitochondrial respiratory chain disorders in children, *Brain 126*, 1905–1912.

6. Schaefer, A. M., Taylor, R. W., Turnbull, D. M., and Chinnery, P. F. (2004) The epidemiology of mitochondrial disorders--past, present and future, *Biochim Biophys Acta 1659*, 115–120.

7. Elliott, H. R., Samuels, D. C., Eden, J. A., Relton, C. L., and Chinnery, P. F. (2008) Pathogenic mitochondrial DNA mutations are common in the general population, *Am J Hum Genet 83*, 254–260.

8. Pagliarini, D. J., Calvo, S. E., Chang, B., Sheth, S. A., Vafai, S. B., Ong, S. E., Walford, G. A., Sugiana, C., Boneh, A., Chen, W. K., Hill, D. E., Vidal, M., Evans, J. G., Thorburn, D. R., Carr, S. A., and Mootha, V. K. (2008) A mitochondrial protein compendium elucidates complex I disease biology, *Cell 134*, 112–123.

9. Wong, L. J., Scaglia, F., Graham, B. H., and Craigen, W. J. (2010) Current molecular diagnostic algorithm for mitochondrial disorders, *Mol Genet Metab 100*, 111–117.

10. Scaglia, F., Towbin, J. A., Craigen, W. J., Belmont, J. W., Smith, E. O., Neish, S. R., Ware, S. M., Hunter, J. V., Fernbach, S. D., Vladutiu, G. D., Wong, L. J., and Vogel, H. (2004) Clinical spectrum, morbidity, and mortality in 113 pediatric patients with mitochondrial disease, *Pediatrics 114*, 925–931.

11. Jaksch, M., Hofmann, S., Kleinle, S., Liechti-Gallati, S., Pongratz, D. E., Muller-Hocker, J., Jedele, K. B., Meitinger, T., and Gerbitz, K. D. (1998) A systematic mutation screen of 10 nuclear and 25 mitochondrial candidate genes in 21 patients with cytochrome c oxidase (COX) deficiency shows tRNA(Ser) (UCN) mutations in a subgroup with syndromal encephalopathy, *J Med Genet 35*, 895–900.

12. Jaksch, M., Kleinle, S., Scharfe, C., Klopstock, T., Pongratz, D., Muller-Hocker, J., Gerbitz, K. D., Liechti-Gallati, S., Lochmuller, H., and Horvath, R. (2001) Frequency of mitochondrial transfer RNA mutations and deletions in 225 patients presenting with respiratory chain deficiencies, *J Med Genet 38*, 665–673.

13. Liang, M. H., and Wong, L. J. (1998) Yield of mtDNA mutation analysis in 2,000 patients, *Am J Med Genet 77*, 395–400.

14. Marotta, R., Chin, J., Quigley, A., Katsabanis, S., Kapsa, R., Byrne, E., and Collins, S. (2004) Diagnostic screening of mitochondrial DNA mutations in Australian adults 1990–2001, *Intern Med J 34*, 10–19.

15. Schaefer, A. M., McFarland, R., Blakely, E. L., He, L., Whittaker, R. G., Taylor, R. W., Chinnery, P. F., and Turnbull, D. M. (2008)

Prevalence of mitochondrial DNA disease in adults, *Ann Neurol 63*, 35–39.

16. Wong, L. J., Naviaux, R. K., Brunetti-Pierri, N., Zhang, Q., Schmitt, E. S., Truong, C., Milone, M., Cohen, B. H., Wical, B., Ganesh, J., Basinger, A. A., Burton, B. K., Swoboda, K., Gilbert, D. L., Vanderver, A., Saneto, R. P., Maranda, B., Arnold, G., Abdenur, J. E., Waters, P. J., and Copeland, W. C. (2008) Molecular and clinical genetics of mitochondrial diseases due to POLG mutations, *Hum Mutat 29*, E150-172.

17. Naimi, M., Bannwarth, S., Procaccio, V., Pouget, J., Desnuelle, C., Pellissier, J. F., Rotig, A., Munnich, A., Calvas, P., Richelme, C., Jonveaux, P., Castelnovo, G., Simon, M., Clanet, M., Wallace, D., and Paquis-Flucklinger, V. (2006) Molecular analysis of ANT1, TWINKLE and POLG in patients with multiple deletions or depletion of mitochondrial DNA by a dHPLC-based assay, *Eur J Hum Genet 14*, 917–922.

18. Shoubridge, E. A. (2001) Cytochrome c oxidase deficiency, *Am J Med Genet 106*, 46–52.

19. Bohm, M., Pronicka, E., Karczmarewicz, E., Pronicki, M., Piekutowska-Abramczuk, D., Sykut-Cegielska, J., Mierzewska, H., Hansikova, H., Vesela, K., Tesarova, M., Houstkova, H., Houstek, J., and Zeman, J. (2006) Retrospective, multicentric study of 180 children with cytochrome C oxidase deficiency, *Pediatr Res 59*, 21–26.

20. Bourdon, A., Minai, L., Serre, V., Jais, J. P., Sarzi, E., Aubert, S., Chretien, D., de Lonlay, P., Paquis-Flucklinger, V., Arakawa, H., Nakamura, Y., Munnich, A., and Rotig, A. (2007) Mutation of RRM2B, encoding p53-controlled ribonucleotide reductase (p53R2), causes severe mitochondrial DNA depletion, *Nat Genet 39*, 776–780.

21. Carrozzo, R., Dionisi-Vici, C., Steuerwald, U., Lucioli, S., Deodato, F., Di Giandomenico, S., Bertini, E., Franke, B., Kluijtmans, L. A., Meschini, M. C., Rizzo, C., Piemonte, F., Rodenburg, R., Santer, R., Santorelli, F. M., van Rooij, A., Vermunt-de Koning, D., Morava, E., and Wevers, R. A. (2007) SUCLA2 mutations are associated with mild methylmalonic aciduria, Leigh-like encephalomyopathy, dystonia and deafness, *Brain 130*, 862–874.

22. Dimmock, D. P., Zhang, Q., Dionisi-Vici, C., Carrozzo, R., Shieh, J., Tang, L. Y., Truong, C., Schmitt, E., Sifry-Platt, M., Lucioli, S., Santorelli, F. M., Ficicioglu, C. H., Rodriguez, M., Wierenga, K., Enns, G. M., Longo, N., Lipson, M. H., Vallance, H., Craigen, W. J., Scaglia, F., and Wong, L. J. (2008) Clinical and molecular features of mitochondrial DNA depletion due to mutations in deoxyguanosine kinase, *Hum Mutat 29*, 330–331.

23. Elpeleg, O., Miller, C., Hershkovitz, E., Bitner-Glindzicz, M., Bondi-Rubinstein, G., Rahman, S., Pagnamenta, A., Eshhar, S., and Saada, A. (2005) Deficiency of the ADP-forming succinyl-CoA synthase activity is associated with encephalomyopathy and mitochondrial DNA depletion, *Am J Hum Genet 76*, 1081–1086.

24. Oskoui, M., Davidzon, G., Pascual, J., Erazo, R., Gurgel-Giannetti, J., Krishna, S., Bonilla, E., De Vivo, D. C., Shanske, S., and DiMauro, S. (2006) Clinical spectrum of mitochondrial DNA depletion due to mutations in the thymidine kinase 2 gene, *Arch Neurol 63*, 1122–1126.

25. Ostergaard, E., Christensen, E., Kristensen, E., Mogensen, B., Duno, M., Shoubridge, E. A., and Wibrand, F. (2007) Deficiency of the alpha subunit of succinate-coenzyme A ligase causes fatal infantile lactic acidosis with mitochondrial DNA depletion, *Am J Hum Genet 81*, 383–387.

26. Ostergaard, E., Hansen, F. J., Sorensen, N., Duno, M., Vissing, J., Larsen, P. L., Faeroe, O., Thorgrimsson, S., Wibrand, F., Christensen, E., and Schwartz, M. (2007) Mitochondrial encephalomyopathy with elevated methylmalonic acid is caused by SUCLA2 mutations, *Brain 130*, 853–861.

27. Wong, L. J., Brunetti-Pierri, N., Zhang, Q., Yazigi, N., Bove, K. E., Dahms, B. B., Puchowicz, M. A., Gonzalez-Gomez, I., Schmitt, E. S., Truong, C. K., Hoppel, C. L., Chou, P. C., Wang, J., Baldwin, E. E., Adams, D., Leslie, N., Boles, R. G., Kerr, D. S., and Craigen, W. J. (2007) Mutations in the MPV17 gene are responsible for rapidly progressive liver failure in infancy, *Hepatology 46*, 1218–1227.

28. Saada, A., Shaag, A., Mandel, H., Nevo, Y., Eriksson, S., and Elpeleg, O. (2001) Mutant mitochondrial thymidine kinase in mitochondrial DNA depletion myopathy, *Nat Genet 29*, 342–344.

29. Duncan, A. J., Bitner-Glindzicz, M., Meunier, B., Costello, H., Hargreaves, I. P., Lopez, L. C., Hirano, M., Quinzii, C. M., Sadowski, M. I., Hardy, J., Singleton, A., Clayton, P. T., and Rahman, S. (2009) A nonsense mutation in COQ9 causes autosomal-recessive neonatal-onset primary coenzyme Q10 deficiency: a potentially treatable form of mitochondrial disease, *Am J Hum Genet 84*, 558–566.

30. Lagier-Tourenne, C., Tazir, M., Lopez, L. C., Quinzii, C. M., Assoum, M., Drouot, N., Busso, C., Makri, S., Ali-Pacha, L., Benhassine, T., Anheim, M., Lynch, D. R., Thibault, C., Plewniak, F., Bianchetti, L., Tranchant, C.,

Poch, O., DiMauro, S., Mandel, J. L., Barros, M. H., Hirano, M., and Koenig, M. (2008) ADCK3, an ancestral kinase, is mutated in a form of recessive ataxia associated with coenzyme Q10 deficiency, *Am J Hum Genet 82*, 661–672.

31. Mollet, J., Delahodde, A., Serre, V., Chretien, D., Schlemmer, D., Lombes, A., Boddaert, N., Desguerre, I., de Lonlay, P., de Baulny, H. O., Munnich, A., and Rotig, A. (2008) CABC1 gene mutations cause ubiquinone deficiency with cerebellar ataxia and seizures, *Am J Hum Genet 82*, 623–630.

32. Quinzii, C. M., and Hirano, M. (2010) Coenzyme Q and mitochondrial disease, *Dev Disabil Res Rev 16*, 183–188.

33. Nishino, I., Spinazzola, A., and Hirano, M. (1999) Thymidine phosphorylase gene mutations in MNGIE, a human mitochondrial disorder, *Science 283*, 689–692.

34. Guan, M. X., Fischel-Ghodsian, N., and Attardi, G. (1996) Biochemical evidence for nuclear gene involvement in phenotype of non-syndromic deafness associated with mitochondrial 12S rRNA mutation, *Hum Mol Genet 5*, 963–971.

35. Wong, L. J., Dimmock, D., Geraghty, M. T., Quan, R., Lichter-Konecki, U., Wang, J., Brundage, E. K., Scaglia, F., and Chinault, A. C. (2008) Utility of oligonucleotide array-based comparative genomic hybridization for detection of target gene deletions, *Clin Chem 54*, 1141–1148.

36. Dimmock, D., Tang, L. Y., Schmitt, E. S., and Wong, L. J. (2010) Quantitative evaluation of the mitochondrial DNA depletion syndrome, *Clin Chem 56*, 1119–1127.

37. Wong, L. J., Perng, C. L., Hsu, C. H., Bai, R. K., Schelley, S., Vladutiu, G. D., Vogel, H., and Enns, G. M. (2003) Compensatory amplification of mtDNA in a patient with a novel deletion/duplication and high mutant load, *J Med Genet 40*, e125.

38. Wong, L. J., Wladyka, C., and Mardach-Verdon, R. (2004) A mitochondrial DNA mutation in a patient with an extensive family history of Duchenne muscular dystrophy, *Muscle Nerve 30*, 118–122.

# Part II

## Biochemical Analysis of Mitochondrial Disorders

# Chapter 4

# Biochemical Analyses of the Electron Transport Chain Complexes by Spectrophotometry

## Ann E. Frazier and David R. Thorburn

## Abstract

In the diagnostic work-up of patients with suspected mitochondrial disease, evaluating the activity of the individual oxidative phosphorylation (OXPHOS) complexes is crucial. Here, we describe spectrophotometric assays for OXPHOS enzymology that can be applied to both tissue samples and cultured cells. These assays are designed to assess the enzymatic activity of the individual OXPHOS complexes I–V, along with the Krebs cycle enzyme citrate synthase as a mitochondrial control. As well, we include an assay for the coupled energy transfer between complexes II and III. Determining the enzymatic activities can be valuable in defining isolated or multicomplex disorders and may be relevant to the design of future molecular investigations.

**Key words:** Mitochondrial disease, Electron transport chain, Respiratory complex, Enzyme assays, OXPHOS, Spectrophotometry

## 1. Introduction

Disorders of the mitochondrial oxidative phosphorylation (OXPHOS) complexes can primarily affect one tissue/organ or cause multisystemic disorders, all with an onset at any age (1–3). In particular, they frequently affect skeletal and cardiac muscle, and the central nervous system, composed of cells and tissues with a high energy demand (4, 5). These OXPHOS diseases are predicted to affect up to 1 in 5,000 people (6), making them the most common cause of inborn errors of metabolism. They can be caused by mutations in either the mitochondrial (mtDNA) or nuclear genomes. Thus far, mutations in over 100 genes have been implicated in OXPHOS disease, complicating searches for the molecular basis of the disorders (7–10).

The contribution of the OXPHOS complexes to mitochondrial disease varies, as the disorders can either result from an isolated

Lee-Jun C. Wong (ed.), *Mitochondrial Disorders: Biochemical and Molecular Analysis*, Methods in Molecular Biology, vol. 837, DOI 10.1007/978-1-61779-504-6_4, © Springer Science+Business Media, LLC 2012

complex deficiency or defects in multiple complexes. For this reason, individual analysis of OXPHOS complex activity is an important diagnostic assessment of a patient suspected of having a mitochondrial disorder. As well, determining tissue or OXPHOS complex specificity can be important in directing molecular investigations (7, 10). For instance, mitochondrial disorders affecting several complexes often indicate a mutation in genes that affect mtDNA maintenance, transcription, translation, or nucleotide regulation, such as the mitochondrial tRNA genes, mtDNA polymerase γ1 (*POLG*), or deoxyguanosine kinase (*DGUOK*) (3, 11) (Fig. 1a). Alternatively, defects in an isolated complex may indicate a mutation in a specific complex subunit or assembly factor. For example, a patient with a mutation in *FOXRED1* (Patient DT22 in (12)), encoding an assembly factor of complex I (NADH-ubiquinone oxidoreductase), displays an isolated complex I deficiency that can be detected in both muscle as well as cultivated skin fibroblasts (Fig. 1b).

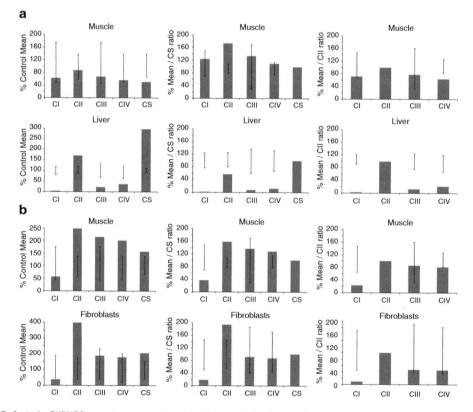

Fig. 1. Defects in OXPHOS complexes may be systemic or affect only specific tissues. They may affect multiple OXPHOS complexes or an isolated complex. Due to mitochondrial proliferation or cell/tissue variability, comparing OXPHOS complex activity to citrate synthase or complex II activity may help in distinguishing defects. (**a**) Patient A, with mutations in *DGUOK* (unpublished), demonstrates liver specific defects in complex I, III, and IV while muscle looks normal. (**b**) Patient B, carrying mutations in the complex I assembly factor *FOXRED1* (12), shows an isolated complex I defect in both muscle and fibroblasts. In muscle, the complex I activity falls within control ranges when expressed as % of control mean, but not when expressed as the ratio of complex I to citrate synthase and complex II. Activity and ratios are expressed as % of the control mean value. *Vertical bars* indicate control ranges.

There is no one correct way to measure the activity of the OXPHOS complexes, and many other protocols are available (13–18). Here, we describe the spectrophotometric assays that we routinely use for the diagnosis of mitochondrial disease in patient tissues and cell lines. In general, these assays are designed to monitor the individual activity of the OXPHOS complexes by following the oxidation/reduction of various substrates or substrate analogues (Fig. 2). An exception is the assay for complex V (ATP synthase) activity, which monitors enzyme function in the reverse. We also include an assay for measuring the coupled transfer of electrons from complex II to complex III, which can be helpful in detecting a deficiency in Coenzyme $Q_{10}$.

**a  Citrate Synthase:**

    *a.* oxaloacetate + acetyl CoA ⟶ citrate + CoA•SH

    *b.* CoA•SH + DTNB ⟶ CoA•S-S-nitrobenzoate + | thionitrobenzoate anion |

**b  Complex I:**

    | NADH | + $H^+$ + $CoQ_1$ ⟶ $NAD^+$ + $CoQ_1$•$H_2$

**c  Complex II:**

    succinate + | $CoQ_1$ | ⟶ fumarate + $CoQ_1$•$H_2$

**d  Complex II + III:**

    succinate + cyt. *c* (oxidized) ⟶ fumarate + | cyt. *c* (reduced) |

**e  Complex III:**

    DB•$H_2$ + cyt. *c* (oxidized) ⟶ DB + | cyt. *c* (reduced) |

**f  Complex IV:**

    | cyt. *c* (reduced) | + $O_2$ ⟶ cyt. *c* (oxidized) + $H_2O$

**g  Complex V:**

    *a.* Mg•ATP $\xrightarrow{ATPase}$ Mg•ADP + $P_i$

    *b.* Mg•ADP + PEP $\xrightarrow{PK}$ Mg•ATP + pyruvate

    *c.* pyruvate + | NADH | + $H^+$ $\xrightarrow{LDH}$ lactate + $NAD^+$

Fig. 2. The enzyme complexes are measured by kinetic spectrophotometric assays. These analyse the change in absorbance at specific wavelengths, indicating the appearance (**a, d, e**) or disappearance (**b, c, f, g**) of specific compounds (indicated by a *box*). The assay for citrate synthase activity measures the generation of free sulfhydryl groups which react with 5,5′-dithio-bis-(2-nitrobenzoic acid) (DTNB) (**a**), while the OXPHOS complexes I–IV are assayed for the transfer of electrons from one intermediate to the next (**b–f**). The activity of complex V is measured in the reverse, linking ATP hydrolysis to the oxidation of NADH using pyruvate kinase (PK) and lactate dehydrogenase (LDH) (**g**). The absorbance of the thionitrobenzoate anion is measured at 412 nm (**a**), NADH at 340 nm (**b, g**), coenzyme $Q_1$ ($CoQ_1$) at 280 nm (**c**), and reduced cytochrome *c* at 550 nm (**d–f**). $CoQ_1$ and DB (decyl-benzylquinone) are soluble analogues of the endogenous hydrophobic $CoQ_{10}$ (**b, c, e**).

Due to variability in collected tissue and isolated cells, we find it helpful to look at both the enzyme activities on their own and in comparison to other mitochondrial enzymes. In our measurements, we use the Krebs tricarboxylic acid (TCA) cycle enzyme citrate synthase (CS), located in the mitochondrial matrix, as a control for mitochondrial OXPHOS complex activity. By expressing the activity of the OXPHOS complexes as a ratio to citrate synthase activity, one can avoid assay variability that may result from differing cell culture conditions (i.e. cell passage number or cell confluence) or mitochondrial proliferation, which may occur in some patients with mitochondrial disease (see Fig. 1, middle panels). Analysing the ratio to complex II activity can also be informative, as it can be indicative of sample integrity since it appears to be the most labile OXPHOS complex in post-mortem samples (19). As well, it is the only OXPHOS complex lacking mtDNA encoded subunits; therefore, its activity should be normal or high in patients carrying mutations in mtDNA (see Fig. 1, right panels).

The assays we describe here are routinely performed on cultured cells (fibroblasts, lymphoblasts, and chorionic villus sampling cells) and frozen tissue samples (skeletal muscle, liver, and heart) since the logistics are prohibitive for obtaining fresh samples from many referred patients. While it is generally thought that fresh tissue samples are ideal in investigating a patient with a suspected OXPHOS defect (20), we find that the activities of OXPHOS complexes I, II, III, and IV and citrate synthase are stable when stored for extended periods at $-70°C$ (19). However, complex V cannot be assayed reliably in frozen tissues (13); therefore, we do not routinely perform the assay. Testing of tissues other than skeletal muscle, such as liver and heart, can be informative as mitochondrial OXPHOS defects are often limited to one or several tissues. For instance, Patient A, carrying mutations in *DGUOK* (unpublished), shows liver-specific defects in complex I, III, and IV that are not observed in muscle (Fig. 1a). And while OXPHOS enzyme defects can only be detected in cultured cells around 50% of the time (13, 21, 22), they can be useful in confirming a defect found in other tissues or in cell biology based assays. Ultimately, results from these assays fulfil a critical part of the diagnostic criteria required for providing a diagnosis of mitochondrial disease. Numerous schemes for providing these diagnoses have been established, which encompass the enzymatic activities, clinical features, laboratory markers, and molecular analyses in their criteria (23–25).

## 2. Materials

All reagents are analytical grade and, unless stated otherwise, are made up in deionised water.

**2.1. Materials and Equipment Required for Sample Preparation**

1. Sonicator with microtip.

2. 1 ml glass–glass homogeniser with cylindrical section clearance of 0.1–0.15 mm.

3. 1 ml smooth surface glass homogeniser with a motor-driven Teflon plunger.

4. Tissue lysis buffer: 5 mM HEPES, 1 mM ethylene glycol-bis-($\beta$-aminoethyl ether)-$N,N,N',N'$-tetraacetic acid (EGTA), 210 mM mannitol, 70 mM sucrose, pH 7.2 (with KOH).

5. MegaFB buffer: 2 mM HEPES, 0.1 mM EGTA, 250 mM sucrose, pH 7.4 (with KOH).

6. Dulbecco's phosphate buffered saline (PBS): 136.89 mM $NaCl_2$, 2.68 mM KCl, 8.1 mM $Na_2HPO_4$, 1.47 mM $KH_2PO_4$, pH 7.3.

7. Hypotonic buffer: 25 mM potassium phosphate buffer (KPi), pH 7.2, 5 mM $MgCl_2$.

8. 5 mg/ml digitonin.

**2.2. Equipment and Reagents for Enzymatic Assays**

1. Spectrophotometer with temperature controller (see Note 1).

2. Two sets of six matched quartz cuvettes (semimicro, 2-mm-wide precision optical cells, 10 mm path length) (see Note 2).

3. Citrate Synthase (CS) assay buffer: 50 mM KPi, pH 7.4, 0.1 mM 5,5'-dithio-bis-(2-nitrobenzoic acid) (DTNB). The buffer is prepared fresh from stock solutions (see Note 3).

4. 10 mM acetyl CoA, trilithium salt, freshly prepared.

5. 10 mM oxaloacetic acid, pH 7.2 (with 2.0 M $KHCO_3$), freshly prepared.

6. Complex I assay buffer: 50 mM KPi (pH 7.4), 50 μM nicotinamide adenine dinucleotide (NADH), 1 mM KCN, 10 μM antimycin A, 0.1% (w/v) BSA (fraction V, fatty acid free), 50 μM coenzymeQ$_1$ (CoQ$_1$). The buffer is prepared fresh from stock solutions (see Note 3).

7. Ethanol (A.R. Grade, 100%, anhydrous).

8. 0.25 mM rotenone (in ethanol).

9. Complex II assay buffer: 50 mM KPi, pH 7.4, 10 mM sodium succinate, 1 mM KCN, 10 μM antimycin A, 2.5 μM rotenone. The buffer is prepared fresh from stock solutions (see Note 3).

10. Complex II + III assay buffer: 50 mM KPi, pH 7.4, 10 mM succinate, 1 mM KCN, 2.5 μM rotenone, 0.1% (w/v) BSA (fraction V, fatty acid-free), and 0.075% ethylene diamine tetraacetic acid (EDTA), pH 7.0. The buffer is prepared fresh from stock solutions (see Note 3).

11. 0.1 M ATP, pH 7.2 (with KOH).

12. 2 mM cytochrome *c* from horse heart (type VI).

13. Complex III assay buffer: 50 mM KPi, pH 7.4, 1 mM $n$-dode-cylmaltoside, 1 mM KCN, 2.5 μM rotenone, and 0.1% (w/v) BSA (fraction V, fatty acid free). The buffer is prepared fresh from stock solutions (see Note 3).

14. 10 mM decylbenzylquinol (reduced DB or DB·H$_2$), freshly prepared in ethanol (see Note 4).

15. L-ascorbic acid (solid).

16. 50 mM potassium phosphate buffer (KPi), pH 7.4.

17. Reduced cytochrome $c$ (see Note 5).

18. 0.1 M K$_3$Fe(CN)$_6$.

19. Complex V assay buffer: 40 mM Tris–HCO$_3$/1 mM EGTA, pH 8.0 (see Note 6), 0.2 mM NADH, 2.5 mM phosphoe-nolpyruvate (PEP), 0.5 μM antimycin A, 5 mM MgCl$_2$, 50 μg/ml lactate dehydrogenase (LDH), and 50 μg/ml pyruvate kinase (PK). The buffer is prepared fresh from stock solutions (see Note 3).

## 3. Methods

In these spectrophotometric assays, the reaction buffers are allowed to equilibrate to 30°C in the spectrophotometer cell changer for 5–10 min, and the reaction is then started by the addition of sample or another component. The reaction mixtures are diluted from the stock solutions such that the final assay volume (0.5 or 1 ml) accounts for the sample and other subsequent additions. For each round of enzyme assays performed, we typically include a normal tissue sample or cell line, which is then compared to established reference ranges to ensure it complies within specified limits (the mean control values should fall within two standard deviations of the reference range). For each sample, we try to assay at least two different sample volumes if enough sample is available. Spectrophotometric traces are generally collected for 2–5 min in order to collect data within a linear range. At the end of the assay, the protein concentration is calculated for each sample and used for calculation of the enzyme rates.

### 3.1. Enriched Mitochondrial Preparation from Cultured Cells

All procedures should be done on ice and reagents chilled prior to use. We typically perform the assays using cultured skin fibroblasts from patients, although other cell types can be used. The preparation below uses fibroblasts harvested from four confluent 175-cm² flasks. Following harvesting and washing in PBS, the cell pellets can be stored at –70°C until preparation of mitochondria. Enriched cell mitochondria are typically assayed for OXPHOS complex activity in 0.5 ml volumes, using 10–40 μl sample per assay (see Note 7).

1. Resuspend cells in 1 ml MegaFb buffer and then transfer to a prechilled glass homogeniser with Teflon plunger. Disrupt cells with 20 strokes of the motor-driven plunger set at 1,000 rpm.

2. Transfer homogenate to an Eppendorf tube and spin at 600 RCF in a chilled centrifuge for 10 min with soft acceleration/braking.

3. Set supernatant aside, then repeat steps 1 and 2 with the pellet using 0.8 ml MegaFb buffer.

4. Combine supernatants from steps 2 and 3 and then spin in a refrigerated tabletop centrifuge at 14,400 RCF for 10 min with soft acceleration/braking.

5. Resuspend the pellet containing enriched mitochondria in 400 μl MegaFb buffer. Retain 60–75 μl for CII + III and CIII assays.

6. The remaining sample from step 5 is treated hypotonically for use in CI, CII, CIV, CV, and CS assays by re-centrifuging sample as in step 4, then resuspending the pellet in 1 ml hypotonic buffer.

7. Centrifuge sample in hypotonic buffer again as in step 4, then resuspend pellet in 300 μl hypotonic buffer.

8. Treat samples from steps 5 and 7 to three freeze/thaw cycles in a dry ice/ethanol slurry.

## 3.2. Tissue Homogenate Preparation

All tissue specimens are received and stored at below –50°C prior to preparation of tissue homogenates, although fresh samples can also be used. The minimum amount of tissue required is 20 mg for skeletal muscle and 10 mg for either cardiac muscle or liver, with all samples kept at 0–10°C throughout the procedure. Tissue homogenates are typically assayed in 1 ml volumes, using 10–40 μl sample per assay (see Note 7).

1. Remove fat and connective tissue from the sample and dice finely with a scalpel.

2. Transfer sample to a chilled glass–glass homogeniser, add tissue lysis buffer to 100 mg (wet weight)/ml, then homogenise with ten strokes until sample is dispersed (see Note 8). Transfer solubilised tissue to a chilled Eppendorf tube.

3. Centrifuge the homogenate at 600 RCF for 10 min with soft acceleration/braking in a refrigerated tabletop centrifuge and then transfer the supernatant to a new tube (see Note 9).

4. Freeze/thaw the homogenate supernatants two times in a dry ice/ethanol slurry and then split them into tubes for batch analysis with the different assays and store at –70°C. Sample aliquots used for CS and CI assays are sonicated on ice just prior to use with 5 × 6 pulses (25% amplitude, 0.3 s on and 0.7 s off).

### 3.3. Citrate Synthase Assay

This reaction gauges the rate of free sulfhydryl group (CoA·SH) production using the thiol reagent 5,5-dithio-bis-(2-nitrobenzoic acid) (DTNB), which reacts with sulfhydryls to produce a free thionitrobenzoate anion that is measured at 412 nm (see Fig. 2a) (26).

1. Prepare CS assay buffer in the cuvettes and equilibrate to 30°C.

2. In the following order, add the sample to the cuvettes, then add acetyl CoA to 0.1 mM, and then start the reaction with the addition of oxaloacetic acid to 0.1 mM. Mix the cuvette and obtain linear reaction rates for ~3 min.

3. A blank reaction containing sample but no oxaloacetic acid should be included with each cuvette run (see Note 10).

4. Calculate results with an extinction coefficient for the thionitrobenzoate anion of 13.6/mM/cm (see Subheading 3.10, step 1).

### 3.4. Complex I Assay

The activity assay for complex I (NADH: $CoQ_1$ oxidoreductase) measures electron transfer from NADH to Coenzyme $Q_1$ ($CoQ_1$), a short chain analogue of $CoQ_{10}$ (also known as ubiquinone) that is more water soluble (see Fig. 2b). The absorbance of each sample is measured at 340 nm with and without the complex I inhibitor rotenone to account for rotenone-insensitive NADH oxidation.

1. For each sample, prepare two cuvettes containing CI assay buffer and equilibrate to 30°C. One cuvette should contain 2.5 μM rotenone, while the other should include a corresponding volume of ethanol.

2. The reactions are started by the addition of sample and read for 3–5 min.

3. Calculate the rotenone-sensitive rate (see Fig. 3a) with an extinction coefficient for NADH of 6.81/mM/cm (see Note 11) (see Subheading 3.10, step 1).

### 3.5. Complex II Assay

The activity of complex II (succinate: $CoQ_1$ oxidoreductase) is assayed by following the reduction of the ubiquinone analogue $CoQ_1$ upon the oxidation of succinate to fumarate (see Fig. 2c), monitoring the absorbance at 280 nm to detect reduction of $CoQ_1$ (see Note 12).

1. Add CII assay buffer to the cuvettes and equilibrate to 30°C.

2. Add the sample to the assay buffer and then incubate for 10 min to activate complex II.

3. Start the reaction by adding $CoQ_1$ to a concentration of 50 μM and read for 3 min. Include a blank sample without $CoQ_1$ (blank rate should be negligible).

4. Calculate the results with an extinction coefficient for $CoQ_1$ of 12/mM/cm (see Subheading 3.10, step 1).

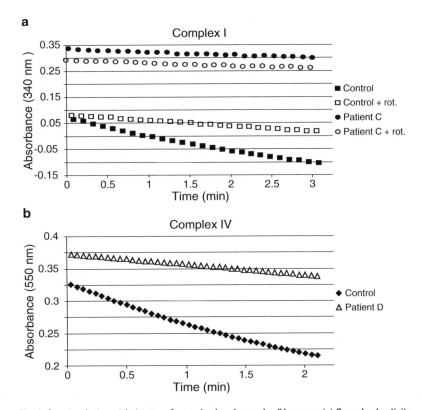

Fig. 3. Spectrophotometric traces of complex I and complex IV assays. (**a**) Complex I activity is assayed by measuring the absorbance at 340 nm to monitor the oxidation of NADH in a control muscle sample versus a patient (Patient C) with a complex I defect of unknown genetic basis. Because of substantial background oxidation of NADH, samples are measured with and without the complex I inhibitor rotenone (rot.), and the rate of rotenone-sensitive activity is calculated. (**b**) The activity of complex IV is measured by monitoring the absorbance at 550 nm to follow the oxidation of reduced cytochrome *c* in a control muscle sample versus a sample from a patient (Patient D) with a mutation in SURF-1 (unpublished). Because complex IV activity decays pseudo-exponentially, first-order rate constants are calculated rather than initial rates.

### 3.6. Complex II + III Assay

This assay (succinate: cytochrome *c* oxidoreductase) assesses the coupled transfer of electrons from complex II to complex III, as electrons from the oxidation of succinate are passed by endogenous ubiquinol to complex III where the reduction of cytochrome *c* occurs and its appearance monitored at 550 nm (see Fig. 2d).

1. Prepare CII + III assay buffer in the cuvettes and equilibrate to 30°C.

2. Add sample to the assay buffer, then add ATP to a final concentration of 1 mM. Incubate for 5 min to activate complex II.

3. Start the reaction by the addition of cytochrome *c* to a final concentration of 50 μM. A reaction lacking cytochrome *c* is used as a blank for muscle, heart, and cultured cells and should be negligible (see Note 13 regarding liver samples).

4. Calculate the results with an extinction coefficient for reduced cytochrome $c$ of 18.7/mM/cm (see Subheading 3.10, step 1).

**3.7. Complex III Assay**

The assay for complex III (decylbenzylquinonol: cytochrome $c$ oxidoreductase) activity utilises the reduced form of decylbenzylquinone (DB), a short chain analogue of endogenous hydrophobic $CoQ_{10}$, which donates electrons for the reduction of cytochrome $c$ (see Fig. 2e). The absorbance of reduced cytochrome $c$ is then measured at 550 nm.

1. Prepare CIII assay buffer in the cuvettes and equilibrate to 30°C.

2. Add sample to the cuvettes and then add $DB \cdot H_2$ to a concentration of 0.1 mM. A blank lacking sample should be included (see Note 14).

3. Start the reaction by adding cytochrome $c$ to 15 µM and then begin reading after a 20-s lag time.

4. At the end of the 3–5-min time course, add a few grains of L-ascorbic acid to the cuvette to fully reduce cytochrome $c$ and make a final reading.

5. The results are calculated as in Subheading 3.10, step 2.

**3.8. Complex IV Assay**

As the terminal enzyme of the electron transport chain, complex IV (cytochrome $c$ oxidase) catalyses the oxidation of cytochrome $c$ along with generation of $H_2O$. By measuring the absorbance at 550 nm, this assay monitors the rate of disappearance of reduced cytochrome $c$ (see Fig. 2f).

1. Prepare reaction cuvettes containing 50 mM KPi (pH 7.4) and equilibrate to 30°C (see Note 15).

2. Add sample, then start reaction by adding reduced cytochrome $c$ to a concentration of 15 µM (see Note 5), mixing as fast as possible and then reading immediately, as the absorbance decreases pseudo-exponentially (see Fig. 3b).

3. At the end of the time course, add $K_3Fe(CN)_6$ to 1 mM to complete oxidation of cytochrome $c$ and perform a final reading after ~1 min.

4. The results are calculated as in Subheading 3.10, step 2.

**3.9. Complex V Assay**

The assay for complex V ($F_1$-ATPase) is measured in the reverse direction, by linking the hydrolysis of ATP to the oxidation of NADH using pyruvate kinase (PK) and lactate dehydrogenase (LDH) (see Fig. 2g). The rate of NADH oxidation is monitored by measuring the absorbance at 340 nm.

1. Prepare CV assay buffer in cuvettes and equilibrate to 30°C.

2. Following equilibration, add ATP to 2.5 mM then leave for 2 min to allow for phosphorylation of any ADP present.

3. Start the reaction by the addition of sample. A blank should be assayed containing 2 μM oligomycin to calculate the oligomycin-sensitive rate (see Note 16).

4. Calculate the results with an extinction coefficient of 6.22/mM/cm for NADH (see Subheading 3.10, step 1).

*3.10. Calculations*

1. For CS, CI, CII, CII + III, and CV assays, spectrophotometric traces are analysed and linear regions are selected for calculating initial rates of enzyme activity in nmol/min/mg protein (see Fig. 3a).

2. The results for the CIII and CIV assays are calculated as first-order rate constants ($K$, /min/mg), as the reactions decay pseudo-exponentially and rapidly become non-linear (see Fig. 3b). The final absorbance readings after the addition of ascorbic acid (CIII) or $K_3Fe(CN)_6$ (CIV) are subtracted from each data point. The data are then plotted as a log plot against time, and the slope is the first-order rate constant.

# 4. Notes

1. For optimal measurements, it is helpful to use a spectrophotometer that can handle several samples at once. We and others use a Varian Cary 300 Bio Spectrophotometer with Cary temperature controller, sample transporter, and multicell holder, driven by Varian WinUV software (13).

2. Keep one set of cuvettes free of antimycin A for use with the CII + III and CIII assays, as residual antimycin A can inhibit complex III.

3. We prepare the assay buffers just before use from stock solutions, taking into account the number, type, and concentration of the samples and the volume of the assays. The following stock solutions are prepared ahead of time in $H_2O$: 0.5 M KPi, pH 7.4; 50 mM KCN; 10% (w/v) BSA; 0.1 M sodium succinate; 10% EDTA, pH 7.0 (with NaOH); 0.1 M phosphoenolpyruvate (PEP); 1 M $MgCl_2$ (monopotassium salt); 10 mg/ml LDH from hog muscle; and 10 mg/ml PK from rabbit muscle. The following stock solutions are prepared ahead of time in ethanol: 0.5 mM antimycin A; 10 mM $CoQ_1$; and 0.25 mM rotenone. The following stock solutions are freshly prepared on the day of use in $H_2O$: 2 mM NADH; 25 mM *n*-dodecylmaltoside; and 1 mM DTNB (prepared in 50 mM KPi, pH 7.4).

4. To prepare DB·$H_2$, add a small amount of $KBH_4$ (~5 mg) to 300 μl 10 mM DB (prepared in ethanol), then add 10 μl 0.1 M

HCl (in ethanol) to acidify. Vortex well until solution becomes clear (1–3 min). To this, add 12 μl 3 M HCl (in ethanol) to stabilise the $DB \cdot H_2$ and allow mixture to bubble to completion. Centrifuge for 30 s to remove excess $KBH_4$ then use clear supernatant kept on ice in CIII assays. If solution becomes yellow, indicating oxidation, a new batch must be made.

5. To prepare reduced cytochrome $c$, add a few grains of ascorbic acid to 0.5 ml oxidised cytochrome $c$ (2 mM) and leave for a few minutes. Then remove the ascorbic acid from the mixture by gel filtration using a PD-10 Sephadex G-25 column (Pharmacia 17-0851-01) that has been equilibrated with 50 mM KPi, 0.1 mM EDTA, pH 7.4. Collect the coloured fraction and store in a dark bottle at –20°C under nitrogen. Just prior to performing assays, estimate the amount of reduced cytochrome $c$ (diluted in 50 mM KPi, pH 7.4) by measuring the absorbance at 550 nm (the extinction coefficient for cytochrome $c$ is 18.7/mM/cm at 550 nm) and 565 nm. For the assay, calculate the amount of reduced cytochrome $c$ needed to give a concentration of 15 μM in the cuvette. If the ratio of $A_{550}/A_{565}$ is <6, then the sample is too oxidised for use, and a new batch must be prepared.

6. Prepare this buffer by mixing Tris Base and EGTA in $H_2O$, then bubble with $CO_2$ until the pH reaches 8.0 (~15 min). This can be stored at 0–10°C but needs to be warmed before use to room temperature and the pH again adjusted to 8.0 with $CO_2$. Bubbling with $CO_2$ can be effected using dry ice placed in a plastic squeeze bottle that has been fitted with tubing capped with a pipette.

7. Smaller sample volumes are recommended for the CIII assay, ranging between 1 and 6 μl for tissue homogenates and 5–10 μl for enriched cell mitochondria. A similar muscle homogenate sample volume range is recommended for the CV assay.

8. If tissue amount is very small, use a minimum of 200 μl tissue lysis buffer and wash homogeniser with a further 50 μl after transferring to a new tube. Assay sample in a 0.5 ml volume.

9. The cell debris pellet left over after tissue homogenisation can be stored for later DNA assays.

10. The blank rates for the CS assay lacking oxaloacetic acid should be negligible. However, liver samples show significant blank rates due to the presence of non-specific thiolase activity. Therefore, an individual blank must be included for each sample and subtracted in the final calculations.

11. The extinction coefficient is adjusted to account for the absorbance of $CoQ_1$ reduction at 340 nM.

12. We find this assay sufficiently sensitive in detecting complex II activity in our samples, but the sensitivity may be improved

using the alternate electron acceptor 2,6-dichlorophenol-indophenol (DCPIP) as either the direct electron acceptor from succinate or in combination with $CoQ_1$ (13). The extinction coefficient for DCPIP is 19.1/mM/cm.

13. Liver samples are not routinely used for complex II + III assays as they have a substantial antimycin A-insensitive rate. Therefore, to measure liver samples in the complex II + III assay, each volume must be assayed in two separate cuvettes, one of which contains 10 μM antimycin A and is used as the blank.

14. The blank for the complex III assay has a significant rate constant. This needs to be subtracted from the sample rate constants when making the final calculations.

15. The complex IV assay can be used with intact cells or mitochondria that have been permeabilised by detergent. For this, include 2.5 mM *n*-dodecylmaltoside in the reaction buffer.

16. Frozen tissues display variable oligomycin sensitivity and cannot be assayed reliably (13).

## Acknowledgements

Special thanks to Simone Tregoning and Wendy Salter for assistance with diagnostic data. This work was supported by grants from the Australian National Health and Medical Research Council (NHMRC) and the Muscular Dystrophy Foundation, an NHMRC Principal Research Fellowship to David Thorburn, and an NHMRC Career Development Award to Ann Frazier.

## References

1. Munnich, A., and Rustin, P. (2001) Clinical spectrum and diagnosis of mitochondrial disorders, *Am. J. Med. Genet.* *106*, 4–17.

2. McFarland, R., Taylor, R. W., and Turnbull, D. M. (2010) A neurological perspective on mitochondrial disease, *Lancet Neurol.* *9*, 829–840.

3. Di Donato, S. (2009) Multisystem manifestations of mitochondrial disorders, *J. Neurol.* *256*, 693–710.

4. Rahman, S., and Hanna, M. G. (2009) Diagnosis and therapy in neuromuscular disorders: diagnosis and new treatments in mitochondrial diseases, *J. Neurol. Neurosurg. Psychiatry 80*, 943–953.

5. Finsterer, J. (2006) Central nervous system manifestations of mitochondrial disorders, *Acta Neurol. Scand. 114*, 217–238.

6. Skladal, D., Halliday, J., and Thorburn, D. R. (2003) Minimum birth prevalence of mitochondrial respiratory chain disorders in children, *Brain 126*, 1905–1912.

7. Wong, L. J., Scaglia, F., Graham, B. H., and Craigen, W. J. (2010) Current molecular diagnostic algorithm for mitochondrial disorders, *Mol. Genet. Metab. 100*, 111–117.

8. Tucker, E. J., Compton, A. G., and Thorburn, D. R. (2010) Recent advances in the genetics of mitochondrial encephalopathies, *Curr. Neurol. Neurosci. Rep. 10*, 277–285.

9. Rotig, A. Genetic bases of mitochondrial respiratory chain disorders, *Diabetes Metab. 36*, 97–107.

10. Kirby, D. M., and Thorburn, D. R. (2008) Approaches to finding the molecular basis of mitochondrial oxidative phosphorylation

disorders, *Twin Res. Hum. Genet. 11*, 395–411.

11. Smits, P., Smeitink, J., and van den Heuvel, L. (2010) Mitochondrial translation and beyond: processes implicated in combined oxidative phosphorylation deficiencies, *J. Biomed. Biotechnol.*, doi:10.1155/2010/737385.

12. Calvo, S. E., Tucker, E. J., Compton, A. G., Kirby, D. M., Crawford, G., Burtt, N. P., Rivas, M., Guiducci, C., Bruno, D. L., Goldberger, O. A., Redman, M. C., Wiltshire, E., Wilson, C. J., Altshuler, D., Gabriel, S. B., Daly, M. J., Thorburn, D. R., and Mootha, V. K. (2010) High-throughput, pooled sequencing identifies mutations in NUBPL and FOXRED1 in human complex I deficiency, *Nat. Genet. 42*, 851–858.

13. Kirby, D. M., Thorburn, D. R., Turnbull, D. M., and Taylor, R. W. (2007) Biochemical assays of respiratory chain complex activity, *Methods Cell Biol. 80*, 93–119.

14. Medja, F., Allouche, S., Frachon, P., Jardel, C., Malgat, M., Mousson de Camaret, B., Slama, A., Lunardi, J., Mazat, J. P., and Lombes, A. (2009) Development and implementation of standardized respiratory chain spectrophotometric assays for clinical diagnosis, *Mitochondrion 9*, 331–339.

15. Benit, P., Goncalves, S., Philippe Dassa, E., Briere, J. J., Martin, G., and Rustin, P. (2006) Three spectrophotometric assays for the measurement of the five respiratory chain complexes in minuscule biological samples, *Clin. Chim. Acta 374*, 81–86.

16. Wibom, R., Hagenfeldt, L., and von Dobeln, U. (2002) Measurement of ATP production and respiratory chain enzyme activities in mitochondria isolated from small muscle biopsy samples, *Anal. Biochem. 311*, 139–151.

17. Zheng, X. X., Shoffner, J. M., Voljavec, A. S., and Wallace, D. C. (1990) Evaluation of procedures for assaying oxidative phosphorylation enzyme activities in mitochondrial myopathy muscle biopsies, *Biochim. Biophys. Acta 1019*, 1–10.

18. Birch-Machin, M. A., Briggs, H. L., Saborido, A. A., Bindoff, L. A., and Turnbull, D. M. (1994) An evaluation of the measurement of the activities of complexes I-IV in the respiratory chain of human skeletal muscle mitochondria, *Biochem. Med. Metab. Biol. 51*, 35–42.

19. Thorburn, D. R., Chow, C. W., and Kirby, D. M. (2004) Respiratory chain enzyme analysis in muscle and liver, *Mitochondrion 4*, 363–375.

20. Thorburn, D. R., and Smeitink, J. (2001) Diagnosis of mitochondrial disorders: clinical and biochemical approach, *J. Inherit. Metab. Dis. 24*, 312–316.

21. Faivre, L., Cormier-Daire, V., Chretien, D., Christoph von Kleist-Retzow, J., Amiel, J., Dommergues, M., Saudubray, J. M., Dumez, Y., Rotig, A., Rustin, P., and Munnich, A. (2000) Determination of enzyme activities for prenatal diagnosis of respiratory chain deficiency, *Prenat. Diagn. 20*, 732–737.

22. Niers, L., van den Heuvel, L., Trijbels, F., Sengers, R., and Smeitink, J. (2003) Prerequisites and strategies for prenatal diagnosis of respiratory chain deficiency in chorionic villi, *J. Inherit. Metab. Dis. 26*, 647–658.

23. Bernier, F. P., Boneh, A., Dennett, X., Chow, C. W., Cleary, M. A., and Thorburn, D. R. (2002) Diagnostic criteria for respiratory chain disorders in adults and children, *Neurology 59*, 1406–1411.

24. Wolf, N. I., and Smeitink, J. A. (2002) Mitochondrial disorders: a proposal for consensus diagnostic criteria in infants and children, *Neurology 59*, 1402–1405.

25. Walker, U. A., Collins, S., and Byrne, E. (1996) Respiratory chain encephalomyopathies: a diagnostic classification, *Eur. Neurol. 36*, 260–267.

26. Srere, P. A. (1969) Citrate Synthase, *Methods Enzymol. 13*, 3–11.

# Chapter 5

# Measurement of Mitochondrial Oxygen Consumption Using a Clark Electrode

## Zhihong Li and Brett H. Graham

## Abstract

Mitochondria require oxygen to produce ATP in sufficient quantities to drive energy-requiring reactions in eukaryotic organisms. The measurement of oxygen consumption rates from isolated mitochondria in vitro is a useful and valuable technique in the research and evaluation of mitochondrial dysfunction and disease since ADP-dependent oxygen consumption directly reflects coupled respiration or oxidative phosphorylation (OXPHOS). This chapter describes the traditional method of mitochondrial polarography using a Clark electrode for measuring coupled respiration in freshly isolated mitochondria from both mammalian tissues and *Drosophila melanogaster*.

**Key words:** Mitochondria, Polarography, Coupled respiration, Oxygen consumption rate, Oxidative phosphorylation

## 1. Introduction

The analysis of oxygen consumption from isolated mitochondria by polarography is a well-established technology that has been utilized for over 50 years (1). This approach is based upon the fundamental biochemical principle of oxidative phosphorylation (OXPHOS) that electron transport along the respiratory chain (with the final step being consumption of molecular oxygen via reduction to $H_2O$ by cytochrome $c$ oxidase) is functionally coupled to the phosphorylation of ADP through the generation and utilization of the proton electrochemical gradient across the mitochondrial inner membrane (2). The ability to measure oxygen consumption of intact mitochondria isolated from human tissues or cells or from animal tissues is an important tool for functional analysis of OXPHOS in the context of mitochondrial dysfunction and disease (2–4). Mitochondrial polarography involves using a Clark electrode to measure soluble oxygen content in a closed system.

Lee-Jun C. Wong (ed.), *Mitochondrial Disorders: Biochemical and Molecular Analysis*, Methods in Molecular Biology, vol. 837, DOI 10.1007/978-1-61779-504-6_5, © Springer Science+Business Media, LLC 2012

In the presence of oxidizable substrates, freshly isolated mitochondria are introduced into the system and oxygen consumption, in the presence of exogenously added ADP and or inhibitors, is measured. In this chapter, the methods for isolating intact mitochondria from tissues and for performing mitochondrial polarography are presented.

## 2. Materials

Prepare all solutions using ultrapure water (prepared by purifying deionized water via reverse osmosis to attain a resistivity of 18 MΩ cm at 25°C). Prepare and store all reagents at room temperature (unless indicated otherwise). All waste disposal regulations should be followed when disposing waste materials.

### 2.1. Stock Solutions (see Note 1)

1. 1 M mannitol: Dissolve 91 g mannitol in 300 mL of ultrapure water and then add water to final volume of 0.5 L (see Note 2).

2. 0.5 M sucrose: Dissolve 85.6 g sucrose in 300 mL ultrapure water and then add water to final volume of 0.5 L.

3. 0.5 M HEPES [4-(2-Hydroxyethyl)piperazine-1-ethanesulfonic acid]: Dissolve 59.6 g HEPES in 300 mL of ultrapure water and adjust pH to 7.2 using 10 M KOH. Add water to final volume of 0.5 L.

4. 0.4 M EGTA (ethylene glycol tetraacetic acid): Add 15.2 g EGTA to 90 mL ultrapure water. Add 3.5 g NaOH pellets initially and then carefully titrate pH to 7.5 by adding individual NaOH pellets as needed (see Note 3). Add water to final volume of 100 mL.

5. 1 M Tris–HCl: Dissolve 60.6 g Tris base in 400 mL of ultrapure water. Carefully adjust pH to 7.2 using concentrated HCl. Add water to final volume of 0.5 L.

6. 1 M KCl: Dissolve 37.3 g KCl in 400 mL of ultrapure water. Add water to final volume of 0.5 L.

7. 1 M $KH_2PO_4$: Dissolve 68 g $KH_2PO_4$ in 400 mL of ultrapure water. Adjust pH to 7.2 with concentrated HCl. Add water to final volume of 0.5 L.

### 2.2. Equipment and Buffer for Mitochondrial Isolation

1. Mitochondrial isolation buffer: 5 mM HEPES (pH 7.2), 210 mM mannitol, 70 mM sucrose, 1 mM EGTA, and 0.5% (w/v) BSA. To make 1 L, combine 210 mL of 1 M mannitol, 140 mL of 0.5 M sucrose, 10 mL of 0.5 M HEPES, 2.5 mL of 0.4 M EGTA, and 5 g of bovine serum albumin (BSA, fraction V, fatty acid free). Add ultrapure water to final volume of 1 L. Filter and sterilize the buffer and store at 4°C.

2. Dounce glass homogenizer (Kontes): It is useful to have at least two sizes of Dounce homogenizers (with accompanying "tight" and "loose" glass pestles) to facilitate homogenization of various amounts of tissues. In our experience, having sizes of 15, 7, and 2 mL provide sufficient capacity and flexibility for homogenizing mammalian tissues, fly tissues, and cell pellets.

3. Refrigerated centrifuge and microfuge (set at 4°C), 50 mL Oak Ridge centrifuge tubes, and 1.5 mL microfuge tubes.

**2.3. Reagents for Polarography**

1. Respiration buffer: 225 mM mannitol, 75 mM sucrose, 10 mM KCl, 10 mM Tris–HCl (pH 7.2), and 5 mM $KH_2PO_4$. To make 0.5 L, combine 112.5 mL of 1 M mannitol, 75 mL of 0.5 M sucrose, 5 mL of 1 M KCl, 5 mL of 1 M Tris–HCl (pH 7.2), and 2.5 mL of 1 M $KH_2PO_4$ (pH 7.2). Add ultrapure water to final volume of 0.5 L. Filter, sterilize, and dispense into 50 mL aliquots and store at –20°C.

2. Malate, glutamate, pyruvate, and succinate stock solutions (0.65 M): Make up 20 mL amounts of 0.65 M stocks of each, adjusting pH to 7.2 with 10 M KOH. Dispense into 150 µL aliquots and store at –80°C. These stock solutions are generally stable for several years when stored at –80°C. All are used at a working concentration of 5 mM (see Note 4).

3. ADP (25 mM): Prepare 20 mL of 25 mM stock solution of ADP (214 mg in 20 mL total volume) and use diluted (1 M) KOH to adjust the pH to within 6.0–6.8 to ensure long-term stability. Immediately dispense into 150 µL aliquots and store at –80°C (see Note 5). Usually 125 nmol of ADP is used to stimulate respiration for 200–300 µg of isolated mitochondria.

4. 6.5 mM DNP (2,4-dinitrophenol): Prepare 20 mL of a 6.5 mM stock by placing 24 mg of DNP into 4 mL of 1 M NaOH in a 15 mL polypropylene tube and fully dissolve by heating at 50°C for 20 min. Adjust the pH to neutral with 1 M HCl by adding ~3 mL. Check the pH until 7.2 is achieved (solution of DNP should have a translucent yellow color). Add ultrapure water to make a 20 mL final volume, dispense into 150 µL aliquots, and store at –80°C (see Note 6).

**2.4. Equipment for Polarography**

1. Clark-type microoxygen probe (YSI LifeSciences, Yellow Springs, OH).

2. Water-jacketed microoxygen chamber (0.65 mL) with magnetic stirrer (YSI LifeSciences, Yellow Springs, OH) and standard circulating water bath.

3. Oxygen monitor (YSI model 5300A, YSI LifeSciences, Yellow Springs, OH).

4. PowerLab 4/30 electronic chart recorder plus LabChart analysis software (ADInstruments, Colorado Springs, CO).

# 3. Methods

## 3.1. Isolation of Mitochondria from Tissue

1. Dissect or collect desired tissue (see Note 7) and wash twice with 2–5 mL of either ice-cold PBS or isolation buffer to remove any associated debris (see Note 8).

2. Mince the tissue with a razor blade and place in a prechilled Dounce homogenizer. Add up to 10 mL of the isolation buffer, then homogenize using eight to ten manual passes of the glass pestle (see Notes 9 and 10).

3. Transfer the homogenate into a 50 mL centrifuge tube (see Note 11).

4. Centrifuge at $1,500 \times g$ for 5 min at 4°C.

5. Carefully decant the supernatant into a fresh tube, avoiding transferring any loose material from the pellet and centrifuge at $8,000 \times g$ for 15 min at 4°C.

6. Discard the supernatant and wash the mitochondrial fraction by suspending the pellet in 25 mL of isolation buffer and repeating centrifugation at $8,000 \times g$ for 15 min at 4°C.

7. Suspend the washed mitochondrial pellet in isolation buffer (use 0.1 mL per gram of tissue used) and keep on ice while preparing the polarographic chamber (see Subheading 3.2 below). Determine the protein concentration using any one of the standard colorimetric methods such as the Bradford assay (5) (see Note 12).

## 3.2. Preparation of Equipment for Polarography

1. Disassemble and clean the microchamber, which consists of a 650-μL plastic chamber with magnetic stirrer, inlet and outlet ports for chamber flushing, a port for the oxygen electrode, a port for introduction of samples/substrates into chamber, and rotating stopcock with transparent viewport that is adhered to the chamber using silicone grease. To facilitate the cleaning of old grease, use a cotton swab saturated with an organic solvent such as xylene or methanol. Once clean, reassemble the chamber apparatus, turn on the circulating water bath to heat and maintain the water-jacketed chamber at 30°C, and turn on the oxygen monitor, electronic chart recorder, and computer.

2. Thaw a 50 mL aliquot of respiration buffer and ensure that it is saturated with room air by shaking in a flask in a 30°C water bath for at least 30 min.

3. Prepare the oxygen probe by covering the probe tip with a drop of fresh KCl solution (approximately half saturated by dissolving 5.25 g KCl in 16 mL ultrapure water) then overlaying the polyethylene membrane and securing with the O-ring as according to the manufacturer's instructions (see Note 13).

4. Calibrate the oxygen monitor and the electronic data recorder by first removing the probe jack from the monitor and setting the zero point for the data recorder. Plug the probe jack back into the monitor, flush the clean chamber with a few milliliters of equilibrated buffer then fill the chamber with buffer, close the chamber cock, and turn on magnetic stirrer (see Note 14). Once the signal has stabilized, calibrate the oxygen monitor for 100% air saturation and set the maximum point on the chart recorder for proper unit conversion of oxygen content (see Note 15).

### 3.3. Performing Mitochondrial Polarography

1. On ice, thaw aliquots of substrates (glutamate, malate, pyruvate, and succinate), ADP, and DNP.

2. Fill the chamber with fresh respiration buffer, and once a stable baseline is observed (usually after 1–3 min), add 5 µL of each of the desired substrates (see Note 4) and observe recording for 2–5 min until baseline restabilizes.

3. Add 200–500 µg of freshly isolated mitochondria to the chamber (see Note 16). There should be a low oxygen consumption rate (state IV or ADP-limiting rate) primarily caused by an endogenous baseline leak of protons across the inner membrane of the isolated mitochondria (4).

4. Add 5 µL (125 nmol) of ADP. The addition of ADP should stimulate a high oxygen consumption rate (state III or ADP-stimulating rate) that will transition back to a baseline state IV rate once the exogenously added ADP is consumed by OXPHOS (see Note 17). For well-coupled mitochondria, the state III rate should be at least threefold greater than the state IV rate (Figs. 1–3) (see Note 18). After state IV respiration has resumed for 1–2 min (allowing sufficient time to accurately measure the rate), a second addition of ADP can be made which should result in a similar state III rate and transition back to state IV (let run for 1–2 min).

5. Add DNP to a final concentration of 50 µM (see Note 6) and measure the uncoupled respiration rate for at least 2–3 min to complete the assay (see Note 19).

### 3.4. Data Analysis

The typical parameters determined from mitochondrial polarography include state III rate, state IV rate, RCR, uncoupled rate, and the ADP/O ratio (Fig. 1). The RCR (respiratory control ratio or state III rate/state IV rate) is a good indicator of the integrity of the inner membrane of the isolated mitochondria and is sensitive for indicating OXPHOS defects, while the ADP/O ratio is a direct reflection of phosphorylation efficiency and can indicate abnormalities of the ATP synthase or coupling (2, 4). The uncoupled rate (UC) reflects the maximal respiratory capacity of the

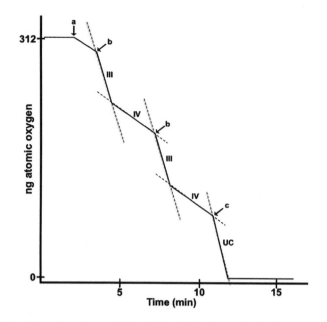

Fig. 1. Idealized trace of a polarographic run. At time "0," substrate (5 μM of glutamate + malate or succinate) is added to chamber full of air-saturated respiration buffer, and a stable baseline is observed. (a) Isolated mitochondria are added to a final concentration of 0.3 mg/mL. (b) 125 nmol of ADP is added, stimulating state III rate which transitions to state IV once exogenous ADP is consumed by OXPHOS. (c) DNP (50 μM) is added to stimulate maximal uncoupled rate (UC).

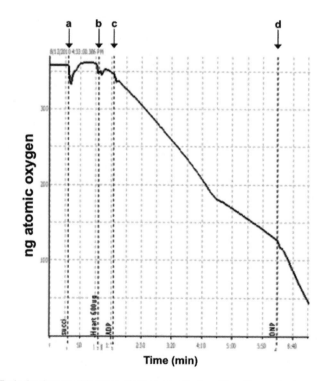

Fig. 2. Typical polarographic trace of rat heart mitochondria with succinate as substrate. A polarographic experiment using mitochondria isolated from wild-type rat heart is shown. (a) Succinate (5 μM) is added. (b) Rat heart mitochondria (0.9 mg/mL) is added. (c) 125 nmol of ADP is added. (d) DNP (50 μM) is added.

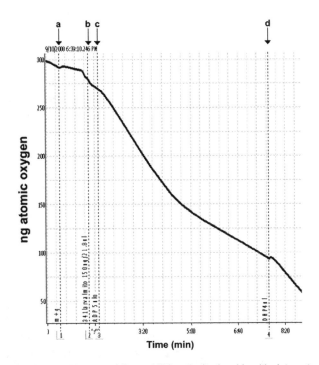

Fig. 3. Typical polarographic trace of *Drosophila* larval mitochondria with glutamate + malate as substrate. A polarographic experiment using mitochondria isolated from wild-type *Drosophila melanogaster* third instar larvae is shown. (**a**) Glutamate + malate (5 μM each) is added. (**b**) Fly larval mitochondria (0.3 mg/mL) is added. (**c**) 125 nmol of ADP is added. (**d**) DNP (50 μM) is added.

mitochondria and in wild-type mitochondria the ratio of state III rate/UC rate is usually between 0.85 and 0.95 (4).

1. Using the LabChart analysis software that accompanies the digital chart recorder, the oxygen consumption rates (i.e., state III, state IV, and UC) are measured from the linear phases and normalized to protein content (rates expressed as ng atomic oxygen/min/mg mitochondrial protein).

2. For the ADP/O ratio, the amount of ADP consumed during state III (125 nmol) is divided by the amount of oxygen consumed during state III (ng atomic oxygen).

## 4. Notes

1. Sterilize all stock solutions by filtration through 0.22-μm vacuum filter and store at 4°C for 1–2 months. Solutions can also be autoclaved, except for those containing sucrose, which will caramelize and turn solution brown.

2. When stored at 4°C, the mannitol stock solution (or other mannitol containing solutions) may precipitate out of solution. When this occurs, simply heat in a 65°C water bath to dissolve crystals immediately prior to using.

3. EGTA will not adequately dissolve without adding base. Once the pH is sufficiently raised, EGTA will rapidly go into solution. Take care when titrating to pH 7.5 so as not to overshoot.

4. The stock solutions are 0.65 M so that adding 5 μL of a stock solution in the microchamber gives a final concentration of 5 mM. The most common substrates utilized include malate, glutamate, pyruvate, and succinate and are oxidized at high rates in multiple mammalian tissues and cell types (3, 4, 6–9). Typically, malate + glutamate or malate + pyruvate are used for assaying complex I-specific respiration, while succinate is used for assaying complex II-specific respiration. Palmitoylcarnitine (0.4 mM) plus malate (1 mM) is a substrate pair useful for assessing fatty acid oxidation. Ascorbate in combination with tetramethylphenylenediamine (TMPD) to reduce intramito-chondrial cytochrome $c$ can be used for testing respiration directly through complex IV (3, 4). It is important to ensure that the pH of substrates and buffer stays between 7.2 and 7.4. Most substrates used for polarography are organic acids and adding substrates that are not pH neutral can uncouple mitochondria.

5. The accuracy of the ADP concentration is important because the calculation of ADP/O ratios is based upon the assumption that each microliter of ADP contains 25 nmol.

6. Uncouplers abolish the functional linkage between the respiratory chain and the ATP synthase by dissipating the proton electrochemical gradient. We routinely use DNP for polarography, but other proton ionophores such as carbonyl cyanide $m$-chlorophenyl hydrazone (CCCP) or carbonyl cyanide-$p$-trifluoromethoxyphenylhydrazone (FCCP) can also be used. The stock solution of DNP is 6.5 mM; therefore, adding 5 μL of DNP to the microchamber full volume will result in a final concentration of 50 μM.

7. In general, 1 g of mammalian tissue will provide a yield of 0.5–2 mg of isolated mitochondria, depending on the specific tissue. For *Drosophila*, 100 adult flies or sufficient third instar larvae to fill a 1.5 mL Eppendorf to 0.5 mL will yield approximately 0.5 mg mitochondria.

8. It is extremely important to keep all reagents and samples at 4°C or on ice at all times.

9. The optimal buffer volume to tissue weight ratio is 10 mL/g tissue. If using smaller amounts of tissue, one can reduce the amount of isolation buffer used proportionally.

10. For liver and brain, the tissues have minimal connective tissue, so eight to ten strokes with pestle B (the "tight" fitting pestle) typically result in complete homogenization. For tissues with more significant connective tissue, such as lung, heart, skeletal muscle and kidney, a razor blade can be used to mince the tissue into small pieces to facilitate homogenization and improve the mitochondrial yield. In addition, for heart and skeletal muscle, pestle A (the "loose" fitting pestle) should be used for five initial passes to ensure complete connective tissue breakdown, followed by five to six additional passes using pestle B to complete homogenization. For lung tissue, which may be still partially inflated after dissection, initially add one-half of the volume of the necessary buffer for the first two strokes due to trapped air contained within the lung tissue sample in order to avoid spillage.

11. For mitochondria isolation from *Drosophila* adult or larvae, we use buffer-saturated cheesecloth to filter either adult exoskeleton debris or larval fat from the homogenate prior to centrifugation.

12. Be sure to account for the 0.5% BSA when determining protein concentration. Either subtract it from the total concentration or wash and resuspend small aliquot of mitochondrial pellet in BSA-free buffer before measuring protein concentration.

13. A polyethylene membrane that allows diffusion of oxygen, but not metabolites, is required. The plastic from any generic sandwich plastic bag actually works very well as the membrane. Avoid using saturated KCl solution as precipitating crystals will disrupt optimal probe functioning. After several days of use, the silver anode will become uneven with AgCl leading to probe instability. Therefore, when the probe exhibits unstable behavior not corrected by changing the membrane, the old chloride layer of the anode should be carefully removed by gentle abrasion using the supplied abrasive pad (avoid abrading the platinum cathode in the center of the probe tip). Once the anode tip appears a bright and clean silver color, add a fresh layer of chloride by electroplating the KCl solution using the "chlorider" supplied with the probe.

14. Avoid introducing air bubbles into the chamber, as bubbles will displace volume, reducing the effective volume of the chamber.

15. Assuming a chamber volume of 650 μL, constant temperature of 30°C, and sea level altitude (1 atm. of pressure), the air-saturated buffer in the chamber will contain 312 ng of atomic oxygen (10).

16. In general, mitochondria should be added to a final protein concentration of 0.3–0.8 mg/mL, and the isolated mitochondrial

suspension should have a minimum concentration of 10 mg/mL in order to minimize the volume added to the chamber.

17. Sufficient mitochondria should be added so that a typical assay takes approximately 10–15 min. Avoid adding too much mitochondria which results in the assay taking less than 5 min and is too fast to make additions of ADP or inhibitors comfortably. If the assay takes longer than 20–25 min, wild-type mitochondria can exhibit functional deterioration. The optimal amount of mitochondria must be determined empirically for each tissue. In our experience, the optimal concentration for mammalian mitochondria is 0.3–0.6 mg/mL and for *Drosophila* mitochondria is 0.45–0.9 mg/mL.

18. The ratio of state III rate to state IV rate is expressed as the respiratory control ratio (RCR) and should be greater than 3.0 for well-coupled wild-type mitochondria (4). An RCR of less than 3.0 for wild-type mitochondria most commonly suggests a suboptimal concentration of mitochondria or a technical problem with the isolation of mitochondria that damages the integrity of the inner membrane.

19. DNP should be added, while there is at least 30% of the total oxygen content remaining in the microchamber. This ensures that maximal uncoupled respiration occurs for a sufficient amount of time to allow an accurate determination of the rate (i.e., at least 2 min).

## References

1. Chance, B., and Williams, G. R. (1955) A simple and rapid assay of oxidative phosphorylation, *Nature 175*, 1120–1121.

2. Brand, M. D., and Nicholls, D. G. (2011) Assessing mitochondrial dysfunction in cells, *Biochem. J. 435*, 297–312.

3. Barrientos, A. (2002) In vivo and in organello assessment of OXPHOS activities, *Methods 26*, 307–316.

4. Trounce, I. A., Kim, Y. L., Jun, A. S., and Wallace, D. C. (1996) Assessment of mitochondrial oxidative phosphorylation in patient muscle biopsies, lymphoblasts, and transmitochondrial cell lines, *Methods Enzymol 264*, 484–509.

5. Bradford, M. M. (1976) A rapid and sensitive method for the quantitation of microgram quantities of protein utilizing the principle of protein-dye binding, *Anal Biochem 72*, 248–254.

6. Puchowicz, M. A., Varnes, M. E., Cohen, B. H., Friedman, N. R., Kerr, D. S., and Hoppel, C. L. (2004) Oxidative phosphorylation analysis: assessing the integrated functional activity of human skeletal muscle mitochondria–case studies, *Mitochondrion 4*, 377–385.

7. Rasmussen, U. F., and Rasmussen, H. N. (2000) Human quadriceps muscle mitochondria: a functional characterization, *Mol Cell Biochem 208*, 37–44.

8. Rossignol, R., Letellier, T., Malgat, M., Rocher, C., and Mazat, J. P. (2000) Tissue variation in the control of oxidative phosphorylation: implication for mitochondrial diseases, *Biochem J 347 Pt 1*, 45–53.

9. Wenchich, L., Drahota, Z., Honzik, T., Hansikova, H., Tesarova, M., Zeman, J., and Houstek, J. (2003) Polarographic evaluation of mitochondrial enzymes activity in isolated mitochondria and in permeabilized human muscle cells with inherited mitochondrial defects, *Physiol Res 52*, 781–788.

10. Truesdale, G. A., and Downing, A. L. (1954) Solubility of Oxygen in Water, *Nature 173*, 1236.

# Chapter 6

# Mitochondrial Respiratory Chain: Biochemical Analysis and Criterion for Deficiency in Diagnosis

## Manuela M. Grazina

## Abstract

Spectrophotometric evaluation of mitochondrial respiratory chain (MRC) enzymatic complexes is the main approach to the biochemical investigation and diagnosis in oxidative phosphorylation disorders (also known as mitochondrial cytopathies). Regular dual beam spectrophotometers may be used, but we describe the protocols for double wavelength devices, allowing the analysis of complex activities from a small amount of tissue, with high sensitivity. An important concern is which tissue should be selected for analysis. Accordingly, we present the results obtained with different tissues and control values to be used. There are no standards available for the determinations and no interlaboratory quality control schemes are implemented. Additionally, different laboratories may use different protocols and comparison of results may be difficult. Currently, there is no consensus in literature for defining a criterion of an MRC deficiency to be used in biochemical diagnosis. There is statistical evidence that the most adequate criterion to define an MRC deficiency is below 40% of the mean control value normalized to citrate synthase activity.

**Key words:** Mitochondrial respiratory chain, Spectrophotometry, Analysis, Lymphocytes, Fibroblasts, Muscle, Liver, Heart

## 1. Introduction

### 1.1. Mitochondrial Respiratory Chain

Mitochondrial respiratory chain (MRC) is a unique system in the cell coded by two genomes, known as the powerhouse system of the cell, where the energy production occurs through the oxidative phosphorylation (OXPHOS) pathway (1, 2). MRC system (Fig. 1) is located in the mitochondrial inner membrane (MIM), organized in five enzymatic complexes (I to V), ubiquinone (or coenzyme Q10, CoQ10), and cytochrome $c$ (2, 3). The electrons are transferred from the reducing equivalents NADH, succinate, or other primary electron donor through flavoproteins, CoQ10, or iron-sulfur (Fe-S) centers, and cytochromes to the final acceptor $O_2$ at

Lee-Jun C. Wong (ed.), *Mitochondrial Disorders: Biochemical and Molecular Analysis*, Methods in Molecular Biology, vol. 837, DOI 10.1007/978-1-61779-504-6_6, © Springer Science+Business Media, LLC 2012

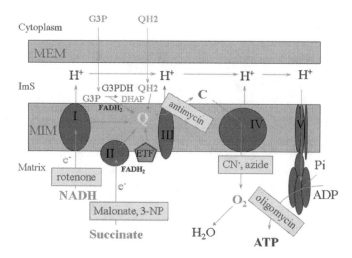

Fig. 1. Mitochondrial respiratory chain system and oxidative phosphorylation pathway. *I* complex I; *II* complex II; *Q* CoQ10; *QH2* quinol; *III* complex III; *C* cytochrome *c*; *IV* complex IV; *V* complex V; *G3P* glycerol-3-phosphate; *G3PDH* glycerol-3-phosphate dehydrogenase; *DHAP* dihidroxyacetone phosphate; *ETF* electron transfer protein; *3-NP* 3-nitropropionate; *CN* cyanide; *MIM* mitochondrial inner membrane; *MEM* mitochondrial external membrane; *ImS* intermembrane space; inhibitors are represented by the *rectangles*.

complex IV level, which is reduced to water. The electron flow is determined by the increasing reducing potential and is confirmed by using specific inhibitors for each complex (rotenone – complex I, NQR; malonate – complex II, SQDR; antimycin A – complex III, QCCR; KCN, azide, cyanide, carbon monoxide – complex IV, COX; oligomycin B – complex V, ATPase). The electric gradient generated by the electron transfer creates a proton chemical gradient from mitochondrial matrix to intermembrane space (ImS) through the enzymatic complexes NQR, QCCR, and COX. The reentrance of protons through complex V channel will allow the synthesis of ATP from ADP and inorganic phosphate, using the energy of the electrochemical gradient. These assumptions are the basis for the biochemical evaluation of MRC enzymes activity.

***1.2. Spectrophoto-metry for Evaluating MRC Complexes Function in Biological Samples***

The basic principle for spectrophotometry comes from the fact that the wavelength ($\lambda$) selected pass through the sample, promoting changes in the electronic, vibrational, and rotational energy of certain molecules, which absorb a part of the light discharged (tungsten or deuterium lamp, for visible range or for ultraviolet ranges, respectively). The enzyme activity is obtained by following the variation in absorption (optical density) in a certain period of time, either for common double beam (mono $\lambda$) or double wavelength spectrophotometry. A double wavelength device usually operates in both options. In general, protein determination is obtained in mono $\lambda$ mode.

**1.3. Analytical
Considerations**

Particular attention needs to be given to collection, transport, and storage of samples. These parameters are determinant for the quality of the result from the enzymatic assay.

Ideally, lymphocytes should be isolated from blood collected in EDTA within a maximum of 1 h after collection. If this is not possible, the time at collection to time the extraction and assay are performed should be recorded. Concerning the tissues, they must be frozen in dry ice immediately at collection and shipped to the laboratory or stored at –80°C (up to 1 week) or in liquid nitrogen (more than 1 week). If longer periods at –80°C are used, the influence of the type of storage in the MRC activities should be tested since it may influence the results. All the steps should be controlled in order to guarantee maximum confidence in results. For analysis, the samples stay on ice for about 30 min before starting the assays and submitted to freeze–thaw cycle to facilitate the access of substrates to the enzymatic complexes.

# 2. Materials

Prepare all analytical grade reagents and solutions using ultrapure water (prepared by purifying deionized water via reverse osmosis to attain a resistivity of 18 M$\Omega$ cm at 25°C). Prepare all reagents at store at –20°C, unless indicated otherwise. All waste disposal regulations should be followed when disposing waste materials.

**2.1. Solutions
and Reagents**

1. PBS (phosphate-buffered saline): 120 mM NaCl, 2.7 mM KCl, 10 mM K-phosphate buffer, pH 7.4 (storage 4°C).

2. 0.05% Trypsin solution: Dissolve 0.5 g of porcine trypsin powder and 0.2 g of Na$_2$EDTA in 1 L of commercially available Hanks' solution (nutritive solution for culturing animal cells) free of calcium and magnesium.

3. Buffer N: 50 mM Tris–HCl, 5 mg/ml BSA, pH 8.0 (see Note 1).

4. Buffer S: 2 mM EDTA, 10 mM KH$_2$PO$_4$, 1 mg/ml BSA, pH 7.8 (see Note 1).

5. Buffer C: 10 mM KH$_2$PO$_4$, 300 mM sucrose, 5 mg/ml BSA, pH 6.5 (see Note 1).

6. Buffer M: 20 mM Tris–HCl, 250 mM sucrose, 40 mM KCl, 2 mM EGTA, 1 mg/ml BSA, pH 7.2 (see Note 1).

7. Phosphate buffer: 0.1 M KH$_2$PO$_4$ and 0.01 M KH$_2$PO$_4$, pH 7.0.

8. 0.4 mM KCN.

9. 0.2 mM and 0.8 mM NADH.

10. 0.08 mM ubiquinone.

11. 4 and 8 μM rotenone.

12. 5 mM succinate.

13. 80 μM dichlorophenolindophenol (DCPIP).

14. 5 mM malonate.

15. 0.2 and 0.5 mM ATP.

16. 60 μM cytochrome $c$.

17. 80 μM ubiquinol.

18. 40 μM antimycin A.

19. 1.25 mM $n$-Dodecyl-β-$D$-maltoside.

20. 10 μM reduced cytochrome $c$: Dissolve 10 mg of cytochrome $c$ (oxidized form) in 1 ml of 0.1 M phosphate buffer. Add 50 μg of L-ascorbate and gently mix with magnetic stirrer at $7 \times 100$ rpm. Transfer the solution to a dialysis bag (diameter 6.3 cm, section of 25 mm), seal, and immerse in 5 L of 0.01 M phosphate buffer at 4°C for 19 h with gentle magnetic stirring ($7 \times 100$ rpm) and buffer change once. 50 μl of the prepared solution is removed and added to 960 μl of 0.1 M phosphate buffer in a cuvette for the measurement of the reduced cytochrome $c$ concentration spectrophotometrically by the absorbance at 550 nm. For the reference cuvette, 50 μl of the prepared solution, 950 μl of 0.1 M phosphate buffer, and 10 μl of saturated KFeCN (to oxidize cytochrome $c$) were used. Concentration ($C$) is determined by using Beer-Lambert equation (absorbance $= \varepsilon l C$), $\varepsilon_{(\text{cytochrome } c)} = 19.2$ units/mM/cm and $l$ is the length of the light path, which is 1 cm for the cuvettes we used.

21. 2 mM $MgCl_2$.

22. 2 mM phosphoenolpyruvate (PEP), storage 4°C.

23. 1,000 U/ml lactate dehydrogenase (LDH), storage 4°C.

24. 1,000 U/ml pyruvate kinase (PK), storage 4°C.

25. 0.2 mM carbamoyl cyanide $m$-chlorophenylhydrazone ($m$-CCP).

26. 0.1 mM oligomycin B.

27. 0.1 and 2% Triton X-100, storage 4°C.

28. 0.2 mM acetyl CoA.

29. 8 mM oxaloacetate.

30. 2 mM 5,5′-dithiobis-(2-nitrobenzoic acid) (DTNB).

31. Commercially available Ficoll-Paque solution for lymphocyte isolation from blood (storage 4°C).

32. 5% Percoll (storage 4°C).

33. 4% bovine serum albumin (BSA), storage 4°C.

34. Commercially available Bradford reagent, storage 4°C.

35. Commercially available protein assay kit (storage room temperature).

36. Commercially available hemoglobin assay kit (storage 4°C).

37. 10 mg/ml cytochrome *c*.

38. 50 µg/ml L-ascorbate.

39. Saturated KFeCN.

40. 1 mg/ml and 5 mg/ml BSA (storage 4°C).

### 2.2. Instrumentation

1. Double wavelength spectrophotometer and plotter.

2. Water bath (37°C).

3. Centrifuge.

4. Mechanical homogenizer.

5. Tissue glass grinder with Teflon pestle.

6. Tissue grinder with glass pestle.

7. Dialysis membrane (diameter 6.3 cm, section of 25 mm).

8. Nylon mesh (pore of 190 µm).

### 2.3. Preparation of Biological Samples for MRC Activity Measurement

#### 2.3.1. Blood Cells (see Note 2)

1. Collect whole blood in EDTA tube; 10 ml for adults and 5 ml in children.

2. Dilute (1:1) blood with PBS in a conic tube.

3. Centrifuge the mixture through 10/5 ml of Ficoll-Paque at $800 \times g$ for 10 min at room temperature.

4. Collect the *buffy coat* with Pasteur-type pipette.

5. Wash the buffy coat with 2 volumes of PBS and centrifugation at $800 \times g$ for 20 min.

6. Suspend the pellet in 50–100 µl of PBS and subject the cell suspension to two cycles of freeze–thaw in liquid nitrogen and start the measurements keeping the samples on ice.

#### 2.3.2. Skin Fibroblasts (see Note 3)

1. Decant medium from six T75 flasks with confluent cells.

2. To each flask, add 1 ml per flask of trypsin solution.

3. Trypsinized cells are collected by centrifugation.

4. Wash cell pellet twice with 20 ml of PBS (room temperature). Collect cell each time by centrifugation at $800 \times g$ for 10 min. The final cell pellet is suspended in 100 µl of PBS and submitted to two freeze–thaw cycles in liquid nitrogen and start the measurements keeping the samples on ice.

#### 2.3.3. Fresh and Frozen Tissues (see Note 4)

1. The frozen tissue is left on ice for about 20 min prior to homogenization.

2. Prepare the homogenate on ice at ~4°C using a tissue grinder with pestle of clearance 0.004–0.006 in. (0.1016–0.1524 mm), adding 100–500 µl of buffer M and grind manually.

3. Filter through a nylon mesh (pore of 190 µm) to a cryotube of 2 ml. The homogenate is ready for use.

*2.3.4. Isolation of Mitochondria from Fresh Muscle Biopsy (see Note 5)*

1. The isolation procedure starts with mechanical homogenization in a tissue glass grinder with Teflon pestle using 2 ml of buffer M at 500 rpm on ice.

2. Filter the homogenate through a nylon mesh (pore of 190 μm) to a clean glass grinder containing 1 ml of cold buffer M. This homogenate can either be analyzed for MRC activities or be used to prepare the mitochondrial fraction. In the second cases, an aliquot of homogenate (HM1), 100–200 μl, is used for control quality.

3. Perform a second mechanical homogenization at 500 rpm followed by centrifugation at $2,410 \times g$ for 10 min at 4°C in conical tubes of 2 ml.

4. Transfer the supernatant to clean tubes and centrifuge at $11,840 \times g$ for 10 min at 4°C. The pellet is suspended in 50–100 μl of buffer M with 5% Percoll.

**2.4. Determination of Protein Concentration by Bradford Method (see Note 6)**

1. Sample is diluted 1:6 in $H_2O$ (1:2), and 2% Triton X-100 (1:3) and 2–6 μl (triplicate) are used with 1 ml of commercially available Bradford reagent.

2. Read absorbance at $\lambda = 595$ nm.

3. Protein concentration is obtained by extrapolation of the standard calibration curve based on the analysis of BSA 0.4% with application of Beer-Lambert principle.

# 3. Methods

The main advantages of double wavelength approach include the reduced amount of biological sample (use of single cuvette) and minimization of interferences caused by some components of the reactions, such as detergents. In double wavelength mode of operation, the emission of two monochromatic beams (two different $\lambda$) occurs alternately through the cuvette containing sample in reaction buffer, neutralizing the effects of dispersion and sedimentation, which, by conventional spectrophotometry do impede the detection of small variations in absorbance. One monochromator is defined with the $\lambda$ associated to isosbestic point (sample absorbance is the same for the two chemical states of the chromophore responsible for the alteration of the registered absorbance; in general, it may correspond to substrate), corresponding to the reference $\lambda$. The other monochromator is defined with the sample $\lambda$, close to reference $\lambda$ in the absorbance spectrum of the chromophore. The absorbance value at the reference $\lambda$ remains constant and a compensation effect occurs with attenuation of the highest intensity beam, allowing the registration of the chemical modification occurring in the assay. This is a high sensitivity method, particularly important in the analysis of pediatric samples and needle biopsies (particularly liver and heart).

The MRC enzymatic activity determination presented in this chapter is described according to modified protocols reported earlier (10, 11), in a double wavelength spectrophotometer connected to a register under controlled temperature conditions (all the assays are performed at 37°C).

The protocols below include determination of activity of the five complexes mediating OXPHOS pathway (complex I: NADH dehydrogenase or NADH-coenzyme Q reductase, NQR – EC 1.6.5.3; complex II: succinate dehydrogenase or succinate-coenzyme Q reductase, SQDR – EC 1.3.5.1; complex III: ubiquinol-cytochrome *c* reductase, QCCR – EC 1.10.2.2; complex IV: cytochrome oxidase, COX – EC 1.9.3.1; complex V: ATP synthase, ATPase – EC 3.6.1.34) and two mobile electron transporters, ubiquinone (or coenzyme Q10, CoQ10) and cytochrome *c* (2, 3).

Citrate synthase (CS – EC 4.1.3.7) activity was also determined on the basis that it can be used as an indicator of mitochondrial proliferation (12) of a certain tissue or cell preparation. The normalization of results to CS activity may allow the minimization of differences due to mitochondria number in the biological sample. This question will be discussed in more detail in Subheading 3.10.

Spectrophotometric studies for evaluation of MRC enzymatic complexes comprise a series of assays, isolated or in combination (13), using specific electron donors and acceptors (11). Regular dual beam spectrophotometry may be used, but dual wavelength is preferable since smaller quantities of biological sample are needed. This is particularly critical in small children. Accordingly, 1–20 mg of tissue is usually sufficient, and this quantity may be easily obtained from a needle biopsy (liver, kidney, myocardium), a pellet of lymphocytes isolated from peripheral blood, or from cultured fibroblasts derived from skin (11), or other tissue (even postmortem if the collection is performed up to 3 h after death and immediately frozen in dry ice). All the samples must be immediately frozen without any type of buffer (if they are used also for DNA extraction, the tube is sterile) and maintained in liquid nitrogen or at –80°C (11, 14–18).

The question of which tissue should be investigated needs particular attention (13). In case of muscle weakness, the skeletal muscle (usually deltoid) is the most indicated. If the hematopoietic system is affected, such as in Pearson syndrome (19), the analysis should be performed in lymphocytes from peripheral blood (13). In liver or heart disease, hepatic (20) or endomyocardial (11) biopsy is recommended (respectively). When the organ affected is difficult to access, such as brain, retina, endocrine gland, and smooth muscle, peripheral tissues should be extensively investigated, including skeletal muscle, cultured fibroblasts, and lymphocytes.

Independently of the organ affected, it is mandatory to perform a skin biopsy (even postmortem) for obtaining cell-cultured fibroblasts and subsequent investigation (13).

In general, lymphocytes isolated from peripheral blood are not widely used, but they are important in the specific cases abovementioned and also in small children, especially if the parents do not authorize a biopsy. In our experience (evaluation in 99 cases with MRC disease in whom several tissues were analyzed for the same patient), the analysis of lymphocytes is quite specific for detecting an MRC deficiency (71.7%), that is, when a deficiency is found, it is present in other tissue/tissues of the same patient although presents low sensitivity (25.3%) when compared to other tissues including the muscle.

Additionally, the result in blood may give clues to the genetic study (e.g., SURF-1 mutations in isolated complex IV deficiencies).

In the protocols described below, all the assays are performed at 37°C under controlled pH, but the biological samples stay on ice during the measurements.

The velocity for register enzymatic activities is 2 cm/min, with acquisition time of 2–10 min for each assay. Final volume is ~1 ml (1 ml of buffer plus 15–100 µl of reagents according to the specifications of each assay, as described in detail below).

### 3.1. Measurement of Complex I Activity

1. Setup parameters: $\lambda = 340-380$ nm, absorbance scale 0.5–0.9.
2. Mix 800 µl of $H_2O$ and 200 µl of buffer N in the cuvette.
3. Add 100 µg of protein (make up final volume up to 20 µl with the suspension buffer) and acquire the baseline at the spectrophotometer.
4. Incubate 1 min, at 37°C.
5. Add 5 µl of 80 mM KCN, 5 µl of 40 mM NADH, and 4 µl of 20 mM ubiquinone to initiate the reaction (see Note 7) and record changes (decrease) in absorbance ($\Delta A$) versus time per 1 cm during at least 2 min.
6. Add 2 µl of 2 mM rotenone to inhibit complex I.
7. The same cuvette may be used for complex V determination (see Subheading 3.6).

### 3.2. Measurement of Complex II Activity

1. Setup parameters: $\lambda = 600-750$ nm, absorbance scale 0.7–1.7.
2. Add 50 µg of protein (make up final volume up to 20 µl with the suspension buffer) to 1 ml of buffer S in the cuvette.
3. Acquire the baseline at the spectrophotometer.
4. Add 2 µl of 80 mM KCN, 4 µl of 2 mM rotenone, 4 µl of 50 mM ATP, 10 µl of 500 mM succinate, 4 µl of 20 mM ubiquinone, and 2 µl of 20 mM DCPIP to initiate the reaction

(see Note 8) and register changes (decrease) in absorbance ($\Delta A$) versus time per 1 cm during at least 2 min.

5. Add 10 µl of 500 mM malonate to inhibit the reaction.

### 3.3. Measurement of Complex II + III Activity

1. Setup parameters: $\lambda = 550$–540 nm, absorbance scale –0.1 to 0.3.

2. Add 50 µg of protein (make up final volume up to 20 µl with the suspension buffer) to 1 ml of buffer S in the cuvette.

3. Acquire the baseline at the spectrophotometer.

4. Add 2 µl of 80 mM KCN, 4 µl of 2 mM rotenone, 4 µl of 50 mM ATP, 30 µl of 2 mM cytochrome $c$, and 10 µl of 500 mM succinate to initiate the reaction (see Note 9) and register changes (increase) in absorbance ($\Delta A$) versus time per 1 cm during at least 2 min.

5. Add 10 µl of 500 mM malonate to inhibit complex II and continue to complex III evaluation (see Subheading 3.4).

### 3.4. Measurement of Complex III Activity

1. Add 10 µl of 20 mM ubiquinol to the cuvette of the assay described in Subheading 3.3 to initiate the reaction.

2. Register changes (increase) in absorbance ($\Delta A$) versus time per 1 cm during at least 2 min.

3. Add 10 µl of 2 mM antimycin A to inhibit complex III.

### 3.5. Measurement of Complex IV Activity

1. Setup parameters: $\lambda = 550$–540 nm, absorbance scale –0.1 to 0.2.

2. Add 50 µg of protein (make up final volume up to 20 µl with the suspension buffer) to 1 ml of buffer C in the cuvette.

3. Acquire the baseline at the spectrophotometer.

4. Add 10 µl of 125 mM $n$-Dodecyl-β-D-maltoside detergent (see Note 10).

5. Add 10 µM of reduced cytochrome $c$ (volume depends on the concentration which is determined after dialysis, but it is advisable to use 2–5 µl).

6. Register changes (decrease) in absorbance ($\Delta A$) versus time per 1 cm during at least 2 min.

7. Add 2 µl of 80 mM KCN to inhibit complex IV.

### 3.6. Measurement of Complex V Activity

1. Setup parameters: $\lambda = 340$–380 nm, absorbance scale 0.5–0.9.

2. Add 20 µl of 200 mM $MgCl_2$, 10 µl of 50 mM ATP, 20 µl of 200 mM phosphoenolpyruvate (PEP), 91 U of lactate dehydrogenase (LDH, EC 1.1.1.27), 91 U of pyruvate kinase (PK, EC 2.7.1.40), 1 µl of 200 mM $m$-CCP to the cuvette of the Subheading 3.1 (see Note 11).

3. Register changes (decrease) in absorbance ($\Delta A$) versus time per 1 cm during at least 2 min.

4. Add 5 µl of oligomycin B 2.5 mM to inhibit complex V and obtain specific activity.

**3.7. Measurement of Segment I + III Activity**

1. Setup parameters: $\lambda = 550$–540 nm, absorbance scale –0.1 to 0.4.

2. Mix 800 µl of $H_2O$ and 200 µl of buffer N in the cuvette.

3. Add 50 µg of protein (make up final volume up to 20 µl with the suspension buffer) and get the baseline at the spectrophotometer.

4. Incubate 1 min at 37°C.

5. Add 2 µl of 80 mM KCN, 20 µl of 40 mM NADH, and 30 µl of 2 mM cytochrome $c$ to initiate the reaction (see Note 12) and register $\Delta A$ (increase).

6. Add 4 µl of 2 mM rotenone to inhibit the reaction.

**3.8. Measurement of Citrate Synthase Activity**

1. Setup parameters: $\lambda = 412$–600 nm, absorbance scale 0–1.

2. Add 50 µg of protein (make up final volume up to 20 µl with the suspension buffer) to 1 ml of buffer S in the cuvette.

3. Acquire the baseline at the spectrophotometer.

4. Add 5 µl of 20% Triton X-100, 10 µl of 5 mM acetyl CoA, 10 µl of 10 mM 5,5′-dithiobis-(2-nitrobenzoic acid) and 25 µl of 320 mM oxaloacetate to initiate the reaction (see Note 13).

5. Register $\Delta A$ (increase).

**3.9. Calculation of Activity**

The activities are calculated in nmol/min/mg of protein. In order to convert absorbance in specific activity, we used an equation similar to the one published by Miró et al. (15), where "$\Delta A$" is the alteration in absorbance, "$\Delta t$" is the time of the assay, "$V$" is the volume of reaction (liters), "Pt" is the quantity of protein used (mg), "$b$" is the length of cuvette (cm), and "$\varepsilon$" is the molar extinction coefficient (units/mM/cm): $(\Delta A/\Delta t) \times V \times 10^6/(Pt \times b \times \varepsilon)$.

**3.10. Data Analysis: How to Define MRC Deficiency for Diagnosis**

The deficiencies observed vary from isolated complex impairment to alterations in multiple MRC enzymes. The most frequent deficits observed affect complex I and IV (21), also in our series (unpublished data).

It is important to be aware that in vitro investigation of OXPHOS is difficult and complex; results may be difficult to interpret in some cases and ambiguities may be found (13) that should be taken into account:

1. Normal MRC enzyme activities may be found to be normal in one tissue/organ that does not express the disease, such as in Friedreich ataxia (22);

2. Normal MRC results do not exclude an OXPHOS disease even if the tissue tested is clinically affected due to kinetic mutants, tissue heterogeneity, or heteroplasmy. In these cases, histology analysis may be helpful to guide more extensive genetic analysis;

3. Normal activities of isolated complexes I, II, and III with decreased I + III and II + III are usually indicative of CoQ10 deficiency (see Chapter 10). On the other hand, the incorrect storage of samples may result in loss of ubiquinone or affecting assembling of each complex. Parafin-embedded tissues are not recommended for MRC enzymatic assays (11);

4. Measuring complex I in lymphocytes or cultured cells may be difficult due to other NADH dehydrogenases resistant to rotenone, but it is possible (23);

5. The phenotypic expression of MRC defects in cultured cells, unstable and active, tends to disappear with cell proliferation. It is necessary to add pyruvate (10 μM) and uridine (200 μM) to the cell culture media to preserve deficiencies and the mutant phenotype (5);

6. Discrepancies in control values may indicate problems in the experimental assay. Relative activities should be consistent when tested under nonlimiting conditions. For example, normal succinate-cytochrome $c$ reductase (SCCR) is expected to be twofold higher than succinate-quinone DCPIP reductase (SQDR) because only 1 electron is required to reduce cytochrome $c$, but 2 electrons are necessary to reduce DCPIP (11);

7. The dispersion of the reference values brings additional difficulties (Fig. 2) to the identification of MRC deficiencies in patients, given the overlap between the value ranges in the two groups (Fig. 3). The ratio between complexes activities may help (16) since normal OXPHOS requires a balanced activity of the several complexes, but this is not a criteria allowing identification of deficiency for all cases, particularly in multiple MRC defects.

The biochemical classification was proposed in 1997 (24), but there is a lack of consensus for defining an MRC deficiency to be considered in diagnosis.

Unfortunately, there is a considerable heterogeneity between laboratories performing MRC deficits screening, concerning both experimental protocols and criteria for MRC defect classification.

The complexity of information for MRC disease diagnosis may contribute to difficulties in interpretation of the MRC results, especially when the patient does not present a well-defined syndrome, as it is the case of the majority of patients.

If we look into the biochemical criteria for MRC deficiency (Table 1) from four different groups (11, 25–27), we consider that

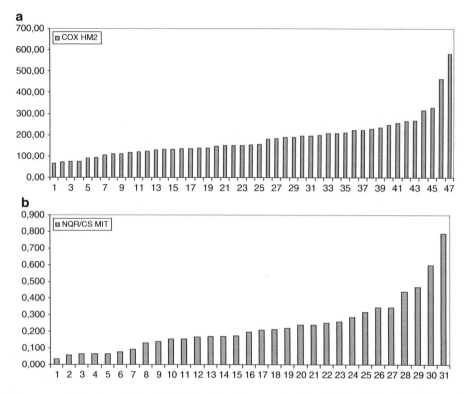

Fig. 2. MRC enzyme activities in control subjects (absolute value – **a**; normalized to CS activity – **b**). *NQR* complex I; *COX* complex IV; *MIT* fresh isolated mitochondrial fraction from skeletal muscle; *HM2* homogenate of frozen skeletal muscle.

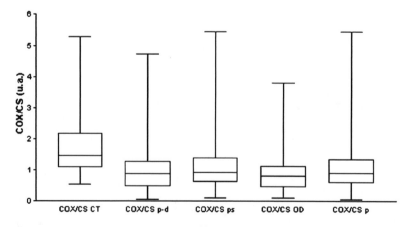

Fig. 3. MRC enzyme activities in lymphocytes of control subjects and patients normalized to CS activity. *COX* complex IV; *CS* citrate synthase; *CT* control subjects; *p-d* patients with MRC disease "probable or definitive"; *ps* patients with MRC disease "possible"; *p* patients with MRC disease "possible, probable, or definitive"; *OD* patients with other neurological diseases; *u.a.* arbitrary units.

**Table 1**
**Criteria for mitochondrial respiratory chain deficiency in diagnosis of OXPHOS disorders**

| References | Major | Minor |
|---|---|---|
| (11) | Activity below minimum reference absolute value; ratio of complexes activities | |
| (25) | Detection of MRC activity decrease: <20% activity of any complex in one or more tissues (compared to the mean corrected for age) | Indicators of decreased MRC function: 20–30% activity of any complex in one or more tissues (compared to the mean corrected for age) |
| (26) | <20% activity[a] of any complex in one tissue <30% activity[a] of any complex in a cell line <30% activity[a] of the same complex in more than 2 tissues | Demonstration of alteration in MRC complexes expression, based on antibody detection 20–30% activity[a] of any complex in one tissue 30–40% of any complex in a cell line 30–40% activity[a] of the same complex in more than 2 tissues |
| (27) | Activity below minimum reference absolute value (normalized to CS activity) | |

[a]Mean control value corrected for a reference enzyme, CS or complex II
*CS* citrate synthase; *MRC* mitochondrial respiratory chain

it is worth to investigate further the comparison with 40% of the mean reference value. The graphical representation of MRC values in different ways (Fig. 4) shows that the lowest dispersion of MRC patients' data is obtained for % of mean reference value, especially if corrected to CS activity.

A significant portion (47–59%) of the patients' values is higher than the reference minimum value. Considering the representation by % of the mean reference activity corrected to CS, about 80% of the patients' values are below control. These results, together with the data published by Bernier et al. (26) and statistical analysis for sensitivity, specificity, negative predictive value (NPV), and positive predictive value, reveal that the criteria with the best scores for sensitivity (39%) and NPV (40%) are obtained when the data of the patients are compared with 40% of the mean reference, corrected to CS activity. The tests with higher sensitivity (capacity to detect the disease) and NPV (probability of a negative result being associated to the absence of disease) are indicated for screening of severe phenotypes. Accordingly, it is logical to suggest that the uniformity required for multicenter studies on MRC analysis must be developed by the starting point of defining sensitivity, specificity, negative predictive value (NPV), and positive predictive value for the biochemical testing of MRC function to be considered in detection, screening, and diagnosis of OXPHOS diseases.

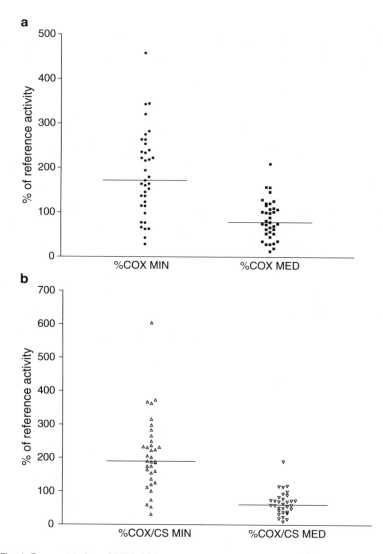

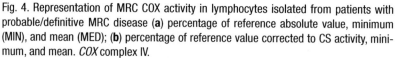

Fig. 4. Representation of MRC COX activity in lymphocytes isolated from patients with probable/definitive MRC disease (**a**) percentage of reference absolute value, minimum (MIN), and mean (MED); (**b**) percentage of reference value corrected to CS activity, minimum, and mean. *COX* complex IV.

Note that it is not possible to differentiate a case of MRC disease from other disorders only by biochemical testing of enzymatic activities. All the other data and criteria must be considered, as described, for example, by Bernier et al. (26), and clinical information is mandatory to guide investigation in a significant portion of cases.

The control values used in our laboratory are presented in Table 2.

## Table 2
## Reference values for the five (I–V) mitochondrial respiratory chain enzymatic complexes activities, corrected to citrate synthase

| | I/CS | II/CS | III/CS | IV/CS | V/CS | I + III/CS | II + III/CS |
|---|---|---|---|---|---|---|---|
| Lymphocytes isolated from peripheral blood ($n=50$) | | | | | | | |
| Mean | | 0.33 | 1.02 | 1.54 | | | 0.51 |
| 40% mean | | 0.132 | 0.408 | 0.616 | | | 0.204 |
| Cultured fibroblasts ($n=42$) | | | | | | | |
| Mean | | 0.26 | 1.15 | 1.34 | | | 0.46 |
| 40% mean | | 0.104 | 0.460 | 0.536 | | | 0.184 |
| Mitochondria isolated from fresh muscle biopsy ($n=35$) | | | | | | | |
| Mean | 0.23 | 0.27 | 2.16 | 2.41 | 0.91 | 0.75 | 0.66 |
| 40% mean | 0.092 | 0.110 | 0.862 | 0.963 | 0.365 | 0.301 | 0.262 |
| Fresh muscle homogenate ($n=12$) | | | | | | | |
| Mean | 0.06 | 0.17 | 1.48 | 1.65 | 0.36 | 0.32 | 0.18 |
| 40% mean | 0.024 | 0.068 | 0.590 | 0.660 | 0.142 | 0.128 | 0.070 |
| Frozen muscle homogenate ($n=47$) | | | | | | | |
| Mean | 0.16 | 0.32 | 1.21 | 1.72 | 0.84 | 0.31 | 0.28 |
| 40% mean | 0.065 | 0.127 | 0.484 | 0.689 | 0.338 | 0.123 | 0.112 |
| Liver ($n=45$) | | | | | | | |
| Mean | 0.34 | 1.88 | 2.36 | 2.70 | 1.22 | 0.81 | 0.97 |
| 40% mean | 0.136 | 0.750 | 0.944 | 1.080 | 0.490 | 0.326 | 0.389 |
| Myocardium ($n=26$) | | | | | | | |
| Mean | 0.46 | 0.77 | 3.01 | 3.50 | 0.63 | 0.70 | 0.94 |
| 40% mean | 0.185 | 0.308 | 1.203 | 1.398 | 0.250 | 0.278 | 0.376 |

## 4. Notes

1. BSA is added to buffers immediately before use.

2. The blood samples are collected in EDTA; 10 ml for adults and 5 ml in children allow investigation of complexes II, III, IV, II + III (SCCR, EC 1.3.99.1), and CS activities. Usually complexes I, V, and I + III (NCCR, EC 1.6.99.3) are investigated only in tissue homogenates. In general, higher amount of cells are required (4) due to the low specific activity of complex I, relatively to other complexes. This process allows isolation of $58–62 \times 10^6$ cells from 10 ml of EDTA blood (90% of lymphocytes) with 98% of cell viability.

3. The most commonly used cultured cells are the fibroblasts derived from skin biopsy (5). However, other types, such as transformed lymphocytes, may be used. Cells are always cultured in the presence of pyruvate and uridine. Fibroblasts

are very useful since these cells share the embryonic origin with the central nervous system and may allow prenatal diagnosis (6). The possibility of cryopreservation for later analysis is also useful. This procedure allows the investigation of complexes II, III, IV, segment II + III (SCCR, EC 1.3.99.1), and CS activities. Usually complexes I, V, and segment I + III (NCCR, EC 1.6.99.3) are investigated only in tissue homogenates. In tissue, in general, higher amounts of cells are required, due to the low specific activity of complex I, relatively to other complexes.

4. The homogenates may be prepared either from fresh or frozen tissue sample. It is important to establish control values for each case since the results are not comparable. In general, the tissue is immediately frozen after collection (in postmortem, note that collection has to be performed up to 3 h after deceased). If possible, it is advisable that tissue collection (cryotubes, sterile and dry) includes more than one fragment in order to prepare a second homogenate, when necessary (e.g., quality control, such as storage/freezing effects), or to DNA extraction.

5. Mitochondrial fraction is isolated from fresh muscle biopsy (100–500 mg) that arrives in buffer M on ice (~4°C) to the laboratory up to 2 h after collection. Processing of the sample in the laboratory should not exceed 1 h, in order to preserve MRC enzymes activity. In theory, analysis of mitochondrial fraction is more specific than muscle homogenate, but the results are different for the two types of sample according to the evidences of our laboratory and Casademont et al. (7).

6. Protein content is determined in all biological preparations and buffers, according to Bradford method (8), allowing detection of 2–50 μg of protein. When the protein content is higher than 50 μg, Lowry method (9) is performed at 660 nm, allowing detection of 10–300 μg of protein with commercially available protein assay kit following the manufacturer's instructions. In the case of lymphocytes, if the pellet is contaminated with hemoglobin, it is possible to determine its content using a commercially available hemoglobin assay kit following the manufacturer's instructions and discount this contribution to the protein content in order to obtain a more exact determination for enzyme activity results.

7. Ubiquinone is the electron acceptor for NADH oxidation. It transfers the electrons to cytochrome $c$. KCN is used to inhibit cytochrome $c$ oxidase (COX) activity, which would affect NADH oxidation rate.

8. Ubiquinone is the electron acceptor for DCPIP oxidation. Rotenone is used to maintain NADH concentration high (regulator of Krebs cycle enzymes) and prevents the formation

of oxaloacetate (inhibits SDH) from malate by malate dehydrogenase in Krebs cycle. ATP is used to stimulate SDH and avoid oxaloacetate production. KCN prevents oxidation of reduced cytochrome $c$ by blocking COX activity, allowing a more precise detection of reduction of cytochrome $c$ by SDH activity.

9. Succinate is oxidized to fumarate by SDH. Effects of rotenone, ATP, and KCN are mentioned above.

10. Dodecyl maltoside is a detergent that promotes the formation of pores in mitochondrial external membrane, allowing the entrance of reduced cytochrome $c$ to transfer its electrons to MRC cytochrome $c$ in inner mitochondrial membrane without compromising integrity. In the analysis of mitochondrial fraction, it is possible to evaluate the quality of the preparation by evaluating the ratio of COX activities in the absence and in the presence of detergent.

11. The oxidation of NADH depends on the conversion of pyruvate to lactate by LDH. Pyruvate is formed from phosphoenolpyruvate by pyruvate kinase activity, which is dependent on ADP formed by ATPase (complex V) from exogenous ATP. The uncoupler $m$-CCP guarantees that complex V activity is uncoupled from electron transport, allowing the maximal specific detection of its function.

12. NADH is the electron donor to cytochrome $c$, and the reaction is monitored by the detection of reduced cytochrome $c$.

13. This reaction corresponds to the first reaction of Krebs cycle, the condensation of oxaloacetate with acetyl coenzyme. The reagent DTNB reacts with coenzyme A formed in the reaction, allowing detection at 412 nm.

## Acknowledgements

The author is grateful to Pierre Rustin, Dominic Chrétien, Arnold Münnich, and Agnès Rötig for the training and teaching at INSERM U-393 in 1994, particularly to Pierre Rustin for the invaluable scientific guidance on the biochemical analysis of MRC, which allowed the implementation of these techniques and the development of a Reference Center in our Laboratory.

## References

1. Hatefi Y. The Mitochondrial electron transport and oxidative phosphorylation system. Ann Rev Biochem 1985, 54: 1015–1069.

2. Nelson DL, Cox MM. Oxidative Phosphorylation and Photophosphorylation, in: Lehninger Principles of Biochemistry, 3rd edition, New York, Worth Publishers, 2000, Ch 19: pp 659–721.

3. Schon EA, Dimauro S, Hirano M, Gilkerson RW. Therapeutic prospects for mitochondrial disease. Trends Mol Med. 2010, 16(6): 268–276.

4. Grazina M, Silva F, Santana I, Santiago B, Mendes C, Simões M, Oliveira M, Cunha L, Oliveira C. Frontotemporal Dementia and Mitochondrial DNA Transitions. Neurobiol Dis 2004, 15: 6–11.

5. Bourgeron T, Chretien D, Amati P, Rötig A, Munnich A, Rustin P. Expression of respiratory chain deficiencies in human cultured cells. Neuromusc Disord 1993, 3 (5/6): 605–608.

6. Smeitink J, Heuvel B, Trijbels F, Ruitenbeek W, Sengers R. Mitochondrial Energy Metabolism, in: Physician's Guide to the Laboratory Diagnosis of Metabolic Diseases. Blau N, Duran M, Blaskovics ME, Gibson KM (eds), 2nd edition, Berlin Heidelberg, Springer-Verlag, 2003, Cap 27:pp 519–536.

7. Casademont J, Perea M, López S, Beato A, Miró O, Cardellach F. Enzymatic diagnosis of oxidative phosphorylation defects on muscle biopsy: Better on tissue homogenate or on a mitochondria-enriched suspension? Med Sci Monit 2004, 10(9): CS49-CS53.

8. Bradford MM. A rapid and sensitive method for the quantitation of microgram quantities of protein utilizing the principle of protein-dye binding. Anal Biochem 1976, 72: 248–254.

9. Lowry OH, Rosebrough NJ, Farr AL, Randall RJ. Protein measurement with the Folin phenol reagent. J Biol Chem 1951, 193(1): 265–275.

10. Rickwood MD, Wilson MT, Darley-Usmar V. Isolation and characteristics of intact mitochondria, in: Mitochondria, a Practical Approach. Darley-Usmar VM, Rickwood D, Wilson MT (eds), Oxford, IRL Press, 1987, ch1, pp 1–16.

11. Rustin P, Chrétien D, Bourgeron T, Gérard B, Rötig A, Saudubray JM Münnich A. Biochemical and Molecular investigations in Respiratory Chain Deficiencies. Clin Chim Acta 1994, 228: 35–51.

12. Robinson BH, De Meirleir L, Glerum M, Sherwood G, Becker L. Clinical presentation of mitochondrial respiratory chain defects in NADH-coenzyme Q reductase and cytochrome oxidase: clues to pathogenesis of Leigh disease. J Pediatr. 1987;110(2):216–222.

13. Munnich A, Rötig A, Cormier-Daire V, Rustin P. Clinical Presentation of Respiratory Chain Deficiency, in: The Metabolic & Molecular Basis of Inherited Disease. Scriver CR, Beaudet AL, Sly WS, Valle D (eds), 8th edition, New York, McGraw-Hill, 2001, Vol II, Cap 99: pp 2261–2274.

14. Taylor RW, Birch-Machin MA, Bartlett K, Turnbull DM. Succinate-cytochrome c reductase: assessment of its value in the investigation of defects of the respiratory chain. Biochim Biophys Acta 1993, 1181(3): 261–265.

15. Birch-Machin MA, Briggs HL, Saborido AA, Bindoff LA, Turnbull DM. An evaluation of the measurement of the activities of complexes I-IV in the respiratory chain of human skeletal muscle mitochondria. Biochem Med Metab Biol 1994, 51(1): 35–42.

16. Chrétien D, Rustin P, Bourgeron T, Rötig A, Saudubray JM, Münnich A. Reference charts for respiratory chain activities in human tissues. Clin Chim Acta 1994, 228: 53–70.

17. Miró O, Cardellach F, Barrientos A, Casademont J, Rötig A, Rustin P. Cytochrome c oxidase assay in minute amounts of human skeletal muscle using single wavelength spectrophotometers. J Neurosci Methods 1998, 80(1):107–111.

18. Chrétien D, Gallego J, Barrientos A, Casademont J, Cardellach F, Münnich A, Rötig A, Rustin P. Biochemical parameters for the diagnosis of mitochondrial respiratory chain deficiency in humans, and their lack of age-related changes. Biochem J 1998, 329(Pt 2): 249–254.

19. Rötig A, Cormier V, Blanche S, Bonnefont JP, Ledeist F, Romero N, Schmitz J, Rustin P, Fischer A, Saudubray JM, Munnich A. Pearson's marrow-pancreas syndrome. A multisystem mitochondrial disorder in infancy. J Clin Invest 1990, 86(5): 1601–1608.

20. Cormier-Daire V, Chrétien D, Rustin P, Rötig A, Dubuisson C, Jacquemin E, Hadchouel M, Bernard O, Munnich A. Neonatal and delayed-onset liver involvement in disorders of oxidative phosphorylation. J Pediatr 1997, 130(5): 817–822.

21. Smeitink JAM, Heuvel L, DiMauro S. The Genetics and Pathology of Oxidative Phosphorylation. Nat Gen Rev 2001, 2:342–352.

22. Rötig A, de Lonlay P, Chrétien D, Foury F, Koenig M, Sidi D, Munnich A, Rustin P. Aconitase and mitochondrial iron-sulphur

protein deficiency in Friedreich ataxia. Nat Genet 1997, 17(2): 215–217.

23. Chrétien D, Benit P, Chol M, Lebon S, Rötig A, Munnich A, Rustin P. Assay of mitochondrial respiratory chain complex I in human lymphocytes and cultured skin fibroblasts. Biochem Biophys Res Commun 2003, 301(1): 222–224.

24. DiMauro S, Bonilla E. Mitochondrial encephalomyopathies, *in*: The Molecular and Genetic Basis of Neurological Disease. DiMauro S, Barchi RL (eds), Boston, Butterworth-Heinemann, 1997, pp 201–235.

25. Walker UA, Collins S, Byrne E. Respiratory Chain Encephalomyopathies: A Diagnostic Classification. Eur Neurol 1996, 36(5): 260–267.

26. Bernier FP, Boneh A, Dennett X, Chow CW, Cleary MA, Thorburn DR. Diagnostic criteria for respiratory chain disorders in adults and children. Neurology 2002, 59(9): 1406–1411.

27. Wolf NI, Smeitink JAM. Mitochondrial disorders: a proposal for consensus diagnostic in infant and children. Neurol 2002, 59(1): 1402–1405.

# Chapter 7

# Assays of Pyruvate Dehydrogenase Complex and Pyruvate Carboxylase Activity

## Douglas Kerr, George Grahame, and Ghunwa Nakouzi

## Abstract

Pyruvate dehydrogenase complex (PDC) and pyruvate carboxylase (PC) are mitochondrial enzymes that provide the initial steps of the two main alternatives for pyruvate metabolism: oxidative decarboxylation vs. anaplerotic carboxylation, gluconeogenesis, and glycerogenesis. Assays of the enzymatic activity of these two enzymes in cells and tissues are described in this chapter, based on evolution or fixation of $^{14}CO_2$. These assays are both suitable for use in crude homogenates of cultured skin fibroblasts, lymphocytes, and frozen muscle (PDC) or liver (PC). Activities of these two enzymes are related to spectrophotometric assays of two other mitochondrial enzymes, dihydrolipoamide dehydrogenase (E3) and citrate synthase (CS), providing initial indices of sample integrity and mitochondrial content. These parameters have proven useful for initial detection of inherited human disorders due to deficiencies of these enzymes, and in combination with available genetic analyses can lead to confirmation of specific diagnoses.

Key words: Pyruvate dehydrogenase complex, Pyruvate carboxylase, Dihydrolipoamide dehydrogenase, Citrate synthase, Pyruvate, Radiometric assays, Spectrophotometric assays

## 1. Introduction

### 1.1. Pyruvate Dehydrogenase Complex

Assay of overall activity of pyruvate dehydrogenase complex (PDC) in human samples (skin fibroblasts, lymphocytes, or tissues) has been the primary method for detecting deficiencies of any of the component proteins of PDC. PDC oxidatively decarboxylates pyruvate in the presence of NAD and coenzyme A, forming carbon dioxide, NADH, and acetyl-CoA (AcCoA). PDC is a mitochondrial enzyme and is the regulated "gateway" for oxidation of energy derived from all carbohydrate sources, producing NADH for oxidation by the electron transport chain and AcCoA for further oxidation in the tricarboxylic acid cycle. Regulation depends on the state of phosphorylation of the pyruvate dehydrogenase component of the complex ($E_1$) which is inactive in the phosphorylated

Lee-Jun C. Wong (ed.), *Mitochondrial Disorders: Biochemical and Molecular Analysis*, Methods in Molecular Biology, vol. 837, DOI 10.1007/978-1-61779-504-6_7, © Springer Science+Business Media, LLC 2012

form and active in the dephosphorylated form. These changes are catalyzed by specific kinases and phosphatases, which are naturally inhibited or stimulated by physiological mediators, and these can be manipulated to achieve full activation (or inactivation) in vitro by pretreatment of the sample to be assayed.

Several methods are available for measuring overall PDC activity (1–3). The method most widely used for clinical diagnosis is the production of $^{14}CO_2$ from $1$-$^{14}C$-pyruvate in the presence of thiamine pyrophosphate (TPP) and coenzyme A (CoASH). This relatively difficult method has the advantage that it can be used in crude homogenates of cells and tissues, not requiring separation of mitochondria or partial enzyme purification. Activity of "total" PDC is measured after activation (dephosphorylation by endogenous E1 phosphatase) by preincubation of cells in the presence of dichloroacetate (an inhibitor of the E1 kinase) or by preincubation of frozen tissues after addition of exogenous phosphatase (as the endogenous phosphatase may dissociate with freezing and thawing). Additionally, some laboratories assay the "native" state of PDC activity without preincubation, which may better reflect partial inadequacy of endogenous E1 phosphatases. To confer additional specificity to the overall assay, the inactivated (phosphorylated) state of PDC is assayed after preincubation of whole cells with fluoride (an inhibitor of the phosphatase) or tissue homogenates with fluoride and ATP (a substrate for the kinase).

Coupled spectrophotometric methods generally are not useful in crude cell or tissue homogenates, due to the presence of NADH diaphorases or AcCoA acetylases, but these can be used after immunocapture of the overall complex that partially purifies the enzyme (4). An alternate assay method for PDC to that described here depends on capture of $^{14}C$-acetylcarnitine on an ion exchange column after incubation with $2$-$^{14}C$-labeled pyruvate (5). Immunoblotting assays for the specific protein components of the complex are available. Immunohistochemistry is potentially very useful for detection of partial deficiency of the $E_1$-alpha protein in heterozygous females or males with somatic mosaicism in mixed cell populations, which may be missed by assay of overall PDC activity (6). Since mutations of the *PDHA1* gene are found in about two-thirds of PDC deficient patients, direct exon sequencing of *PDHA1* recently has been used directly for diagnosis without enzymatic assay of PDC or to follow up in those cases with strong suspicion of PDC deficiency that have normal PDC activity.

## 1.2. Dihydrolipoamide Dehydrogenase ($E_3$) (and Other Components of PDC)

PDC catalyses three enzymatic steps: (1) decarboxylation of pyruvate in the presence of TPP by pyruvate dehydrogenase ($E_1$), including the alpha and beta subunits encoded by the *PDHA1* (X-linked) and *PDHB* genes respectively; (2) transfer of the hydroxyethyl moiety from TPP to CoASH via lipoic acid, covalently bound to dihydrolipoamide transacetylase ($E_2$), encoded by the *DLAT* gene; and (3) oxidation of lipoic acid and reduction of NAD to NADH by $E_3$, encoded by the *DLD* gene (7). Activities of these PDC component enzymes can be assayed separately. However, the $E_1$ and $E_2$ assay methods are not physiological, and the overall rate of the complex is most useful for initial determination of possible PDC deficiency. Only the assay for $E_3$ is described here, as this is used both as a mitochondrial reference enzyme in the PDC assay and simultaneously for direct detection of $E_3$ deficiency. This is possible because $E_3$ deficiency is relatively rare, $E_3$ is a component of not only PDC but also the other α-ketoacid dehydrogenases, and if low $E_3$ activity is found, then an additional assay of CS is performed (as an alternative mitochondrial reference enzyme). The spectrophotometric assay method for $E_3$ described here is based on free lipoamide-dependent oxidation of NADH to NAD, measured by the change of absorbance at 340 nm. Lipoamide is used as a surrogate substrate for the natural substrate, covalently bound lipoamide as part of the $E_2$ component of α-ketoacid dehydrogenases (8, 9).

### *1.3. Pyruvate Carboxylase*

PC is a mitochondrial enzyme that catalyses energy-dependent carboxylation of pyruvate to oxaloacetate and dephosphorylation of ATP to ADP. PC is covalently bound to biotin and requires AcCoA for active conformation. This important metabolic reaction is the initial step for pyruvate entry into what are known as anaplerotic pathways, including gluconeogenesis, glycerogenesis, replenishment of intermediates of the tricarboxylic acid and urea cycles, and biosynthesis of nonessential amino acids (7). Activity of PC is greatest in liver and kidney but is also expressed in many other tissues, including skin fibroblasts, adipose tissue, and the central nervous system. PC is a homodimeric protein, encoded by the *PC* gene.

The clinical assay of PC in crude homogenates, described here, is based on pyruvate-dependent fixation of $^{14}CO_2$, coupled with conversion of the oxaloacetate product to citrate in the presence of exogenous CS and AcCoA (3, 10). Linearity is established with respect to time and amount of sample protein. In homogenates of cells and tissue samples, this method has the advantage of avoiding interference from competing reactions for pyruvate and oxaloacetate or other $CO_2$ fixing reactions. Assay of endogenous CS is used as an internal indicator of mitochondrial content (see below). The disadvantages of using volatile radioactive $^{14}CO_2$ are significant but manageable with standard precautions. Several spectrophotometric assays are available for partially purified PC (11). Sequencing of the *PC* gene is also available in clinical laboratories.

### *1.4. Citrate Synthase*

Assay of CS is included in this chapter because it is used as a primary mitochondrial reference enzyme for PC and as a secondary mitochondrial reference enzyme for PDC. The spectrophotometric assay employed is based on oxaloacetate-dependent generation of free CoASH, measured as the reduction of dithionitrobenzoate (DTNB; increase of absorbance at 412 nm) due to release of the free thiol group of CoASH by CS (12).

## 2. Materials

All reagents are prepared using ultrapure water (purified deionized water to 18 MΩ cm at 25°C) and analytical grade reagents, unless noted otherwise.

### 2.1. Pyruvate Dehydrogenase Complex

#### 2.1.1. Reagents

1. Dulbecco's phosphate buffered saline (PBS; 1× liquid) with 1 mM calcium and 1 mM magnesium.
2. KCl-MOPS buffer, pH 7.4: 80 mM KCl, 50 mM MOPS, 2 mM $MgCl_2$, 0.5 mM EDTA. Dissolve 592 g of KCl, 1.04 g of MOPS, 40.6 mg of $MgCl_2$, and 18.6 mg of EDTA in 50 ml water. Adjust pH to 7.4 with 1N NaOH. Bring up to final volume 100 ml. Store at 4°C, good for 6 months.
3. Protease inhibitors (PI):

   0.1 mg/ml leupeptin hemisulfate solution. Store at –20°C, good for 1 year.

   0.1 M phenylmethanesulfonyl fluoride (PMSF): Add 17.4 mg PMSF to 1 ml isopropyl alcohol. Store at 25°C, good for 1 year.

4. PBS + PI (for fibroblast and lymphocyte sample preparation): Dilute each inhibitor 1/200 in PBS to a final concentration of 0.5 μg/ml leupeptin and 0.5 mM PMSF. For example, dilute 20 μl leupeptin + 20 μl PMSF in 4.0 ml of PBS. Prepare fresh.
5. KCl-MOPS + PI (for tissue sample preparation): Dilute each inhibitor 1/200 in KCl-MOPS to a final concentration of 0.5 μg/ml leupeptin and 0.5 mM PMSF. Prepare fresh.
6. Fetal calf serum (FCS): Freeze FCS in 10 ml aliquots in –80°C freezer. Then thaw an aliquot as needed and store at –20°C freezer in 400 μl aliquots.
7. 125 mM dichloroacetate (DCA), pH 6–7: Weigh out 16.1 g of DCA and dissolve in 80 ml of water. Adjust pH with 1N NaOH to 6.5 ± 0.5. Add water to 100 ml. Store at 4°C, good for 1 year.
8. 1.5 M sodium fluoride (NaF): Weigh out 189 mg and dissolve in 3 ml of water. Store at 4°C, good for 6 months.
9. Stopping solution: Dissolve 105 mg NaF, 930 mg EDTA, and 62 mg DTT in 40 ml of water. Adjust to pH 7.4 with 1N NaOH.

Add 40 ml absolute ethanol and bring to a final volume of 100 ml with water. Store at 4°C, good for 1 year.

10. Stopping mix: Dilute FCS 1/8 in stopping solution, prepare fresh.

11. PBS + PI/stopping solution/FCS solution: (additive for fibroblast or lymphocyte assays) Combine 800 µl of PBS + PI, 175 µl of stopping solution, and 25 µl of FCS. Prepare fresh (volume sufficient for 100 reaction tubes).

12. 50 mM $CaCl_2$: 7.4 mg/ml. Store at 4°C, good for 6 months.

13. 0.5 M $MgCl_2$: 100 mg/ml. Store at 4°C, good for 6 months.

14. 100 mM $MgCl_2$: 20 mg/ml. Store at 4°C, good for 6 months.

15. 300 mM ATP: 165 mg/ml. Store at –20°C in 0.2 ml aliquots, good for 1 year.

16. 200 mM potassium phosphate ($KPO_4$) buffer, pH 8: Dissolve 544 mg of $KH_2PO_4$ in 20 ml water. In a separate container, dissolve 6.96 g of $K_2HPO_4$ in 200 ml water. Adjust $K_2HPO_4$ solution to pH 8.0 by addition of $KH_2PO_4$ solution. Store at 4°C, good for 6 months.

17. 60 mM potassium oxalate (KOx): Dissolve 1.1 g of KOx into 100 ml of water. Store at 4°C, good for 6 months.

18. NAD: Store in desiccator at –20°C.

19. 1,000 U/ml phosphotransacetylase (PTA, in 50% glycerol) (≥3,000 U/mg). Store at 4°C, good for 1 year.

20. 20 mM dithiothreitol (DTT): Dissolve 30.8 mg in 10 ml of water. Prepare DTT fresh for each PDC assay. Store remainder at –20°C for E3 assay, good for 2 weeks.

21. 10 mM thiamine pyrophosphate (TPP): Dissolve 46.1 mg in 10 ml of water. Store at 4°C, good for 6 months.

22. Coenzyme A (CoASH): Store in desiccator at –20°C.

23. 1 mg/ml NADH solution: Dissolve 10 mg NADH in 10 ml of water. Store at 4°C, good for 2 months.

24. Trizma base (solution): Store at 4°C.

25. Approximately 1,000 U/ml lactate dehydrogenase (LDH, use as supplied). Store at 4°C.

26. Approximately 15 mCi/mmol Na-1-$^{14}$C-pyruvate stock: Total 250 µCi. Store at 4°C until ready to prepare (see below).

27. 50 mM pyruvic acid: Dissolve 55 mg in 10 ml of 30 mM HCl. Prepare fresh.

28. Na-1-$^{14}$C-pyruvate working solution: Dilute 250 µCi of Na-1-$^{14}$C-pyruvate to approximately 2,000 dpm/nmol in 30 mM HCl (final prep = 250 µCi/125 µmol/2.5 ml): Add 2.5 ml of

50 mM pyruvic acid to 250 µCi of Na-1-$^{14}$C-pyruvate stock solution. Store at –80°C, good for 3 months.

(a) Assay concentration and specific activity of Na-1-$^{14}$C-pyruvate working solution: In four 1 ml spectrophotometric cuvettes (in duplicate for sample and blank), add 590 µl of water, 150 µl of NADH solution, 250 µl of Trizma base solution, and 1 µl of Na-1-$^{14}$C-pyruvate working solution (omit in blanks). Mix cuvettes. Read baseline absorbance ($E_1$) in a spectrophotometer at 340 nm. Add 10 µl of LDH to each cuvette (total volume equals 1.0 ml) and mix again. Allow 5 min for reaction to go to completion and then take endpoint reading ($E_2$). Average the duplicate readings of $E_1$ and $E_2$ for the sample and blank.

(b) Calculation of Na-1-$^{14}$C-pyruvate concentration and specific activity: Pyruvate concentration (nmol/µl) = $[(E_2 - E_1)_{sample} - (E_2 - E_1)_{blank}]/0.00622/\mu M$      (NADH extinction coefficient).

(c) Pyruvate radioactivity (total counts in dpm): Count 25 µl from each reaction cuvette in scintillation counter (1/40th of total sample); dpm × 40 = total counts.

(d) Pyruvate specific activity (dpm/nmol) = pyruvate radioactivity/pyruvate concentration; approximately 2,000 (see Note 1).

29. Pyruvate dehydrogenase phosphatase, from bovine kidney (available from Sigma), approximately 850 U/mg protein (20 µg protein/ml). Store at –80°C, good for 3 months.

30. Phosphate buffered saline with 0.5 mM EGTA, without Ca and Mg (PBS with EGTA, 10×): Dissolve 0.40 g KCl, 16 g NaCl, 0.40 g $KH_2PO_4$, 4.32 g $Na_2HPO_4 \cdot 7H_2O$, and 0.38 g EGTA in 140 ml water. Adjust pH to 7.4 with 1N NaOH. Bring volume up to 200 ml. Dilute 1/10 for working solution. Store both at 4°C, good for 6 months (used for harvesting cultured fibroblasts).

31. Trypsin solution: Combine 200 mg trypsin, 160 mg KCl, 3.6 g NaCl, 400 mg dextrose, 232 mg $NaHCO_3$, 48 mg EDTA, and 0.4 ml of phenol red (2 mg/ml) in 280 ml water. Adjust pH to 7.4 with 0.1N HCl. Bring volume up to 400 ml. Store at 4°C, good for 2 months (used for harvesting cultured fibroblasts).

32. Hank's balanced salt solution (HBSS): Store at 25°C, expiration date on bottle (used for harvesting cultured fibroblasts).

33. Ficoll-Paque: Store at 4°C, expiration date on bottle (used for separation of lymphocytes from blood).

34. 0.2% NaCl: Dissolve 500 mg NaCl in 250 ml of water. Store at 25°C, good for 1 month (used for separation of lymphocytes from blood).

35. 1.6% NaCl: Dissolve 4.0 g in NaCl in 250 ml of water. Store at 25°C, good for 1 month (used for separation of lymphocytes from blood).

36. Tissue homogenate activating solution: On day of assay, for each tissue sample plus one (multiply volumes by $n+1$), mix 1 μl of PDH phosphatase, 44 μl of KCl-MOPS+PI, 10 μl of 125 mM DCA, 5 μl of 0.5 M $MgCl_2$, and 5 μl of 50 mM $CaCl_2$ (volume of mix to be added to each sample is 65 μl).

37. Tissue homogenate inactivating solution: On day of assay, for each tissue sample plus one (multiply volumes by $n+1$), mix 7.5 μl of 1.5 M NaF, 7.5 μl of ATP, and 50 μl of KCl-MOPS+PI (volume of mix to be added to each sample is 65 μl).

38. PDC assay mixture: (example volumes sufficient for 100 total tubes) On day of assay, mix 0.20 ml of 100 mM $MgCl_2$, 14 mg of NAD, 0.01 ml of 1,000 U/ml PTA, 2.5 ml of 200 mM $KPO_4$ pH 8.0, 2.5 ml of 60 mM KOxalate, 0.25 ml of FCS, and 1.44 ml water (final volume = 6.9 ml or 69 μl/tube).

39. Solution A: (example volumes sufficient for 100 reaction tubes) On day of assay, mix 500 μl water + 400 μl DTT + 100 μl TPP + 5 mg CoASH.

40. Solution B: (example volume sufficient for 100 blank tubes) On day of assay, mix 600 μl water + 400 μl DTT.

41. 1.0 M hyamine hydroxide in methanol. Store at 4°C.

42. Bovine serum albumin (BSA): Lyophilized powder, ≥96% albumin.

43. 0.1–1 U/mg protein PDC enzyme standard: Dilute an aliquot to approximately 0.1 U/ml (=0.001 U/10 μl) with KCl-MOPS buffer containing 10 mg/ml BSA. Store at –80°C, good for 3 months for diluted enzyme.

44. Trichloroacetic acid (TCA) + pyruvate solution: 20% TCA solution, 30 mM Na pyruvate. Dissolve 20 g of TCA and 33 mg of Na pyruvate in a total volume of 100 ml. Store at 4°C, good for 6 months.

45. Scintillation fluid (compatible with methanol): Store at 25°C.

46. Dry ice, acetone, and container for mixture.

47. Protein assay reagents for Lowry method (see Note 2).

*2.1.2. Reaction Tubes and $CO_2$ Traps*

1. Polypropylene tubes (conical, 12×75 mm) with caps, two per sample (activated and inactivated preparations for each sample).

2. Polystyrene tubes (12×75 mm) with caps.

3. Glass tubes (round bottom, 16×100 mm), 12 needed per sample, including blanks, standards, and samples.

4. Stopper top (Kimble Chase Kontes), one per glass tube.

5. Stemmed center wells (Kimble Chase Kontes), one per glass tube.

6. Filter paper (Grade 1) cut into small square (~1 cm²), one per center well.

7. Scintillation vials: polyethylene midi-vial, 8 ml and caps (see Note 3).

*2.1.3. Equipment*

1. Shaking water bath, 37°C, set for slow, gentle oscillation.

2. Liquid scintillation counter (see Note 4).

3. Repeating glass micro syringe/dispensers (e.g., Hamilton), 1,000, 500, and 50 μl capacity (set for 20, 10, and 1 μl/injection, respectively).

4. Tissue homogenizer (see Note 5).

5. Centrifuge compatible with 4- to 50-ml conical tubes at $1,000 \times g$ and with a swinging-bucket rotor set for 10- and 50-ml tubes.

6. UV–Vis spectrophotometer for kinetic studies, with temperature-controlled cuvette holder.

*2.1.4. Preparation of Cells and Tissue Samples*

Cultured skin fibroblasts: (see Notes 6 and 7)

1. One confluent T-75 flask is used to assay PDC for each unknown subject and for each control. This typically will provide 1–2 mg of total cell protein.

2. On day of assay, harvest cells with trypsin solution, wash twice with PBS. Resuspend cell pellet in 500 μl PBS + PI (see Note 8).

3. Assay 10 μl aliquots, in triplicate, for protein concentration (see Note 2).

Blood lymphocytes:

1. Whole blood should be collected in ACD (adenosine/citrate/dextrose) tubes (6 ml for an infant, 10 ml for a child up to 3 years, and 20 ml for an older child or adult). A concurrent control sample should be obtained from an unrelated adult volunteer (see Note 9).

2. On day of assay, isolate lymphocyte fraction with Ficoll-Paque, following manufacturer's instructions (see Note 10).

3. Resuspend cells in 375 μl of PBS + PI. Remove 10 μl aliquots in triplicate for assay of protein concentration (see Note 2).

Tissue samples (muscle, heart, liver, etc.):

1. Tissue sample should be stored at –80°C until ready for assay (see Note 11). On day of the assay, an aliquot of approximately 30 mg frozen tissue is homogenized on ice using 200 μl KCl-MOPS + PI (see Note 5). Homogenize thoroughly for at least

30 s. Dilute the homogenate to protein concentration of 50 mg/ml with KCl-MOPS + PI solution.

2. Dilute a 50-μl aliquot with 50 μl KCl-MOPS + PI. Divide into 3 × 10 μl aliquots for protein assay (1/2 dilution) (see Note 2).

***2.2. E3***

1. 30 mM EDTA: Store at 4°C, good for 6 months.

2. 10% Triton X-100 solution: Store at 4°C.

3. 50% Ethanol solution: Store at 4°C.

4. Reagents for preparation of $LAH_2$ powder stock.

   (a) Lipoamide (thiotic amide).

   (b) Methanol.

   (c) Sodium borohydride ($NaBH_4$).

5. Preparation of $LAH_2$ powder stock.

   (a) Weigh 400 mg of lipoamide into a 50 ml glass beaker with a small stirring bar. Add 8 ml of methanol and 2 ml of GDW. Put in ice tray on a magnetic stirrer to keep the particles in suspension. Mix 400 mg of $NaBH^4$ in 2 ml GDW, mixing gently. Add $NaBH_4$ solution to the cold lipoamide suspension.

   (b) Stir on ice until solution is clear and colorless (can take up to 90 min) (see Note 12).

   (c) Carefully adjust pH to about 2 with 1N HCl.

   (d) Transfer to two 30-ml glass conical stoppered centrifuge tubes. In a ventilated fume hood, add 20 ml chloroform to each tube and shake well to extract $LAH_2$ into chloroform. Lift stopper often to release pressure.

   (e) Centrifuge for 5 min at $1,000 \times g$ to separate layers. In a ventilated hood, remove aqueous (top) layer and discard. Transfer chloroform (bottom) layer to a 100-ml glass beaker. Dry overnight in fume hood. White precipitate on beaker is $LAH_2$.

   (f) When completely dry, scrape $LAH_2$ off beaker, weigh, and transfer to capped 12 × 75 polypropylene conical tube. Store in desiccator at −20°C.

6. Dihydrolipoamide ($LAH_2$) solution: 8.3 mg/ml in 50% ethanol, prepare fresh each reaction.

***2.3. Pyruvate Carboxylase***

*2.3.1. Reagents*

1. 200 mM $KPO_4$ solution: Dissolve 136 mg of $KH_2PO_4$ in 5 ml water. Dissolve 870 mg $K_2PO_4$ in 25 ml water. Adjust pH of $K_2PO_4$ solution to 7.4 with $KH_2PO_4$. Store at 4°C, good for 1 year.

2. Homogenizing buffer for tissues (HB): 1.5 M sucrose, 50 mM $KPO_4$ solution, 0.5 M EDTA. To ~50 ml water, add 51 g of sucrose, 18.6 mg of EDTA, and 25 ml of 200 mM $KPO_4$ solution. Bring to final volume of 100 ml. Store at 4°C, good for 1 year.

3. Protease inhibitors (same as for PDC).

4. MOPS + PI: (for fibroblast sample preparation) Dilute each inhibitor 1/200 in MOPS buffer. Final concentration = 0.5 μg/ml leupeptin, 0.5 mM PMSF. Prepare fresh each assay.

5. HB + PI: (for liver preparation) Dilute each inhibitor 1/200 in PC-HB. Final concentration = 0.5 μg/ml leupeptin, 0.5 mM PMSF. Prepare fresh each assay.

6. 0.5 mM Tris–HCl buffer, pH 7.8: Dissolve 12.1 g Tris in water, adjust to pH 7.8 with HCl 5N, final volume 200 ml. Store at 4°C, good for 1 year.

7. 100 mM $MgCl_2$ (same as for PDC).

8. 100 mM ATP: 55 mg/ml. Adjust to pH 6 with 1N NaOH before bringing to final volume. Store at –20°C, good for 1 year.

9. 10% Triton X-100: (same as for E3).

10. 1 M $KHCO_3$: 100 mg/ml.

11. Approximately 1 mCi/18 μmol Na-$^{14}$C-bicarbonate (NaH$^{14}$CO$_3$) stock: (≈53 mCi/mmol in 0.5 ml). Store in desiccator at 4°C.

12. NaH$^{14}$CO$_3$ diluted working solution: Combine 0.5 ml of NaH$^{14}$CO$_3$ stock, 8.3 ml of 1 M $KHCO_3$, and 1.19 ml water. Final concentration = 0.1 mCi/833 μmol/ml (1 μCi/ 8,330 nmol/10 μl). Check specific activity in PC reaction mixture, as follows:

    (a) Determine total $^{14}CO_2$ specific activity: Transfer 25 μl of PC reaction mixture (see Subheading 2.3.1, step 25 below) directly to two scintillation vials, add 4 ml scintillation fluid, and count immediately for 5 min in scintillation counter. Specific activity (dpm/nmol) = (dpm × dilution factor)/ nmol, where dpm = radioactivity counted in vial, dilution factor = 440/25, and nmol = nmol of $CO_2$ in 440 μl PC reaction mix. This corresponds to 10 μl of NaH$^{14}$CO$_3$ working solution and 5 μl of 1 M $KHCO_3$ (8,330 + 5,000 nmol = 13,330 total nmol) in 440 μl of reaction mixture. Calculated specific activity, for example, would be [102,733 dpm × (440/25)]/ 13,330 nmol = 136 dpm/nmol $HCO_3^-$.

    (b) Determine nonvolatile radioactivity (reagent blank): In a fume hood, to each of two 16 × 100 round bottom glass tubes, add 440 μl of PC reaction mixture, 60 μl MOPS + PI, and 250 μl 20% TCA. Bubble $CO_2$ through solution for 10 min. Transfer 0.5 ml to scintillation vial, add 4 ml scintillation fluid, and count for 5 min in scintillation counter.

13. Approximately 40 U/ml CS stock (>100 U/mg): (use as supplied). Store at 4°C.

14. 0.5 M pyruvic acid: Dissolve 55 mg pyruvic acid in 1 ml 0.01N HCl. Prepare fresh each assay, keep on ice (stock reagent stored in desiccator at 4°C).

15. 10% TCA: 10 g/100 ml. Store at 4°C, good for 6 months.

16. Scintillation fluid (same as for PDC).

17. PC enzyme standard: (see Note 13) May use frozen rat liver or purified PC from bovine liver (e.g., Sigma #P-7173, approximately 20–25 U/mg, 50 U/ml), store PC stock at –20°C in desiccator. Dilute 1/400 with HB on day of assay.

18. 1 mM dithionitrobenzoic acid (DTNB): 3.96 mg/ml in Trizma base solution. Prepare fresh (used in AcCoA preparation).

19. 0.05 M oxaloacetic acid (OAA): 6.6 mg/ml. Store at –20°C in ~2 ml aliquots, good for 1 year (used in AcCoA preparation).

20. 0.1 M $KHCO_3$: 10 mg/ml. Prepare fresh (used for AcCoA preparation).

21. 3 M $KHCO_3$: 300 mg/ml. Store at 25°C (used in AcCoA preparation).

22. Acetic anhydride, store in safety cabinet under hood (used in AcCoA preparation).

23. Coenzyme A (CoASH) (same as for PDC; used in AcCoA preparation).

24. Preparation of 7 mM AcCoA working solution:

    (a) Dissolve 200 mg CoASH in 8 ml fresh cold 0.1 M $KHCO_3$.

    (b) Add 100 μl of acetic anhydride to form AcCoA.

    (c) Adjust pH to 7 with 3 M $KHCO_3$. Keep on ice.

    (d) Dilute DTNB 1/10 with water. Add 1 μl AcCoA to 200 μl of this diluted DTNB solution. Check for yellow color (there should be none, indicating that all of the free CoASH sulfhydryl group has been acetylated). Adjust pH to 7 with 3 M $KHCO_3$.

    (e) Determine concentration of AcCoA: In four 1 ml cuvettes, add 675 μl water, 290 μl of 0.5 M Tris–HCl buffer (pH 7.8), 10 μl of 1 mM DTNB, and 5 μl CS. To two of the four cuvettes, add 10 μl water (blank). To the two remaining cuvettes, add 10-μl diluted (1/10) AcCoA working solution (sample). Mix and read absorbance at 412 nm (E1). Add 10 μl 0.05 M OAA to all cuvettes, mix and incubate for 1 min. Read endpoint absorbance (E2). Average the duplicate readings of E1 and E2 for the sample and blank. AcCoA concentration (μmol/ml): $=[(E_2 - E_1)_{sample} - (E_2 - E_1)_{blank}] \times 10/0.01$ ml $\times 13.6$ (DTNB extinction coefficient).

    Dilute to final concentration of 7 μmol/ml (7 mM). Store at –20°C in 1 ml aliquots, good for 1 year (see Note 14).

25. PC reaction mixture: Prepare on day of assay (see Note 15 before preparing this for the first time after diluting a new batch of $NaH^{14}CO_3$). The following example is sufficient for 50 tubes: Mix 2.5 ml of 0.5 M Tris–HCL buffer pH 7.8, 1.0 ml of 0.1 M $MgCl_2$, 0.5 ml of 0.1 M ATP, 1.0 ml of 7 mM AcCoA, 0.62 ml of 10 mg/ml Triton X-100, 0.5 ml of $NaH^{14}CO_3$ diluted working solution, 0.25 ml of 1 M $KHCO_3$, 50 µl of CS, and 15 ml of water (final volume 22.0 ml). Before each assay, check specific activity of $NaH^{14}CO_3$ of reaction mix (see step 13).

*2.3.2. Preparation of Cells and Tissue Samples*

Cultured skin fibroblasts:

1. One confluent T-75 flask is used for each unknown subject and control (see Notes 6 and 7).

2. On day of assay, harvest cells with trypsin solution, wash with PBS. Resuspend cell pellet in 500 µl MOPS + PI (see Note 8).

3. Remove 10 µl aliquots of cell suspension in triplicate to assay protein (see Note 2).

4. Dilute remaining cell suspension with MOPS + PI to make final protein concentration = 2 mg/ml (50 µg/25 µl, 100 µg/50 µl, etc.).

5. Freeze/thaw cells twice prior to assay (freeze in dry ice/acetone bath, thaw in 37°C water bath). Keep samples frozen during second freeze, thaw just prior to assay, and then keep on ice during assay.

Tissue samples (liver):

1. Tissue samples should be stored at −80°C until ready for assay. On day of the assay, a 5- to 10-mg aliquot of frozen liver is homogenized on ice using 200 µl HB + PI.

2. Remove 20 µl aliquots of homogenate in triplicate to assay protein (see Note 2).

3. Freeze/thaw tissue suspension twice prior to assay and keep samples frozen during second freeze, until ready to start assay incubation, then keep on ice during assay.

*2.4. Citrate Synthase*

1. 0.5 mM Tris–HCl buffer, pH 7.8 (same as for PC assay).

2. 1 mM dithionitrobenzoic acid (DTNB) (same as for PC assay).

3. 7 mM AcCoA (same as PC assay).

4. 10% Triton X-100 (same as PC assay).

5. 0.05 M Oxaloacetic acid (OAA) (same as for PC assay).

6. CS reaction mixture: (prepare on day of assay) Example sufficient for 50 cuvettes: Mix 10.0 ml of 0.5 M Tris–HCL,

pH 7.8; 0.50 ml of 10 mM DTNB, 0.72 ml of 7 mM AcCoA, 2.50 ml of 10 mg/ml Triton X-100, and 34.3 ml of water (total volume = 48.0 ml). Warm to 37°C in water bath, and maintain at 37°C during assay.

7. CS enzyme standard solution: Dilute CS stock (same as for PC) 1/16,000 before each assay as follows: Add 25 μl of CS stock to 975 μl Trizma base solution. This solution is then diluted 1/400 in CS reaction mixture.

# 3. Methods

## 3.1. PDC Assay

1. Activation of cells and tissue samples (see Note 16).

   (a) Fibroblasts: Aliquot 200 μl of fibroblast cell suspension into each of two 12×75 mm polypropylene conical tubes, labeled "activated" and "inactivated"

   - Add 10 μl 125 mM DCA to "activated" tube.

   - Add 10 μl 1.5 M NaF to "inactivated" tube.

   - Incubate both tubes for 15 min at 37°C. Add 50 μl stopping mix (final volume per sample = 260 μl). Freeze/thaw cells twice before assay (freeze in dry ice/acetone bath, thaw in 37°C water bath). Keep samples frozen at second freeze until all tubes and reagents are prepared. Thaw samples when ready to start assay incubation.

   (b) Blood lymphocytes: Aliquot 160 μl of lymphocyte cell suspension into each of two 12×75 mm polypropylene conical tubes labeled "activated" and "inactivated"

   - Add 8 μl of 125 mM DCA to each "activated" tube.

   - Add 8 μl of 1.5 M NaF to each "inactivated" tube.

   - Incubate both tubes for 15 min at 37°C. Add 40 μl stopping mix (final volume per sample = 208 μl). Freeze/thaw cells twice and keep samples frozen at second freeze until ready to assay.

   (c) Tissue samples (muscle, heart, liver, etc.) (see Note 17)

   - Aliquot 200 μl of diluted homogenate into each of two 12×75 mm polypropylene conical tubes, labeled "activated" and "inactivated". Freeze/thaw these tubes twice to complete tissue disruption. Keep samples at second freeze until ready to add activating and inactivating solutions.

   - Add 65 ml of tissue homogenate activating solution to each tube labeled "activated."

- Add 65 μl of tissue homogenate inactivating solution to each tube labeled "inactivated". Incubate both tubes for 15 min at 37°C (final volume per sample = 265 μl). Freeze these tubes again and keep frozen until ready to assay.

2. Preincubate Na-1-¹⁴C-pyruvate: Measure an aliquot of Na-1-¹⁴C-pyruvate working solution equivalent in microliter to the number of assay tubes + 20% (i.e., for an assay of 80 tubes, 100 μl of Na-1-¹⁴C-pyruvate is used). Pipette 100 μl of 1 M hyamine into the bottom of a 12×75-mm polystyrene tube. Pipette the Na-1-¹⁴C-pyruvate into a 0.25-ml microtube and place the uncapped microtube into the 12×75 tube containing hyamine. Cap the 12×75 tube and incubate at 37°C for 1 h (to remove any volatile radioactivity). After incubation, the capped 12×75 tube containing the Na-1-¹⁴C-pyruvate is kept on ice throughout the assay and discarded as radioactive waste at the end of the assay. Measure the specific activity per aliquot at the end of the assay (see preparation of Na-1-¹⁴C-pyruvate).

3. Prepare Kontes stemmed center wells with stoppers: one for each assay or blank tube. Place stem of center well through stopper, put small square (about 1 cm²) of filter paper in well, and add 100 μl hyamine hydroxide to the well to soak the filter paper (see Fig. 1).

4. Prepare reaction tubes (activated and inactivated):

   (a) For activated: Label eight 16×100-mm glass tubes (4 blanks and 4 reactions) with a unique identifier for each sample (patient, control, and PDC standard). Two protein concentrations at two incubation periods (5 and 10 min) are run for the sample blanks and reactions.

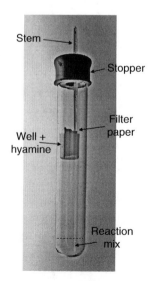

Fig. 1. PDC reaction tube (Kerr, Grahame, and Nakouzi, 04/21/2011).

**Table 1**
**Plan for assay of PDC (one sample)**

| Tube | Type of solution | Sample volume (μl) | Time | Function |
|------|------------------|--------------------|------|----------|
| #1 | B | 10 | 5′ | Blank |
| #2 | A | 10 | 5′ | Reaction |
| #3 | B | 20 | 5′ | Blank |
| #4 | A | 20 | 5′ | Reaction |
| #5 | B | 10 | 10′ | Blank |
| #6 | A | 10 | 10′ | Reaction |
| #7 | B | 20 | 10′ | Blank |
| #8 | A | 20 | 10′ | Reaction |

(b) For inactivated: Label four 16×100-mm glass tubes (2 blanks and 2 reactions) with a unique identifier for each sample (patient and control). Two protein concentrations at one incubation period (10 min) are run for the sample blanks and assays. Perform the activated PDC reaction first followed by the inactivated PDC reaction.

(c) Add reagents for preincubation (see Table 1):
 • To each 16×100-mm glass tube, add 69 μl of PDC assay mixture.
 • To tubes which will be getting only 10 μl of cell/tissue sample, add 10 μl PBS + PI/stopping solution/FCS (for fibroblast and lymphocyte assays) or 10 μl KCl-MOPS + PI (for tissue assays).
 • To blank tubes, add 10 μl of solution B; to reaction tubes, add 10 μl of solution A.

(d) Preincubation: (see Note 16) Place all unstoppered tubes in 37°C shaking water bath. Thaw sample and draw up entire volume into a Hamilton repeating dispenser (10 μl/injection, ×1 or ×2) (see Table 2). Time exactly, starting at 0 s, add specific sample volume (see table above) to each of the eight designated tubes, adding sample to consecutive tubes at 10-s intervals. Place tip of syringe directly in reaction mixture when adding sample (save remaining sample on dry ice or at −80°C for E3 assay).

(e) Start reactions: Exactly 90 s after adding sample, add 1 μl of Na-1-$^{14}$C-pyruvate to each tube, using the 50 μl glass repeating syringe/dispenser (set for 1 μl/injection). Place tip of syringe directly in reaction mixture when adding 1-$^{14}$C-pyruvate. Immediately insert a stopper fitted with

**Table 2**
**Incubation time table for PDC assay (for one sample set)[a]**

| Tube # | (Add) Sample | (Start) [14]C-Pyr | (Stop) TCA[b] | (Reaction time) |
|---|---|---|---|---|
| (volume, μl) | (10, 20 μl) | (1 μl) | (20 μl) | |
| 1–10 | 00″ | 1′ 30″ | 6′ 30″ | 5′ |
| 2–10 | 10″ | 1′ 40″ | 6′ 40″ | 5′ |
| 3–20 | 20″ | 1′ 50″ | 6′ 50″ | 5′ |
| 4–20 | 30″ | 2′ 00″ | 7′ 00″ | 5′ |
| 5–10 | 40″ | 2′ 10″ | 12′ 10″ | 10′ |
| 6–10 | 50″ | 2′ 20″ | 12′ 20″ | 10′ |
| 7–20 | 1′ 00″ | 2′ 30″ | 12′ 30″ | 10′ |
| 8–20 | 1′ 10″ | 2′ 40″ | 12′ 40″ | 10′ |

[a]′ = minute, ″ = second
[b]See Section 2.1.1, step 44

stemmed center well and hyamine saturated filter paper and seal the skirt of the stopper over the top of the tube and then replace the tube into the water bath.

(f) Stop reactions: (Tubes # 1–4) Exactly 5 min after adding Na-1-[14]C-pyruvate, add 20 μl of TCA+pyruvate solution using a 1-ml repeating dispenser/syringe (20 μl/injection), inserting the needle through the center of the stopper and down the side of the tube (do not allow needle tip to touch the stemmed well containing the hyamine saturated filter paper). (Tubes # 5–8) Exactly 10 min after adding Na-1-[14]C-pyruvate, add 20 μl of TCA+pyruvate as above.

(g) Complete absorption of $^{14}CO_2$: Keep samples in the 37°C water bath for at least 30 min (no more than 2 h) to allow time to collect evolved $^{14}CO_2$ in the hyamine.

(h) Count $^{14}CO_2$: Remove tubes from the water bath, remove the serum stopper with the center well, and cut the stem of the center well dropping it and the filter paper into a midi-scintillation vial. Add 4 ml scintillation fluid to the midi-vial, cap tightly, mix well, and place in the liquid scintillation counter. Samples are dark-adapted for 10 h then radiation is measured for 5 min per sample.

(i) Calculation of results: dpm in assay tube – dpm in blank tubes (mean of 4 blanks)/specific activity of 1-[14]C-pyruvate

$(\text{dpm}/\text{nmol}) \times \text{sample protein}(\text{mg}) \times \text{assay time}(\text{min}) = \text{nmol}$ of $CO_2/\text{mg protein}/\text{min}$

Calculate results separately for each of four replicates and average the replicates. If one of the four replicates is an outlier (>2 SDs from mean of other 3), express the average of the other samples.

(j) Interpretation: Report results of "activated" ("total" or dephosphorylated) PDC and "inactivated" (phosphorylated) PDC (see Note 18). Compare activity of unknown samples with concurrent controls and reference range of prior controls (3rd to 97th percentile, mean, and number of prior controls; see Note 19). Report PDC activity result relative to both total cell protein concentration and to the activity of the mitochondrial reference enzyme E3 (or CS activity, if E3 is low).

*3.2. E3 Assay*

1. Prepare E3 reaction mixture: Mix 25 ml of 200 mM KPO₄ pH 8, 5 ml of 30 mM EDTA, 5 ml of 20 mM DTT, 5 ml of Triton X-100, 100 mg of NAD, and 56 ml water (total volume = 96 ml). Warm to 37°C in water bath. Maintain at 37°C during assay.

2. For each sample, prepare six spectrophotometric cuvettes (see Table 3).

3. Mix cuvettes and read at 340 nm, preferably on a kinetics capable spectrophotometer with temperature-controlled cuvette holder at 37°C (see Note 20).

4. Calculation of E3 enzyme activity: (Sample $\Delta OD/\text{min}$ – Blank $\Delta OD/\text{min})/0.00622 \times$ sample protein (mg) = nm/mg protein/min. Report the average of the two duplicate samples.

5. Interpretation: E3 is used as an internal mitochondrial reference enzyme. PDC activity is reported as a ratio relative to E3 activity. Two caveats apply to interpretation of these ratios. First, E3 is relatively more stable than total PDC activity, thus

## Table 3
## Cuvette setup for assay of E3 (one sample)

| Cuvette | Reaction (µl) | Sample mix (µl) | Ethanol (µl) | LAH₂ (µl) |
|---|---|---|---|---|
| 1-blank | 960 | 10 | 25 | |
| 2-reaction | 960 | 10 | | 25 |
| 3-reaction | 960 | 10 | | 25 |
| 4-blank | 960 | 20 | 25 | |
| 5-reaction | 960 | 20 | | 25 |
| 6-reaction | 960 | 20 | | 25 |

a low PDC/E3 ratio may suggest sample deterioration that is only reflected by the decreased activity of the less stable PDC enzyme. Therefore, low PDC/E3 ratios should be confirmed by assay of another relatively labile enzyme in the same sample (such as PC in fibroblasts or cytochrome oxidase in muscle). Second, although rare, E3 deficiency must be considered if both PDC and E3 activities are low, and another mitochondrial reference enzyme, such as CS should be assayed to determine if the low activity of E3 is specific (see Note 21 and Table 6 below for reference values for E3 and PDC activities).

### 3.3. PC Assay (see Table 4)

1. Prepare assay tubes: Label five $12 \times 75$-mm glass tubes (1 pyruvate blank + 4 reaction tubes) with a unique identifier for each sample (patient, control, and PC standard). Two protein concentrations and two incubation periods (30 and 60 min for skin fibroblasts or 5 and 10 min for liver sample) are run for each sample. For every assay, label an additional two $12 \times 75$ glass tubes as "sample blank". Pipette 440 μl of PC reaction mixture into each tube. For tubes which will contain only 25 μl of fibroblast sample, also add 25 μl of MOPS + PI to maintain consistent volumes. Likewise, for tubes that contain 25 μl of tissue sample, add 25 μl of HB + PI.

2. Preincubate: Place all assay tubes in a 37°C water bath in a fume hood. Thaw sample, and time exactly, starting at 0 s, add specific sample volume (see table above) to each of the five designated tubes at 15-s intervals. Preincubate for 4 min (save remaining sample on dry ice or at –80°C for CS assay; see Subheading 3.4).

3. Start reactions: Exactly 4 min after adding sample, add 10 μl of 0.5 M pyruvic acid to each reaction tube, vortex, incubate for 30 or 60 min (5 or 10 min for liver samples).

4. Stop reaction: (see Note 22) Exactly 30 or 60 (5 or 10 for liver) min after adding pyruvic acid, add 250 μl of 10% TCA solution (TCA solution also added to sample blank tubes at 60 min), cap tube, and vortex (final volume = 750 μl). Centrifuge tubes for 10 min at $1,000 \times g$.

5. Remove volatile $^{14}CO_2$: Working in a fume hood, transfer the supernatant from each tube (using a Pasteur pipette) into $12 \times 75$-mm glass tubes. Bubble $CO_2$ through all of the solutions for 10 min. (After this step, samples can be taken out of the fume hood since volatile $^{14}CO_2$ has been removed).

6. Count $^{14}CO_2$: Transfer 0.5 ml from each tube into a scintillation midi-vial. Add 4 ml scintillation fluid, cap tightly, mix well, and place in liquid scintillation counter. Allow chemiluminescense to subside for an hour and then measure radiation for 5 min per vial.

**Table 4**
**Plan for assay of PC (for one sample)**

| Tube # | Additions (µl) | | | | | | Time schedule (minute = ' second = ") | | | |
|---|---|---|---|---|---|---|---|---|---|---|
| | PC mix | MOPS buffer | GDW | Sample | Pyruvate | | Add sample | Add pyruvate | Add TCA[a] | Reaction time |
| Sample blank | 440 | 25 | 35 | | | | | | 60' | 60' (10')[b] |
| Pyruvate blank | 440 | 25 | 10 | 25 | | | 0 | | 60' | 60' (10')[b] |
| 1 | 440 | 25 | | 25 | 10 | | 15" | 4' 15" | 34' 15" | 30' (5')[b] |
| 2 | 440 | | | 50 | 10 | | 30" | 4' 30" | 34' 30" | 30' (5')[b] |
| 3 | 440 | 25 | | 25 | 10 | | 45" | 4' 45" | 64' 45" | 60' (10')[b] |
| 4 | 440 | | | 50 | 10 | | 60" | 5' | 65' | 60' (10')[b] |

[a]See Note 4.22 re radioactivity safety precautions upon addition of TCA.
[b] First number is reaction time for fibroblasts samples and number in parentheses is time for liver samples.

7. Calculation of results: Average dpm for sample and pyruvate blanks together and subtract from dpm of samples. Multiply dpm $\times 1.5$ (counted 0.5 ml out of 0.75 ml total final assay volume): (sample dpm – blank dpm) $\times 1.5$/specific activity of $NaH^{14}CO_3$ (dpm/nmol) $\times$ time (min) $\times$ mg protein = nmol $CO_2$ incorporated/mg protein/min

8. Interpretation: Report mean of replicate PC assay results expressed as PC activity relative to total cell protein, as well as relative to CS activity (see Note 23). These results should be compared to both concurrent controls and previous controls (mean, range, and $\pm$ SD; see Note 24 and Table 7).

**3.4. Citrate Synthase**

1. For each sample (including patients, CS calibration standard, and controls), prepare six spectrophotometric cuvettes (see Table 5).

2. Mix cuvettes and read in a kinetic capable spectrophotometer at $\lambda = 412$ nm while incubating in a temperature-controlled cuvette holder at 37°C (see Note 20 for kinetic settings).

3. Calculations: Sample $\Delta OD_{412}$/min – blank $\Delta OD_{412}$/min/ $0.136 \times$ mg protein = nmol/min/mg protein/

   The specific activities of the sample replicates are averaged and the standard deviation calculated. If one of the four replicates differs by more than $\pm 2$ SD of the mean, this sample is considered an outlier and discarded. The overall mean of 4 or at least 3 replicates is reported as the final result.

4. Interpretation: CS activity is interpreted as a reference mitochondrial protein in relation to the total sample protein content and to the activity of PC, to determine if the sample aliquot was sufficient and whether the sample may have deteriorated, resulting in nonspecific lowering of activities of mitochondrial enzymes (see Note 25 and Table 7). Since CS is not as labile as PC (or PDC), combined low activity is suggestive of sample

**Table 5**
**Spectrophotometric assay of CS (for one sample)**

| Cuvette | Type | Reaction mix (μl) | Sample (μl) | OAA (μl) |
|---------|------|-------------------|-------------|----------|
| 1 | Blank | 980 | 20 | |
| 2 | Reaction | 970 | 20 | 10 |
| 3 | Reaction | 970 | 20 | 10 |
| 4 | Blank | 960 | 40 | |
| 5 | Reaction | 950 | 40 | 10 |
| 6 | Reaction | 950 | 40 | 10 |

deterioration or mitochondrial depletion (requiring further investigation). An increase of CS activity may suggest increased mitochondrial content of the sample but should be corroborated with other data.

## 4. Notes

1. If specific activity is >2,000 dpm/nm, dilute solution with additional 50 mM pyruvic acid and recheck specific activity. For example, if specific activity of Na-1-$^{14}$C-pyruvate = 3,200 dpm/nm and volume = 2.5 ml, 3,200/2,000 = 1.6 × desired concentration; therefore, mix 2.5 ml of Na-1-$^{14}$C-pyruvate and 1.5 ml of 50 mM pyruvic acid and recheck specific activity.

2. Protein assay: By Lowry method, following reagent manufacturer's instructions (13). For lymphocyte and fibroblast samples, prepare protein standard curve tubes and sample protein assay tubes during the last centrifugation step of sample preparation. Set up standard curve and sample protein assay tubes prior to homogenization of tissue samples.

3. Glass scintillation vials will also work but since the solution is radioactive waste, it is easier to dispose of the entire vial

4. Recommend scintillation counter capable of direct calculation of disintegrations per minute (DPM) for single label pure beta nuclides, self-normalized and calibrated via sealed $^{14}$C reference.

5. Kontes Duall glass homogenizers can be used with good results or, alternatively, an electric mechanical shear tissue homogenizer.

6. Recommended skin fibroblast conditions: Use 37°C incubator, humidified, with 10% $CO_2$. Modified cell culture media: to a 500-ml bottle of Dulbecco's Modified Eagle Media (D-MEM), add 1 ml of 1 µg/ml fibroblast growth factor, 1 ml of 5 mg/ml insulin/transferrin/sodium selenite (2.5 mg/ml, 2.5 mg/ml, 2.5 µg/ml respectively), 5 ml of L-glutamine/penicillin/streptomycin solution (200 mM, 10,000 U/ml, and 10 mg/ml solution respectively), and 100 ml fetal bovine serum.

7. Mycoplasma contains PDC activity and can result in either false positive, negative, or altered test results. Therefore, skin fibroblasts must be pretested for mycoplasma and established as free of contamination before they are cultured in the test facility. If obtained without such documentation, the cells should be completely isolated until such testing has been done. Skin fibroblast "plugs" stored in liquid nitrogen are suitable after culture, but mycoplasma also is viable in liquid nitrogen, and such cell lines should be tested again while in culture.

8. Procedure for harvesting fibroblasts:

   (a) Warm PBS with EGTA, and trypsin solution to 37°C in water bath. Pour media off cells, add 5 ml PBS with EGTA, mix gently over cells, and pour off. Add another 5 ml PBS with EGTA and let stand on cells for 5 min.

   (b) Pour off PBS with EGTA. Add 5 ml trypsin solution and allow to remain on cells for 5 min while rocking flasks. After 5 min, add 10 ml HBSS solution. Scrape cells off flask with a cell scraper. Transfer cell suspension from the same patient to a 50-ml centrifuge tube. Centrifuge for 10 min at $400 \times g$ at 25°C.

   (c) Discard supernatant and resuspend cells in 10 ml PBS. Transfer to a 14-ml conical centrifuge tube. Centrifuge for 10 min at $400 \times g$, at 25°C.

   (d) Repeat wash with PBS. Using a Pasteur pipette, remove supernatant, leaving only cell pellet.

   (e) For PDC assay, resuspend cells in 500 µl PBS + PI and continue with PDC method. For PC assay, resuspend cells in 500 µl MOPS + PI and continue with PC method.

9. These blood samples can be kept or shipped at room temperature for as long as 3 days prior to isolation of the lymphocyte (mononuclear) fraction. Do not freeze, refrigerate, or separate plasma.

10. Method for separation of blood lymphocyte fraction with Ficoll-Paque:

   (a) Collect blood in 10-ml tubes (e.g., yellow-top Vacutainer) (containing acid-citrate-dextrose, ACD) and mix well by inverting. Blood in ACD tubes is stable for 48–72 h when held at 25°C.

   (b) Centrifuge blood for 15 min at $400 \times g$. Mark the level of the plasma on the tube then discard the plasma, leaving 1–2 ml on top of the cells (a layer of white cells, the buffy coat, will be visible on top of the red cells). Add PBS to the mark to replace the plasma and mix the tubes gently to resuspend the cells.

   (c) Put 3 ml of Ficoll-Paque into a 12-ml conical centrifuge tube. Layer 4 ml of the cell suspension on top of the Ficoll-Paque by touching the pipette tip to the side of the tube above the Ficoll-Paque and slowly allowing the cells to run down to the interface. Centrifuge for 40 min at $400 \times g$.

   (d) After centrifuging, remove the upper layer of PBS/plasma and discard, leaving 1–2 ml on top of the lymphocytes. With a Pasteur pipette, transfer all of the lymphocyte layer to a 50-ml centrifuge tube, combining all the cells from the same donor (use one 50-ml centrifuge tube for every 10 ml of blood). It is important to remove the entire lymphocyte layer, but a minimum of the Ficoll-Paque.

(e) Dilute the lymphocytes to 15 ml with PBS, mix gently, and centrifuge for 10 min at $400 \times g$. Pour off the supernatant and lyse the red cells as follows: using a stopwatch, add 15 ml of 0.2% NaCl to one tube, resuspend the lymphocytes gently with a Pasteur pipette. Exactly 1 min later, add 15 ml of 1.6% NaCl, cap, and mix gently. Centrifuge for 10 min at $400 \times g$.

(f) Pour off the supernatant. Add 10 ml of PBS to one tube from each patient, resuspending the cells gently with a Pasteur pipette, and then serially combining all the cells from the same patient into a 12-ml centrifuge tube. Centrifuge for 10 min at $400 \times g$.

(g) Remove the supernatant with a Pasteur pipette, taking off as much as possible. Resuspend the cells in 500 μl PBS + PI and continue with instructions for PDC assay.

11. Thawing of tissue samples to obtain aliquots should be avoided as deterioration can occur quickly. By placing the sample in a plastic weigh dish on top of a small flat block of dry ice, samples can stay frozen while cutting.

12. If yellow color remains after 90 min, prepare another 100 mg Na borohydride in 0.5 ml and add to the lipoamide suspension.

13. For the PC assay, frozen liver (from rat or other species) provides a reproducible standard. Our experience with use of purified bovine liver PC for this assay is more limited; purified PC should be diluted in HB buffer (containing a high-concentration sucrose) for stability since the purified enzyme is otherwise cold labile. The stock solution of PC should be diluted to approximately 0.5 U/ml (to add 0.001 U in 25 μl per standard assay tube.

14. AcCoA stock solution is usually around 25 mM; therefore, each milliliter of AcCoA should be diluted to around 3.5 ml with water.

15. Before preparing a full amount of PC reaction mix needed for an assay for the first time after diluting a new batch of $NaH^{14}CO_3$, make a small batch (e.g., enough for three tubes), using proportionate volumes of each ingredient, and check the specific activity of this trial batch (see Subheading 2.3.1, step 12a), to be sure the specific activity is within the desired range of 100–150 dpm/nmol. If the specific activity is too high, increase the volume of 1 M $KHCO_3$ added to the PC reaction mix sufficiently to lower the specific activity below 150 dpm/nmol,

taking into account the total concentration of $NaHCO_3$ and $KHCO_3$ added to the modified mix.

16. It is strongly recommended to create a step by step worksheet before starting this complex assay (e.g., see Tables 1 and 2). It is recommended that a printout of the raw counter data be included with the worksheet.

17. In tissue homogenates, exogenous phosphatase is added for pretreatment of the sample under similar conditions. Inactivation (phosphorylation) depends on keeping E1 kinase active, providing ATP if cells are disrupted, and inhibiting E1 phosphatase with fluoride.

18. Some laboratories routinely assay PDC in cells (fibroblasts) in the untreated (or "native") state, as well as activated PDC dependant on endogenous E1 phosphatases. This approach may be more sensitive for detection of potential partial $E_1$ phosphatase deficiency (14).

19. Reference values for PDC activity (see Table 6)

20. If kinetics programmable, set for run time, 300 s; start time, 30 s; cycle time, 13 s; and zero order, calculation time, 30–300 s.

21. Reference values for E3 activity and PDC/E3 ratios (see Table 6 below)

22. Since $^{14}CO_2$ is evolved during this assay, the assay is performed in a well-vented fume hood, with the sash set at an established safe level. Venting this relatively small quantity of $^{14}CO_2$ into the outside atmosphere is considered safe and acceptable. Caution should be used especially when the assay is stopped with TCA, since most of the $^{14}CO_2$ is released at that point. Tubes must be tightly capped before they are taken out of the fume hood for centrifugation.

23. In this situation, citrate synthase CS activity serves as another reference mitochondrial protein, which indirectly reflects overall mitochondrial protein, for example, in cultured skin fibroblasts. However, CS is relatively more stable than PC and is not a sensitive indicator of potential loss of sample biochemical integrity, which is likely to be a concern in postmortem tissue samples or samples whose integrity may have not been rapidly or consistently maintained. In such cases, assay of another relatively sensitive enzyme activity, such as PDC, may be necessary to determine if low activity of PC is specific or secondary.

24. Reference values for PC assay and PC/CS (see Table 7).

25. Reference values for CS assay and PC/CS ratios (see Table 7).

**Table 6**
**Control reference values for PDC and E3 activities**

| Assay | Sample type | Mean activity | Units | SD | 3rd percentile | 97th percentile | N (controls)[a] |
|---|---|---|---|---|---|---|---|
| PDC (activated) | Skin fibroblasts | 2.42 | nmol/min/mg protein | 0.88 | 1.26 | 4.42 | 329 |
| PDC (inactivated) | Skin fibroblasts | 0.92 | nmol/min/mg protein | 0.63 | 0.19 | 2.30 | 322 |
| PDC (activated) | Blood lymphocytes | 1.63 | nmol/min/mg protein | 0.48 | 0.98 | 2.72 | 596 |
| PDC (inactivated) | Blood lymphocytes | 0.53 | nmol/min/mg protein | 0.23 | 0.22 | 1.09 | 524 |
| PDC (activated) | Skeletal muscle | 3.17 | nmol/min/mg protein | 1.49 | 1.20 | 6.52 | 340 |
| PDC (inactivated) | Skeletal muscle | 0.48 | nmol/min/mg protein | 0.50 | 0.06 | 1.39 | 336 |
| E3 | Skin fibroblasts | 60 | nmol/min/mg protein | 20 | 25 | 98 | 267 |
| E3 | Blood lymphocytes | 70 | nmol/min/mg protein | 16 | 45 | 103 | 596 |
| E3 | Skeletal muscle | 128 | nmol/min/mg protein | 39 | 72 | 222 | 440 |
| PDC/E3 (activated) | Skin fibroblasts | 3.7 | ×100 | 1.2 | 2.2 | 6.6 | 198 |
| PDC/E3 (activated) | Blood lymphocytes | 2.3 | ×100 | 0.6 | 1.4 | 3.6 | 596 |
| PDC/E3 (activated) | Skeletal muscle | 2.3 | ×100 | 1.0 | 0.8 | 4.5 | 311 |

[a] $N$ = number of separate samples assayed from control subjects without known metabolic or other genetic disorder

## Table 7
## Control reference values for PC and CS activities

| Assay | Sample type | Mean activity | Units | SD | 3rd percentile | 97th percentile | N[a] (controls) |
|-------|-------------|---------------|-------|-----|----------------|-----------------|-----------------|
| PC | Skin fibroblasts | 1.42 | nmol/min/mg protein | 0.79 | 0.56 | 3.22 | 338 |
| PC | Liver (frozen) | 68 | nmol/min/mg protein | 38 | 15 | 38 | 74 |
| CS | Skin fibroblasts | 38 | nmol/min/mg protein | 11 | 22 | 58 | 254 |
| CS | Liver (frozen) | 54 | nmol/min/mg protein | 21 | 28 | 104 | 112 |
| PC/CS | Skin fibroblasts | 4.4 | ×100 | 2.3 | 2.0 | 9.0 | 195 |
| PC/CS | Liver | 138 | ×100 | 72 | 45 | 301 | 27 |

[a] N= number of separate samples assayed from controls subjects without known metabolic or other genetic disorder

## Acknowledgments

The authors are indebted to the late Merton Utter and his associates, whose biochemistry laboratory at Case Western Reserve University was responsible for initial development of these assays; to Mulchand Patel and his associates who continued this work; and to Marilyn Lusk, who trained many members of the CIDEM staff to perform and refine these methods for their clinical application.

## References

1. Sheu, K. F. R., Hu, C. W. C., and Utter, M. F. (1981) Pyruvate dehydrogenase complex activity in normal and deficient fibroblasts. J. Clin. Invest. 67, 1463–1471.

2. Robinson, B. H., Taylor, J., and Sherwood, W. G. (1980) The genetic heterogeneity of lactic acidosis: occurrence of recognizable inborn errors of metabolism in pediatric population with lactic acidosis. Pediatr. Res. 14, 956–962.

3. Kerr, D. S., Ho, L., Berlin, C. M., Lanoue, K. F., et al. (1987) Systemic deficiency of the first component of the pyruvate dehydrogenase complex. Pediatr. Res. Pediatr. Res. 22, 312–318.

4. Lib, M., Rodriguez-Mari, A., Marusich, M. F., et al (2003) Immunocapture and microplate-based activity measurement of mammalian pyruvate dehydrogenase complex. Anal. Biochem. 314, 121–127.

5. Sterk, J. P., Stanley, W. C., Hoppel, C. L., et al. (2003) A radiochemical pyruvate dehydrogenase assay: activity in heart. Anal. Biochem. 313, 179–182.

6. Lib, M. Y., Brown, R. M., Brown, G. K., et al. (2002) Detection of pyruvate dehydrogenase E1 alpha-subunit deficiencies in females by immunohistochemical demonstration of mosaicism in cultured fibroblasts. J. Histochem. Cytochem. 50, 877–884.

7. Kerr, D. S. and Zinn, A. B. (2009) Disorders of Pyruvate Metabolism and the Tricarboxylic Acid Cycle. In: Sarafoglu, K. (ed) Essentials of Pediatric Endocrinology and Metabolism. McGraw-Hill.

8. Chuang, D. T., Niu, W. L., and Cox, R. P. (1981) Activities of branched-chain 2-oxo acid dehydrogenase and its components in skin fibroblasts from normal and classical-maple- syrup-urine-disease subjects. Biochem. J. 200, 59–67.

9. Patel, M. S., Hong, Y. S., and Kerr, D. S. (2000) Genetic defects in E3 component of alpha-keto acid dehydrogenase complexes. Methods Enzymol. 324, 453–464.

10. Atkin, B. M., Utter, M. F., and Weinberg, M. B. (1979) Pyruvate carboxylase and

phosphoenolpyruvate carboxykinase activity in leukocytes and fibroblasts from a patient with pyruvate carboxylase deficiency. Pediatr. Res. 13, 38–43.

11. Scrutton, M. C., Olmsted, M. R., and Utter, M. F. (1969) Pyruvate carboxyalse. Methods Enzymol. Volume 13, 235–249.

12. Srere, P. A. (1969) Citrate synthase: [EC 4.1.3.7. Citrate oxaloacetate-lyase (CoA-acetylating)]. *Methods Enzymol. Volume* 13, 3–11.

13. Lowry, O. H., Rosebrough, N. J., Farr, A. L., et al (1951) Protein measurement with the folin phenol reagent. J. Biol. Chem. 193, 265–275.

14. Maj, M. C., MacKay, N., Levandovskiy, V., et al. (2005) Pyruvate dehydrogenase phosphatase deficiency: identification of the first mutation in two brothers and restoration of activity by protein complementation. J. Clin. Endocrinol. Metab 90, 4101–4107.

# Chapter 8

# Assessment of Thymidine Phosphorylase Function: Measurement of Plasma Thymidine (and Deoxyuridine) and Thymidine Phosphorylase Activity

**Ramon Martí, Luis C. López, and Michio Hirano**

## Abstract

We describe detailed methods to measure thymidine (dThd) and deoxyuridine (dUrd) concentrations and thymidine phosphorylase (TP) activity in biological samples. These protocols allow the detection of TP dysfunction in patients with mitochondrial neurogastrointestinal encephalomyopathy (MNGIE). Since the identification of mutations in *TYMP*, the gene encoding TP, as the cause of MNGIE (Nishino et al. Science 283:689–692, 1999), the assessment of TP dysfunction has become the best screening method to rule out or confirm MNGIE in patients. *TYMP* sequencing, to find the causative mutations, is only needed when TP dysfunction is detected. dThd and dUrd are measured by resolving these compounds with high-performance liquid chromatography (HPLC) followed by the spectrophotometric monitoring of the eluate absorbance at 267 nm (HPLC-UV). TP activity can be measured by an endpoint determination of the thymine formed after 1 h incubation of the buffy coat homogenate in the presence of a large excess of its substrate dThd, either spectrophotometrically or by HPLC-UV.

**Key words:** Thymidine, Deoxyuridine, Thymidine phosphorylase, Mitochondrial neurogastrointestinal encephalomyopathy, MNGIE, HPLC, Biochemical diagnosis

## 1. Introduction

The assessment of thymidine phosphorylase (TP) function is the reference method to diagnose mitochondrial neurogastrointestinal encephalomyopathy (MNGIE) (1) in patients with clinical features suggestive of this disorder. TP initiates the catabolism of thymidine (dThd) and deoxyuridine (dUrd) in humans (2) by catalyzing the phosphorolysis of these nucleosides to deoxyribose phosphate and the corresponding bases (Fig. 1). In MNGIE patients, TP dysfunction caused by *TYMP* gene mutations results in the systemic accumulation

Lee-Jun C. Wong (ed.), *Mitochondrial Disorders: Biochemical and Molecular Analysis*, Methods in Molecular Biology, vol. 837,
DOI 10.1007/978-1-61779-504-6_8, © Springer Science+Business Media, LLC 2012

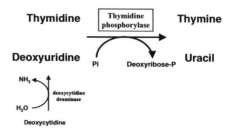

Fig. 1. Role of TP in the catabolism of pyrimidine deoxyribonucleosides in humans. Thymidine and deoxyuridine are the TP substrates; indirectly, TP also contributes to the degradation of the third pyrimidine deoxyribonucleoside, deoxycytidine, because it is first converted in deoxyuridine via deamination.

of these nucleosides (3, 4) and increased urinary excretion of dThd and dUrd (5, 6).

Analyzing plasma dThd and dUrd concentrations is the easiest way to test for TP dysfunction. Screening of urine is also useful, but precautions should be taken to prevent bacterial growth in the sample; otherwise, in vitro catabolism of dThd and dUrd via bacterial TP can lead to underestimation of urinary excretion. For both plasma and urine samples, dThd and dUrd concentrations can be easily assessed by high-performance liquid chromatography coupled to ultraviolet spectrophotometric detection (HPLC-UV) or to tandem mass spectrometry (HPLC-MS/MS). UV detection is less sensitive and selective than MS/MS but simpler to set up and more accessible to many nonspecialized laboratories that may not have access to HPLC-MS/MS. In addition, the sensitivity of the UV detection (typically 0.05 μM in our hands (7)) is sufficient to detect and accurately quantify dThd and dUrd levels typically found in MNGIE patients, including those with residual TP activity (late-onset MNGIE patients (8, 9)), in contrast to plasma and urine levels that are undetectable by UV in healthy subjects and *TYMP* mutation carriers.

In our experience, there is complete concordance between nucleoside accumulation and reduced TP activity in buffy coat. Therefore, testing only nucleoside levels is generally sufficient for diagnostic purposes. Nevertheless, we prefer to measure TP activity in buffy coat in all cases, including those with undetectable dThd and dUrd plasma concentrations, because several late-onset MNGIE, with moderately elevated plasma levels of nucleosides as low as 0.4 μM, have been reported (8, 9). In such cases, TP activity can be around 15–20% of normal in contrast to most MNGIE cases with little or no detectable TP activity. If blood samples are not kept in cold and immediately centrifuged to separate the plasma from the cellular fraction, residual TP activity in vitro can artifactually reduce dThd and dUrd to levels near or below the detection limit, thus masking the moderate nucleoside accumulation in late-onset MNGIE patients.

## 2. Materials

Prepare all solutions using HPLC grade water (e.g., obtained by deionization to a resistivity of 18.2 MΩ-cm).

### 2.1. Reagents

1. Hemolysis buffer: 10 mM $NH_4HCO_3$, 144 mM $NH_4Cl$. Dissolve 0.791 g of $NH_4HCO_3$ ($M = 79.06$ g/mol) and 7.70 g of $NH_4Cl$ ($M = 53.49$ g/mol) in 1 L of water. Sterilize by filtering through a 0.22-μm membrane and keep it refrigerated for less than 6 months.

2. Eluent A: 20 mM potassium phosphate monobasic ($KH_2PO_4$), pH 5.6.

   Dissolve 2.72 g of $KH_2PO_4$ (anhydrous salt, $M = 136.09$ g/mol) in approximately 0.9 L of water, adjust to pH 5.6 with a concentrated solution of KOH (e.g., 0.5 M), and add water up to 1 L. Filter eluent A through a 0.2-μm nylon or PVDF membrane before use.

3. Methanol, gradient grade (see Note 1).

4. Thymidine phosphorylase from *Escherichia coli*, recombinant (TP, SIGMA). Specific activity ~1,000 U/ml, the exact value varies according to lot.

5. Concentrated 70–72% (approximate 11.7 M) perchloric acid (PCA; $H_3ClO_4$).

6. 0.55 M PCA: Dilute 11.7 M PCA 1:21 in HPLC grade water.

7. Phosphate buffered saline (PBS): 140 mM NaCl, 10 mM $Na_2HPO_4$, pH 7.4 (see Note 2). This reagent is only needed for the dilution of urine samples. Dissolve 8.2 g of NaCl ($M = 58.5$ g/mol) and 1.42 g of $Na_2HPO_4$ ($M = 141.96$ g/mol) in nearly 1 L of water. Adjust to pH 7.4 with HCl. Then, complete the volume to 1 L with water.

8. Lysis buffer: 50 mM Tris–HCl, pH 7.2, 1% ($w/v$) triton X-100, 2 mM phenylmethylsulfonyl fluoride (PMSF), 0.02% ($v/v$) 2-mercaptoethanol. Dissolve 0.606 g of Tris base ($M = 121.14$ g/mol) in 80 ml of water and adjust the pH to 7.2 with HCl. Add 1 g of triton X-100, 20 μl of 2-mercaptoethanol and 2 ml of stock PMSF solution (stock PMSF solution: 17.4 mg/ml PMSF in isopropanol). Add water to reach 100 ml final volume.

9. 5× TP reaction buffer: 0.5 M Tris–arsenate, pH 6.5 (see Note 3). Dissolve 6.06 g of Tris base ($M = 121.14$ g/mol) and 15.6 g of disodium arsenate heptathydrate ($Na_2HAsO_4 \cdot 7H_2O$, $M = 312.0$) to nearly 100 ml of water, adjust to pH 6.5 with HCl, and complete to 100 ml with water. Store at room temperature.

10. 167 mM Thymidine: Dissolve 0.404 g of thymidine ($M = 242.2$ g/mol) in 10 ml of water by sustained stirring,

with slight heating to favor dissolution. Once completely dissolved, keep in frozen aliquots at –20°C until used. Upon thawing before its use, make sure that dThd eventually precipitated is completely redissolved (warming at 37°C and several vortexings will help to complete resolubilization).

11. 0.3 M NaOH: Dissolve 1.2 g of NaOH ($M=40$ g/mol) in 100 ml of water (exercise caution with this exothermic dissolution). Keep the solution in a well-closed plastic bottle at room temperature and discard when signs of carbonate appear in the surroundings of the cap.

### 2.2. Standards for HPLC-UV Quantification

1. To make the standard curve (see Note 4), prepare 10 mM aqueous solutions of:

    (a) dThd ($M=242.2$ g/mol): dissolve 0.242 g of dThd in 100 ml of water.

    (b) dUrd ($M=228.2$ g/mol): dissolve 0.228 g of dUrd in 100 ml of water.

    (c) Thymine ($M=126.1$ g/mol): dissolve 0.126 g of thymine in 100 ml of water (see Note 5).

2. Prepare a 100 μM multistandard (100 μM dThd, 100 μM dUrd, 100 μM thymine) by diluting 1 ml of 10 mM dThd + 1 ml of 10 mM dUrd + 1 ml of 10 mM thymine up to 100 ml final volume with water).

3. Use water to serially dilute the 100 μM multistandard to obtain the following concentrations: 50, 25, 10, 8, 5, 2, 1, 0.5, 0.1 and 0.05 μM.

4. Once prepared, the aqueous multistandards can be frozen at –20°C in several aliquots and used as needed.

### 2.3. Other Materials and Equipment

1. HPLC apparatus. The procedure described here has been set for a HPLC apparatus with at least three independent eluent lines and mixer (see Note 6).

2. Vials appropriate for the injection device included in your HPLC apparatus. Vials with reduced dead volume are preferable in cases of limited volume of sample.

3. Column: The method, retention times, and overall runtimes described here have been setup and obtained with a column *Alltima C18 NUC*, 100Å pore size; 5 μm particle size, $250\times4.6$ mm (Alltech). A different C18 column with similar length and particle size can be used. In this case, retention times and work pressures may change.

4. Spectrophotometer and quartz cuvettes (semimicro size). Although any spectrophotometer that can measure absorbance at wavelength 300 nm can be used, models with multisample capacity, enabling the simultaneous load of several cuvettes for reading, are preferable because they minimize the time of UV

reading. A spectrophotometer is not needed if the final thymine measurement for TP activity determination is performed by HPLC-UV.

5. Nylon or PVDF 0.2 μm pore-size filter membranes, and filtering device for vacuum-driven filtration of aqueous solutions (eluent A and water) are used prior to HPLC (see Note 7).

# 3. Methods

### 3.1. Sample Collection and Preparation

1. Centrifuge 5–10 ml of anticoagulated blood in a 15-ml falcon tube at $1,500 \times g$ for 10 min. Then separate the plasma and keep it –20°C until nucleoside analysis (see Note 8).

2. On the cell fraction (erythrocytes + buffy coat), add approximately two volumes of hemolysis buffer (e.g., 8 ml of buffer on 4 ml of cellular fraction) and shake vigorously 15 s, at room temperature. Then keep the tube on ice 30 min, followed by vigorous shaking.

3. Centrifuge at $1,500 \times g$ for 10 min and discard the supernatant.

4. Resuspend the pellet in 10 ml of hemolysis buffer, shake vigorously, and maintain on ice 15 min.

5. Centrifuge at $1,500 \times g$ for 10 min and discard the supernatant.

6. Resuspend the pellet in 2 or 3 ml of hemolysis buffer and transfer the suspension to two or three 1.5-ml microcentrifuge tubes, vortex 5 s, and keep for 10 min on ice.

7. Centrifuge at $20,000 \times g$ for 10 min and discard the supernatant. Eliminate as much as possible the liquid red crown that covers the surface of the white buffy coat, to minimize hemoglobin contamination.

8. Keep the dry buffy coat at –80°C until TP activity analysis (see Note 9).

9. For the assessment in urine samples, collect fresh random urine and freeze immediately until the analysis (see Note 10).

### 3.2. dThd and dUrd Determination

1. Divide the sample (plasma or urine) (see Note 11) in two aliquots of equal volume (e.g., 200 μl each aliquot). If the sample is urine, because of the presence of many metabolites at high concentrations, 1:20 dilution with PBS will help avoiding interferences.

2. Add *E. coli* TP to one of the aliquots to a final catalytic concentration of approximately 10 U/ml (e.g., 200 μl sample + 2 μl of 1,000 U/ml TP). Vortex and incubate TP treated and untreated aliquots for 10 min at 37°C.

3. To deproteinize the sample, add concentrated perchloric acid (PCA, 11.7 M approximately) up to a final concentration of

0.5 M (e.g., 200 or 202 μl of incubated sample + 9 μl of concentrated PCA) to the TP treated and untreated aliquots. Vortex and keep on ice for 5 min to facilitate complete protein precipitation.

4. Centrifuge at $20,000 \times g$ for 10 min. Discard the pellet and save the clear supernatant, which is now ready to be injected to the HPLC or kept at –20°C until analysis (see Note 12).

5. Thaw the following multistandards: 25, 10, 8, 5, 2, 1, 0.5, 0.1, and 0.05 μM. Once thawed, treat each multistandard with 11.7 M PCA in the same proportion as done for the samples. Because the multistandards will be injected more than once, you should use larger volumes, for example 500 μl of multistandard + 22.5 μl of 11.7 M PCA.

6. Separation of nucleosides is carried out through HPLC with gradient elution and UV detection. The conditions of the separation are as follows:

Column: *Alltima C18 NUC* (see Subheading 2).

Column temperature: 30°C (see Note 13).

Injection volume: 50 μl.

Detection: UV absorbance at wavelength 267 nm.

Eluent A: 20 mM $KH_2PO_4$, pH 5.6; eluent B: methanol gradient grade; eluent C: water.

For each run, the gradient elution should be programmed as follows:

| Time (min) | Flow (ml/min) | %A | %B | %C | Comment |
|---|---|---|---|---|---|
| 0 | 1.5 | 100 | 0 | 0 | 5 min isocratic segment |
| 5 | 1.5 | 100 | 0 | 0 | From 5 to 25 min, linear gradient from 0 to 17.4% MeOH |
| 25 | 1.5 | 82.6 | 17.4 | 0 | |
| 26 | 1.5 | 0 | 0 | 100 | Washout of eluent A to avoid salt precipitation (see Note 14) |
| 30 | 1.5 | 0 | 0 | 100 | |
| 31 | 1.5 | 0 | 100 | 0 | Washout of compounds still retained in the column |
| 35 | 2.0 | 0 | 100 | 0 | |
| 45 | 2.0 | 0 | 100 | 0 | |
| 46 | 1.5 | 0 | 100 | 0 | |
| 47 | 1.5 | 0 | 0 | 100 | Washout of MeOH to avoid salt precipitation (see Note 14) |
| 50 | 1.5 | 0 | 0 | 100 | |
| 51 | 1.5 | 100 | 0 | 0 | Re-equilibration of the column at 100% A for the next run |
| 60 | 1.5 | 100 | 0 | 0 | |

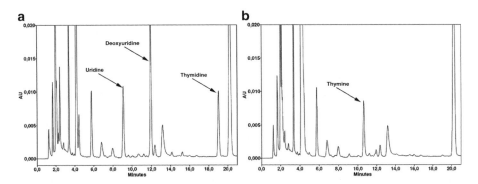

Fig. 2. Chromatograms of plasma from a MNGIE patient showing thymidine and deoxyuridine peaks. Representative chromatograms obtained from plasma of a MNGIE patient (**a**), and the same plasma after the selective elimination of TP substrates by treatment with *Escherichia coli* TP (**b**). The peaks of dThd and dUrd (5.4 and 10.6 μM in this specific case) virtually disappear in the TP-treated aliquot, whereas the thymine observed in the **panel b** is the product of dThd phosphorolysis by TP. Uracil, derived from phosphorolyses of dUrd and uridine, is also present, but it elutes very early in the chromatogram and is not resolved well due to coeluting peaks. Note that the ribonucleoside uridine, which is normally found in plasma of healthy controls and MNGIE patients at similar concentrations, is also degraded by *E. coli* TP.

7. Generate a sample set for analysis. Inject, for every sample, both TP-treated and untreated aliquots. Intercalate the multistandards among the samples. To obtain reliable calibration curves, two or three injections of every multistandard are advisable for every run (see Note 15). Figure 2 shows two typical chromatograms of untreated and TP-treated plasma from a MNGIE patient. Retention times and resolution of the peaks will vary depending on the HPLC apparatus and the column used (see Note 16).

8. Integrate the areas of the standards' peaks using the software of your HPLC apparatus. Generate each a calibration curve for dThd and dUrd by adjusting peak areas ($Y$ axis) versus standards' concentrations ($X$ axis) to a linear regression with no offset (i.e., force the linear regression to pass by the $x=0$, $y=0$).

9. Calculate dThd and dUrd concentrations from the peak areas, using the calibration curves. Because TP from *E. coli* degrades dThd and dUrd, the results from any remaining peaks present in the TP-treated aliquots at the dThd and dUrd retention times should be subtracted from the untreated result, as only the TP-labile portions of these peaks are true dThd or dUrd (see Note 17).

*3.3. TP Activity Assay*

Always work with samples on ice:

1. Resuspend the buffy coat obtained as indicated in the Subheading 3.1 in approximately 300 μl of lysis buffer and homogenize the pellet by passing the suspension through a 27 g × 1 in. needle several times.

2. After homogenization, centrifuge at $20,000 \times g$ for 30 min at 4°C and save the supernatant (homogenate).

3. Determine the protein concentrations of the homogenates (see Note 18).

4. Dilute all the homogenates with lysis buffer up to a protein concentration of 1.35 mg/ml (diluted homogenate). At this concentration, the reaction mixture will contain 100 µg of protein (we will add 74 µl of this extract to a final volume of 100 µl of the reaction mixture). If the undiluted homogenate has a protein concentration lower than 1.35 mg/ml, the assay can be performed with undiluted homogenate, but this will result in a slight overestimation of the activity (see Note 19).

5. For every sample, make two mixtures:

|  | Reaction mixture (µl) | Blank mixture (µl) |
| --- | --- | --- |
| 5× TP reaction buffer[a] | 20 | 20 |
| Diluted homogenate (1.35 mg protein/ml) | 74 | 74 |
| 167 mM thymidine | 6 | – |
| Final volume | 100 | 94 |

[a]Final concentrations in the reaction mixture: 0.1 M Tris–arsenate; 10 mM thymidine; 1 mg protein/ml.

6. Incubate both blank and reaction mixtures for 1 h at 37°C. At the end of this incubation, the reaction must be stopped and the thymine concentration measured, either spectrophotometrically or by HPLC-UV. From here on, the protocol varies according to methods of detection of thymine.

7. Spectrophotometric determination of thymine:

    1. After incubation for 1 h at 37°C, terminate the reaction by the addition of 1 ml of 0.3 M NaOH. Then, add 6 µl of 167 mM thymidine to all the blank mixtures and vortex (see Note 20).

    2. Measure the absorbance of the reaction and blank mixtures at 300 nm, subtract the blank result from that obtained for the reaction mixture and calculate from this difference the amount of thymine formed, based on the $3.4 \times 10^3$ L$^{-1}$ mol cm$^{-1}$ difference in molar absorptivity at 300 nm between dThd and thymine at alkaline pH.

    3. Express enzyme activity as nanomoles of thymine formed per hour and mg of protein.

8. HPLC-UV determination of thymine:

    1. Stop the reaction by adding 1 ml of 0.55 M perchloric acid and vortex.

2. Add 6 µl of 167 mM thymidine to all the blank mixtures and vortex (see Note 20).

3. Keep all mixtures on ice for 5 min.

4. Centrifuge at $20,000 \times g$ for 10 min. Save the supernatant, which is now ready to be injected in the HPLC apparatus.

5. Use the HPLC method and conditions described for dThd and dUrd determination to measure the thymine formed in the reaction and blank mixtures. Alternatively, the runtime can be shortened to around 15 min or less because the thymine in this supernatant can be readily separated and quantified using the following isocratic elution eluent A 90% and eluent B 10%. Flow 1.5 ml/min. In these conditions, the retention times for thymine and thymidine are around 4.5 and 9 min, respectively.

6. Quantify the thymine formed from the areas of the peaks using a calibration curve done with aqueous standards. The range of concentrations of your calibration curve should cover concentrations between 0.5 and 100 µM.

7. Calculate TP activity from the thymine formed in the reaction, taking in account that the blank must be subtracted (see Note 20). Express the results referred to protein (nmol of thymine formed/h/mg protein).

## 4. Notes

1. Methanol can be purchased from various suppliers, but it is important to use HPLC "gradient grade" methanol. This will minimize the changes in the UV absorbance due to the increasing methanol proportion in the mobile phase for the HPLC separation.

2. Although the most common and physiological formulation of PBS includes potassium cation, for the purposes of this protocol, both formulations of PBS are acceptable. We have described here a formulation with sodium salts only. Moreover, it is also possible to purchase inexpensive preformulated PBS tablets to be dissolved in an appropriate volume of water to obtain PBS ready to use.

3. The final concentration in the TP reaction mixture is 0.1 M Tris–arsenate pH 6.5.

4. We detail here the procedure to prepare a multistandard of dThd, dUrd (for the quantification of these two compounds), and thymine (for the TP activity determination). Additional UV-absorbing compounds can be resolved using the HPLC

procedure described here, so that more compounds can be included in the multistandard if needed for other purposes. In our hands, the following compounds are well resolved: cytosine, uric acid, cytidine, hypoxanthine, xanthine, uridine, thymine, dUrd, inosine, guanosine, dThd, tryptophan, and adenosine.

5. The dissolution of thymine in water to 10 mM is slow, but it will be easily accelerated by moderate heating or alkalinization with NaOH.

6. Under the conditions described here, the HPLC system must be able to bear pressures slightly above 4,000 pounds per square inches (psi). Alternatively, the column may be set to a higher temperature than that described in this protocol (described in Subheading 3.2 **step 6**), or the flow may be reduced. In this latter case, the gradient will have to be reformulated and the runtime for each injection increased.

7. As a general rule, all eluents and other liquids to be passed through the column and the HPLC apparatus should be sterile and particle free to prevent damage to the column and system. Therefore, water and aqueous solutions must be filtered through a 0.2 μm filter prior to use, using a suitable vacuum-driven filter unit and hydrophilic nylon or PVDF membranes. HPLC grade organic solvents (e.g., methanol) do not need to be filtered because they are free of particles and do not support microbial growing.

8. Blood can be collected in tubes with any anticoagulant because they do not affect the plasma nucleoside concentration (unless the anticoagulant addition involves significant volume addition, which will have a dilution effect). Although the influence of the anticoagulant on the buffy coat TP activity has been only marginally studied (4), in our experience, standard anticoagulants (e.g., EDTA, heparin, and acid citrate) do not interfere with TP assessment.

9. Separating the buffy coat from each patient into two or three aliquots will provide one for TP assessment and the other(s) for additional studies (for example, DNA extraction for *TYMP* sequencing in cases of TP deficiency).

10. The determination of the nucleosides in urine can be useful for diagnostic purposes, but it presents several pitfalls (variability, microbial contamination, sample collection problems, etc.) that commonly confound biochemical determinations with urine samples. A common way to normalize the results and correct from sample-related variability is to refer the results to milligram of creatinine. In addition, great care should be paid to avoid bacterial contamination, especially in 24-h urine samples that may be stored for several hours, often in poorly

controlled conditions. Because of these potential problems, we recommend use of plasma rather than urine to test for dThd and dUrd overload in candidate MNGIE patients, but if urine should be finally tested, we recommend using freshly collected urine (less prone to be contaminated) that is frozen immediately until analysis, and refer the results to milligram of creatinine.

11. Although this protocol refers to plasma and urine samples, this method can be used for other biological fluids or tissue samples of interest (10).

12. Although it is safer to store the supernatant frozen if the sample will not be analyzed in the following days, we have observed that dThd, dUrd, as well as many other nucleosides are very stable in acidified supernatant, and their concentrations remain unchanged after more than 1 week at 4°C.

13. In our experience, the pressure of the HPLC system reaches maximum values of 4,000 or some higher when the column is set at 30°C. If the conditions described here lead to system pressure levels above those allowed by the HPLC apparatus, the best alternative is setting the column temperature above 30°C. Increasing the temperature of the column will reduce significantly the pressure, and the resolution of the peaks will be only slightly modified. Temperatures as high as 50 or 60°C can be tried, but the manufacturer instructions on the column specifications should always be checked to ensure the maximum temperature tolerated by the column.

14. The contact of high percentages of organic phases with phosphate buffers may result in salt precipitation, which may damage the column and the HPLC apparatus. To avoid this potential problem, we recommend including washout with water at the points indicated in the program, but these steps can be eliminated if you determine empirically that no salt precipitation occurs under your conditions. This would shorten the time needed for each injection.

15. In our experience, slopes of the calibration curves obtained for different runs are very reproducible. Typical slope variations are usually below 10%.

16. To find out the exact retention times of each compound in your conditions (e.g., dThd, dUrd, and thymine, as well as additional compounds of interest), separate injections of standards containing one compound only may be needed. Once the retention times for every compound have been established, standards of single compounds are not generally needed because the retention times change very little over different runs.

17. In our experience, one or several TP-resistant small peaks usually coelute or elute close to dUrd. TP-resistant peaks coeluting

with dThd are rarer but sometimes happen (e.g., circulating UV-absorbing drugs). Care should be taken to avoid false-positives due to coeluting peaks.

18. The protein concentration of the homogenate can be determined with the Bradford method (11) by using bovine serum albumin for the preparation of the calibration curve, or with the method routinely used in your laboratory. It is imperative to use a reliable method for the determination of the protein concentration because biases or inaccuracies in this determination largely affect the final result, as the activity is expressed as nmoles of thymine/h/mg protein.

19. We describe here the assay for the enzyme reaction containing 100 μg of protein. Under these conditions, the reference values should be similar to those previously reported (4, 7), around 650–700 nmol of thymine/h/mg protein. The assay can be performed with smaller amounts of protein, but we have observed that the rate of thymine production is not strictly linear with the amount of protein, and smaller amounts of protein in the reaction mixture tend to result in slightly higher enzyme activities when expressed as nmoles of thymine formed per hour per mg of protein. We strongly recommend always working with the same amount of protein in the reaction mixture and to determine the reference values for healthy controls under these conditions.

20. In the blank mixture, the substrate thymidine should be added only at the end of the incubation, after TP protein contained in the homogenate has been inactivated. Therefore, any thymine detected in the blank is not produced by TP activity. We always observe very low but clearly measurable concentrations of thymine in the blank, as a result of the trace thymine contaminant in thymidine. We have estimated the thymine contamination of our thymidine preparations to be around 0.03%, but the substrate thymidine is added to the mixture in such a high concentration (10 mM) that this small contamination should be taken in account. It is negligible when TP activity is normal, but it may be significant when compared with the minute TP activities observed in MNGIE patients.

## Acknowledgment

Dr. Martí is supported by a grant from the Spanish Instituto de Salud Carlos III (PS09/01591). Dr. López is supported by grants from the Marie Curie International Reintegration Grant Programme (COQMITMEL-266691) of the Seventh European Community Framework Programme, from Ministerio de Ciencia e Innovación,

Spain (SAF2009-08315) and from the Consejería de Economía, Innovación y Ciencia, Junta de Andalucía (P10-CTS-6133). Dr. Hirano is supported by NIH grants R01 HD056103 (cofunded by NICHD and the NIH Office of Dietary Supplements), R01 HD057543, RC1 NS070232; MDA grant 115567; and the Marriott Mitochondrial Disorder Clinical Research Fund.

## References

1. Nishino I, Spinazzola A, Hirano M (1999). Thymidine phosphorylase gene mutations in MNGIE, a human mitochondrial disorder. *Science* 283:689–692.

2. Focher F, Spadari S (2001). Thymidine phosphorylase: a two-face Janus in anticancer chemotherapy. *Curr Cancer Drug Targets* 1:141–153.

3. Marti R, Nishigaki Y, Hirano M (2003). Elevated plasma deoxyuridine in patients with thymidine phosphorylase deficiency. *Biochem Biophys Res Commun* 303:14–18.

4. Marti R, Spinazzola A, Tadesse S, Nishino I, Nishigaki Y, Hirano M (2004). Definitive diagnosis of mitochondrial neurogastrointestinal encephalomyopathy by biochemical assays. *Clin Chem* 50:120–124.

5. la Marca G, Malvagia S, Casetta B, Pasquini E, Pela I, Hirano M, et al. (2006). Pre- and post-dialysis quantitative dosage of thymidine in urine and plasma of a MNGIE patient by using HPLC-ESI-MS/MS. *J Mass Spectrom* 41: 586–592.

6. Schupbach WM, Vadday KM, Schaller A, Brekenfeld C, Kappeler L, Benoist JF, et al. (2007). Mitochondrial neurogastrointestinal encephalomyopathy in three siblings: clinical,

genetic and neuroradiological features. *J Neurol* 254:146–153.

7. Spinazzola A, Marti R, Nishino I, Andreu AL, Naini A, Tadesse S, et al. (2002). Altered thymidine metabolism due to defects of thymidine phosphorylase. *J Biol Chem* 277:4128–4133.

8. Marti R, Verschuuren JJ, Buchman A, Hirano I, Tadesse S, van Kuilenburg AB, et al. (2005). Late-onset MNGIE due to partial loss of thymidine phosphorylase activity. *Ann Neurol* 58:649–652.

9. Massa R, Tessa A, Margollicci M, Micheli V, Romigi A, Tozzi G, et al. (2009). Late-onset MNGIE without peripheral neuropathy due to incomplete loss of thymidine phosphorylase activity. *Neuromuscul Disord* 19:837–840.

10. Valentino ML, Marti R, Tadesse S, Lopez LC, Manes JL, Lyzak J, et al. (2007). Thymidine and deoxyuridine accumulate in tissues of patients with mitochondrial neurogastrointestinal encephalomyopathy (MNGIE). *FEBS Lett* 581:3410–3414.

11. Bradford MM (1976). A rapid and sensitive method for the quantitation of microgram quantities of protein utilizing the principle of protein-dye binding. *Anal Biochem* 72:248–254.

# Chapter 9

## Measurement of Mitochondrial dNTP Pools

### Ramon Martí, Beatriz Dorado, and Michio Hirano

### Abstract

Because deoxyribonucleoside triphosphates (dNTPs) are the critical substrates for DNA replication and repair, dNTP pools have been studied in context of multiple basic biochemical processes. Over the last 12 years, interest in dNTPs, and specifically the mitochondrial dNTP pools, has expanded to biomedical science because several mitochondrial diseases have been found to be caused by dysfunctions of several enzymes involved in dNTP catabolism or anabolism. Techniques to reliably measure mitochondrial dNTPs should be sensitive and specific to avoid interference caused by the abundant ribonucleotides. Here, we describe detailed protocols to measure mitochondrial dNTPs from two specific samples, cultured skin fibroblasts and mouse liver. The methods can be easily adapted to other types of samples. The protocol follows a polymerase-based method, which is the most widely used approach to measure dNTP pools. Our description is based on the latest update of the technique, which minimizes the potential interference from ribonucleotides.

**Key words:** dNTP, Mitochondrial DNA depletion, Mitochondrial DNA replication, PCR, Radio-labeled primer extension

## 1. Introduction

The deoxyribonucleoside triphosphates (dNTPs) are the essential building blocks for DNA replication and repair, and, for over four decades, measurements of dNTPs in biological samples have been related to basic cell biology and biochemistry. Within the last decade, the study of dNTPs, especially the mitochondrial dNTP pool, has gained increasing interest due to the identification of mutations that alter dNTP metabolism in several severe mitochondrial disorders (1–4). Recent studies have identified intermediate RNA: DNA hybrid fragments during mitochondrial DNA replication (5, 6) suggesting that ribonucleotides may also be incorporated into newly synthesized mitochondrial DNA. Nevertheless, mutations in four different genes disrupt dNTP pathways causing severe mitochondrial diseases with altered dNTP pools (1–4), while only

Lee-Jun C. Wong (ed.), *Mitochondrial Disorders: Biochemical and Molecular Analysis*, Methods in Molecular Biology, vol. 837,
DOI 10.1007/978-1-61779-504-6_9, © Springer Science+Business Media, LLC 2012

one disease gene has been linked to mitochondrial ribonucleotide metabolism, but not to unbalanced ribonucleotide triphosphate pools (7). Therefore, the studies on the balance of the mtDNA replication substrates have been centered on dNTPs only. These studies have been mainly restricted to experimental animal or cell culture models and aimed at uncovering the molecular bases of mitochondrial dNTP homeostasis and pathomechanisms of genetic diseases with nucleotide pool imbalances.

Methods based on HPLC separation followed by ultraviolet (UV) or mass spectrometry (MS, MS/MS) detection have been employed to resolve dNTPs (dATP, dGTP, dCTP, and dTTP) and ribonucleoside triphosphates (rNTPs; ATP, GTP, CTP, and UTP) (8–11). However, cellular dNTP concentrations are usually between 10- and 1,000-fold lower than those of rNTPs and other partially phosphorylated ribonucleotides, such as ADP and AMP. For this reason, interference from ribonucleotides often impedes reliable quantification of the dNTPs in cell extracts, even when using highly selective methods as HPLC-MS/MS (8, 11). This problem has been often circumvented by chemically removing the ribonucleotides from the extracts (e.g., with boronate or periodate treatment) prior to HPLC injection (11, 12). However, HPLC methods do not achieve the sensitivity needed for the measurement of mitochondrial dNTPs, much less abundant than cytosolic dNTPs.

A different strategy, based on the DNA polymerase-catalyzed incorporation of dNTPs into a template DNA, was described in 1969 (13). Twenty years later, a modified version of this strategy was proposed (14) and adopted, with slight modifications, by many investigators (15–19). The method is based on the incorporation of tritium-labeled dATP (for the determination of dGTP, dCTP, or dTTP) or tritium-labeled dTTP (for the determination of dATP), via primer extension of daughter oligonucleotides from four parental oligonucleotides specifically designed for dATP, dGTP, dCTP, or dTTP determination. The reaction mixture contains radiolabeled dATP (or dTTP) and primed parental oligonucleotide in excess, together with DNA polymerase, such that the limiting reagent for the elongation of the primed oligonucleotide is the endogenous dNTP specie to be measured. After 1 h of reaction, the amount of radioactive nucleotide incorporated depends on the amount of dNTP species in the specimen, which can be calculated from a calibration curve obtained with aqueous standards.

The DNA polymerase most widely used for this assay used to be the Klenow subunit of the *E. coli* polymerase I. However, a recent report from Dr. Vera Biachi's group has shown that, under some conditions, Klenow may incorporate ribonucleotide instead of deoxynucleotide, which may lead to overestimations of some dNTPs, especially dGTP and dCTP (20). For this reason, the method described here uses the enzyme Thermo Sequenase, which has been shown to discriminate well between dNTPs and rNTPs even in the presence of 1,000-fold higher ribonucleotides than deoxynucleotides (20).

# 2. Materials

## 2.1. Reagents

1. Mitochondrial isolation buffer: 210-mM d-mannitol, 70-mM sucrose, 10-mM Tris–HCl pH 7.5, 0.2-mM EGTA, and 0.5% (w/v) bovine serum albumin (BSA).

   Dissolve 0.121 g of Tris-base (M = 121.14 g/mol) in 85 ml of water and adjust the pH to 7.5 by adding concentrated HCl. Add 3.83 g of d-mannitol (M = 182.2 g/mol), 2.40 g of sucrose (M = 342.3 g/mol), and 7.6 mg of EGTA (380.4 g/mol). Once dissolved, adjust the total volume up to 100 ml with water, and sterilize by filtering through a 0.2-μm pore nylon membrane. Store at –20°C in 10-ml aliquots. Before use, add BSA up to 0.5% (w/v) to the volume needed for the experiment.

2. 1-M Tris–HCl, pH 7.5. Dissolve 12.1 g of Tris-base (M = 121.14 g/mol) in 90 ml of water, and adjust the pH to 7.5 by adding HCl. Store in 10-ml aliquots at –20°C.

3. 0.1-M $MgCl_2$. Dissolve 0.203 g of magnesium chloride hexahydrate ($MgCl_2 \cdot 6H_2O$, M = 203.3 g/mol) in 10 ml of water. Store frozen at –20°C in aliquots.

4. 0.5-M dithiothreitol (DTT). Dissolve 0.772 g of DTT (M = 154.3 g/mol) in 10 ml of water. Store frozen at –20°C in aliquots.

5. Oligonucleotides (sequences are written from 5′ to 3′) (see Note 1):

   Oligo A: AAATAAATAAATAAATAAATGGCGGTGGAGGCGG

   Oligo G: TTTCTTTCTTTCTTTCTTTCGGCGGTGGAGGCGG

   Oligo C: TTTGTTTGTTTGTTTGTTTGGGCGGTGGAGGCGG

   Oligo T: TTATTATTATTATTATTAGGCGGTGGAGGCGG

   Common primer: CCGCCTCCACCGCC

   To make stock solutions of the oligonucleotides, reconstitute with sterile water to a final concentration of 100 μM as follows: Add the appropriate volume of sterile water to the vial containing the solid oligonucleotide, allow rehydration at room temperature for 30 min, and then vortex to ensure complete and homogeneous dissolution. Store the stock solutions at –20°C.

6. Primed oligonucleotides: Mix 50 μl of 100-μM oligo A with 50 μl of 100-μM common primer; make the same separate mixtures for oligo G, oligo C, and oligo T. In each mixture, the

```
Primed oligonucleotide A    5'-AAATAAATAAATAAATAAATGGCGGTGGAGGCGG-3'
                                           3'-CCGCCACCTCCGCC-5'


Primed oligonucleotide G    5'-TTTCTTTCTTTCTTTCTTTCGGCGGTGGAGGCGG-3'
                                           3'-CCGCCACCTCCGCC-5'


Primed oligonucleotide C    5'-TTTGTTTGTTTGTTTGTTTGGGCGGTGGAGGCGG-3'
                                           3'-CCGCCACCTCCGCC-5'


Primed oligonucleotide T     5'-TTATTATTATTATTATTAGGCGGTGGAGGCGG-3'
                                           3'-CCGCCACCTCCGCC-5'
```

Fig. 1. Primed oligonucleotides used for measurement of dNTPs. Primed oligonucleotides, as designed by Sherman and Fyfe (14). See the text for the details on preparation.

final concentrations of both oligos (specific and common) are 50 μM. In a beaker, heat 2 L of water to 70°C, and then, incubate the four tubes in this bath while letting cold the water to room temperature (1–2 h), which will allow the oligos to hybridize to obtain the following primed oligonucleotides (Fig. 1).

7. Once the mixtures have reached room temperature, dilute each 50-μM primed oligonucleotide solution (100 μl) by adding 900 μl of sterile water (1:10 dilution) to bring the primed oligonucleotides to 5 μM. Make aliquots and store at –20°C until needed.

8. Tritium-labeled dATP ([8-³H]dATP, tetrasodium salt, at a specific activity of 10–25 Ci/mmol, 1 mCi/ml) for the determination of dGTP, dCTP, and dTTP. This reagent comes in ethanol:water (1:1) solution. Store at –20°C (see Note 2). On the day of the assay, take the volume of radiolabeled reagent that contains the amount of ³H-dATP needed for all the samples to be analyzed (15 pmol of radiolabeled dATP are needed for every reaction tube). Lyophilize by speed vacuum. Once completely dry, add the amount of water needed to achieve a concentration of 15 μM. Let the pellet rehydrate in the tube for 5 min on ice, then vortex and keep the tube on ice until the reagent is used (see Note 3).

9. Tritium-labeled dTTP ([methyl-³H]dTTP, tetrasodium salt, specific activity 10–25 Ci/mmol, 1 mCi/ml) for the determination of dATP. This reagent comes in ethanol:water (1:1) solution. Store at –20°C. On the day of the assay, dry and reconstitute as indicated for tritium-labeled dATP (see above).

10. Thermo Sequenase. Keep as indicated by the manufacturer until use. Use preparations with catalytic concentrations around 30 U/μl.

11. 95% ethanol (see Note 4).

12. 60% methanol (v/v) in water: This reagent can be stored at room temperature but will have to be at –20°C before use.

13. 5% (w/v) $Na_2HPO_4$: Dissolve 100 g of $Na_2HPO_4$ (M = 142.0 g/mol) in 2 L of water.

14. Phosphate buffered saline (PBS): 140-mM NaCl, 10 mM $Na_2HPO_4$, pH 7.4 (see Note 5).

    Dissolve 8.2 g of NaCl (M = 58.5 g/mol) and 1.42 g of $Na_2HPO_4$ (M = 141.96 g/mol) in nearly 1 L of water. Adjust to pH 7.4 with HCl. Then, complete the volume to 1 L with water.

### 2.2. Aqueous dNTP Standards

1. From a stock solution of dATP, dGTP, dCTP, and dTTP, 10 mM each (see Note 6), make serial dilutions with water to obtain the following concentrations for the aqueous standards to be included in the assay: 400, 200, 100, 50, 25, and 5 nM. The standard zero (pure water) should be also included in the calibration curve.

### 2.3. Other Materials and Equipment

1. Syringes and needles (22 gauge × 1¼″) (0.70 × 30 mm) (for homogenization of cells).

2. Glass–glass homogenizer (for homogenization of tissues).

3. Whatman DE81 discs (2.3-cm diameter).

4. Scintillation cocktail.

5. Beta counter.

6. Scintillation vials.

7. Scrappers (for cultured cells).

8. Speed vacuum concentrator.

## 3. Methods (see Note 7)

### 3.1. Isolation of Mitochondria from Cultured Skin Fibroblasts (see Note 8)

1. Culture a minimum of $30 × 10^6$ cells in Petri dishes (see Note 9) for collection and use under testing conditions (see Note 10). If the final results are to be expressed per million of cells, an additional plate should be cultured in the same conditions for counting purposes.

2. Once the cells are in the desired conditions, put the plates on ice, and carry out the following steps in a cold chamber.

3. Remove the culture medium, and wash all the plates with cold PBS three times.

4. After the last wash, eliminate as much PBS as possible, and then, add 2 ml of isolation buffer to the first plate.

5. Collect thoroughly the cells from the surface with a scraper, and carefully transfer the whole volume of the suspension to the second plate of cells using a Pasteur pipet.

6. Repeat the cell collection with scraper and transfer to the third plate of cells. Repeat the same procedure with all the plates needed for the determination.

7. The final suspension should contain all the cells in the 2 ml of isolation buffer initially poured in the first plate (see Note 11).

8. Break the cells by passing the suspension through a 22-gauge 1¼-inch needle ($0.70 \times 30$ mm) in a 5-ml syringe (ten strokes). Further steps can be performed outside of the cold chamber, by keeping the samples always on ice.

9. Centrifuge the homogenate at $20,000 \times g$ at 4°C for 20 min.

10. Remove the supernatant (cytosolic fraction), which can be stored at −80°C for further use (see Note 12)

11. Wash the pellet (combined nuclear + mitochondrial fraction) with 0.5 ml of isolation buffer and centrifuge at $20,000 \times g$ for 20 min.

12. Discard the supernatant and resuspend the mitochondrial pellet in 1 ml of isolation buffer without BSA. Use one portion of this suspension for the determination of protein (see Note 13). Centrifuge the rest of the suspension at $20,000 \times g$ at 4°C for 20 min.

13. Discard the supernatant and save the pellet on ice for dNTP extraction (Subheading 3.3)

### 3.2. Isolation of Mitochondria from Mouse Liver

1. Fresh whole liver, rapidly excised from a recently sacrificed mouse, must be rapidly washed in cold PBS to eliminate the excess of blood and transferred to cold isolation buffer (4-ml/g tissue). All further steps should be carried out on ice in a cold room.

2. Cut the liver into small pieces using dissection scissors and homogenize in a glass–glass homogenizer (four up-and-down strokes).

3. Centrifuge the homogenate at $1,000 \times g$ for 5 min at 4°C to eliminate the nuclei and remains of undisrupted tissue.

4. Transfer the supernatant to a tube of the appropriate volume and centrifuge at $13,000 \times g$ at 4°C for 2 min (see Note 14).

5. Discard the supernatant, wash the pellet with 1 ml of isolation buffer, and then, centrifuge at $13,000 \times g$ at 4°C for 2 min.

6. Resuspend the mitochondrial pellet in 1 ml of isolation buffer without BSA, and save one portion of this suspension for the determination of protein (see Note 13). Centrifuge the rest of the suspension at $13,000 \times g$ at 4°C for 2 min.

7. Discard the supernatant and save the pellet on ice for dNTP extraction (Subheading 3.3).

**3.3. dNTP Extraction from the Mitochondrial Pellets**

1. Add 2 ml of 60% methanol at −20°C to the mitochondrial pellet, vortex and keep at −20°C for 2 h.

2. Centrifuge at 25,000×$g$ for 10 min and transfer the supernatant to a new tube, and discard the pellet.

3. Incubate the supernatant in boiling water for 3 min, and centrifuge at 25,000×$g$ for 10 min. Transfer the supernatant to a new tube, and discard the pellet (see Note 15).

4. Evaporate the solvent (60% methanol) of the supernatant with a speed vacuum concentrator.

5. Add the appropriate volume of sterile distilled water to the dry residue (see Note 16), let redissolve on ice for 10 min followed by vortexing. Keep the extract at −80°C until analysis.

**3.4. DNA Polymerase Assay (20)**

1. Label 0.5-ml (or 0.2-ml) capped tubes according to the number of samples to be processed, including duplicates and different dilutions. In addition, label tubes for the standard curves with duplicate or triplicate of the expected concentrations. The standard zero (water) should be included in the standard curve (see Note 17).

2. Table 1 shows the reaction mixture for each individual tube.

**Table 1**
**Composition of the reaction mixture for the determination of dNTPs (see Note 18)**

| Reagent | Volume (µl) | Final concentration |
|---|---|---|
| 1-M Tris–HCl, pH 7.5 | 0.8 | 40 mM |
| 0.1-M MgCl$_2$ | 2 | 10 mM |
| 0.5-M DTT | 0.2 | 5 mM |
| 5-µM specific primed oligonucleotide | 1 | 0.25 µM |
| 15-µM radiolabeled deoxynucleotide | 1 | 0.75 µM |
| Thermo sequenase (32 U/µl) | 0.016 | 0.025 U/µl |
| Water | 10 | – |
| Extract (or standard) | 5 | 1:4 |
| Total volume | 20 | – |

3. Prepare four master mixes (for determination of dATP, dGTP, dCTP, and dTTP, respectively) containing the following reagents in the proportions indicated in Table 1: Tris–HCl buffer, $MgCl_2$, DTT, the specific primed oligonucleotide, the radiolabeled nucleotide ($^3$H-dATP for dGTP, dCTP, and dTTP determinations, and $^3$H-dTTP for dATP determination), water, and Thermo Sequenase. Keep the master mixes on ice all the time (see Note 19).

4. Put 15 µl of each master mix in the corresponding tubes, previously labeled for samples and standards. Maintain the tubes constantly on ice.

5. Add 5 µl of standards, undiluted extracts, and diluted extracts to the corresponding tubes, and mix with several up-and-downs using the pipette.

6. Incubate at 48°C for 1 h.

7. Label with a pencil as many DE81 Whatman discs as reaction tubes are in process. The identification of the disc should be clearly visible to make easier further steps of the protocol and to minimize the chances of losing the label (see Note 20). Distribute the discs individually on a parafilm surface or aluminum foil sheet (see Note 21).

8. After 1 h of reaction at 48°C, spot 18 µl of the reaction mixtures on the corresponding discs. Air-dry the discs.

9. Wash the discs three times (10 min each) with 5% (w/v) $Na_2HPO_4$, once (10 min) with water, and once (10 min) with 95% ethanol. Distribute the discs on a surface and air-dry (see Note 22).

10. Place the discs in scintillation vials (previously labeled), add the appropriate volume of scintillation cocktail, and measure the dpm in a program set up for $^3$H (see Note 23).

11. Generate a calibration curve for each dNTP by plotting the standard concentration (independent variable) against the dpm obtained for the standards (dependent variable). Adjust to a linear regression $y = Ax + B$, where $y$ represents dpm, $x$ represents the dNTP concentration of the standard (in nmol/L), $A$ is the slope in dpm·L/nmol, and $B$ is the offset in dpm (see Note 24)

12. Calculate the concentration of each dNTP in the extracts using the calibration curves. The values may be expressed as for millions of cells (in the case of cell culture) or per milligram of protein.

# 4. Notes

1. Synthetic oligonucleotides can be purchased from several companies that offer custom designed oligos. We recommend to order HPLC-purified oligos to ensure the accurate length of most molecules received in the material purchased, which will improve the performance of the assay (e.g., it will enhance the sensitivity by minimizing the background signal).

2. There are different preparations of tritium-labeled dNTPs, including different cationic salts and different shipping solvents. The water:ethanol (1:1) solution does not freeze at −20°C and manufacturers indicate that, in these conditions, the radiolabeled compound is more stable than in solid state or in aqueous solution. The chemical concentration (μmol/L) of the solution received should be calculated for every batch from the activity per volume (e.g., 1 mCi/ml) to the specific radioactivity (we usually purchase at 10–25 Ci/mmol). From these particular values, the range of concentrations results to be 40–100 μmol/L.

3. Since the radiolabeled reagents are in ethanol:water (1:1) solution, the time for the solvent evaporation is not expected to be very long (around 1–1.5 h for 50 μl, depending on the speed vacuum device). The temperature of the solution should not exceed 20°C during the drying process, and the compound should not remain in the solid state any longer than necessary. The protocol described here does not involve dilution of the radiolabel; thus, the specific radioactivity remains unchanged. Theoretically, the higher specific radioactivity, the better sensitivity of the assay, but in our experience, acceptable sensitivities for most purposes can be achieved after a radiolabel 1:3 dilution by adding cold dATP (or cold dTTP) to the radiolabeled dATP (or dTTP). If this is done, it should be taken in account that the chemical concentration of the reagent to be used should be 15 μM in any case.

4. The ethanol is only needed for a washing step (Subheading 3.4, step 9 in the protocol). Many descriptions of the method indicate to use absolute ethanol, which is more expensive than the common 95% ethanol. However, the protocol works perfectly with 95% ethanol.

5. Although the most common and physiological formulation of PBS includes potassium cation, for the purposes of this protocol both formulations of PBS are acceptable. We have described here a formulation with sodium salts only. Moreover, it is also possible to purchase inexpensive preformulated PBS tablets to be dissolved in an appropriate volume of water to obtain PBS ready to use.

6. Premade dNTP aqueous solutions containing the four dNTP, usually at 10 mM, can be purchased from several companies. This is the most convenient way to obtain a concentrated stock dNTP solution, and then to prepare the standards at the concentrations needed. Separate 10-mM stock solutions of dATP, dGTP, dCTP, and dTTP are also available. Alternatively, the solid compounds can be purchased to prepare the solutions.

7. The protocols described here for the isolation of mitochondrial dNTPs involve two steps: (1) isolation of a mitochondria-enriched fraction by differential centrifugation and (2) extraction of dNTPs from the mitochondria-enriched fraction by treatment with 60% methanol. The protocol to obtain a mitochondria-enriched fraction may vary depending on the type of cultured cells or the tissue (detailed protocols can be found in refs. 21 and 22). Once this fraction has been obtained, the steps to extract the dNTPs are identical, regardless of what was the original sample. We describe here the protocols for cultured human skin fibroblasts and for mouse liver.

8. This method for isolation of mitochondrial dNTP pools is based on the protocol described in ref. 23. Most protocols to prepare mitochondria-rich fraction use differential centrifugation of the cell homogenates, with a first low speed centrifugation ($\sim 300 \times g$) to pellet and discard the nuclei, and a second high speed centrifugation ($\sim 20,000 \times g$) to pellet the mitochondria-enriched fraction. However, a substantial amount of mitochondria can be present in the nuclear fraction discarded. This may constitute an important loss of material from samples, e.g., cultured cells, which do not usually produce enough mitochondria to allow dNTP determinations. The method established for cells by Pontarin et al. (23) and described here uses a single $20,000 \times g$ centrifugation, which sediments a combined nuclear + mitochondrial fraction. Nuclear dNTPs are eliminated by a single wash because nuclei (but not mitochondria) are freely permeable for nucleotides. The combined mitochondria plus dNTP-free nuclei are pelleted from the wash at $20,000 \times g$ and used for the dNTP extraction.

9. The cells should be cultured on a surface accessible to collection using scraper; cell collection using trypsin is not recommended as even the limited proteolysis produced by this enzyme may enhance the leakage of dNTPs, thus affecting the final result. In our experience, human skin fibroblasts reach high confluence at around 20,000 cells/cm², but this value may present some variability depending on several factors, including the culture conditions and interindividual variations.

10. The amounts of mitochondrial and cytosolic dNTPs depend on the cell cycle status; in cultured cells, it has been shown that cycling cells have up to 50 times more dNTPs than quiescent

cells (24). Moreover, the mitochondria-rich fraction should be prepared rapidly to avoid dNTP leakage from mitochondria and to minimize ATP breakdown, which also influences negatively the dNTP content (25).

11. The number of plates needed for mitochondrial dNTP measurement depends on the particular purposes of the experiment. In our experience, around 1,500 cm² (i.e., ten plates of 13.7-cm diameter) are advisable when mitochondrial pools from confluent skin fibroblasts are to be measured. In this case, 2 ml of isolation buffer initially poured in the first plate should be sufficient to obtain a final suspension of between 2 and 4 ml due to the additional volume of the cells and residual PBS from the plates. Every plate should be carefully scraped, and the cell suspension should be fully recovered and passed successively from every plate to the next one. For experiments with cycling fibroblasts (therefore, contact inhibition should be avoided), the number of plates needed should be estimated taking in account that more surfaces will be needed to get the same number of cells. However, cycling cells contain more dNTPs than quiescent cells, and so, less than 30 million cells may be sufficient.

12. The cytosolic fraction can be used for dNTP measurement or for other purposes, depending on the particular objectives of the specific experiments. If dNTPs are to be measured in this fraction, one aliquot of the cytosolic fraction should be taken, pure methanol added up to a final concentration of 60%, and then follow the protocol as from the Subheading 3.3, step 1. Because the amounts of cytosolic dNTPs per million of cells are expected to be 10- to 50-fold higher than those expected of mitochondria, the volumes and the dilutions used in the final steps (see Subheading 3.3, step 5 and Notes 16 and 17) should be rescaled accordingly.

13. The protein concentration of the mitochondrial suspension can be determined using the Bradford method (26) using bovine serum albumin to prepare the calibration curve or with the method routinely used in your laboratory.

14. Alternatively, the supernatant can be aliquoted in eppendorf tubes, centrifuged as indicated, and then combine all the pellets together in the next wash.

15. Boiling the extract after the first methanol precipitation helps inactivate any remaining enzyme activity in the extracts (23). The pellet after the second precipitation may be barely visible, but most versions of this protocol include this boiling step to minimize dNTP catabolism in the extract.

16. The dry extract should be redissolved in the appropriate volume of water, which will depend on the type of sample and the

initial amount of mitochondria. We recommend 100 μl for the mitochondrial extract obtained from ~30 × 10⁶ cultured cells and 200 μl for the mitochondrial extract obtained from the whole mouse liver. It is expected that the resulting dNTP concentrations, as well as those of a 1:3 dilution, are quantifiable in the conditions of the assay described here. If different amounts of mitochondria are processed, the final volumes will have to be adjusted accordingly to ensure quantifiable concentrations in the final extract. Adding too much water in this step may lead to undetectable or barely quantifiable dNTPs.

17. Each nucleotide is determined in a different reaction. In addition, we recommend processing each sample with a duplicate of the undiluted extract and a duplicate with a dilution of the extract (e.g., 1:3 or 1:5 dilution, trying to ensure that the dNTP concentration in the diluted sample is reliably quantifiable). Because this is a long technique that can be influenced by many factors, it is necessary to include as many controls as possible to ensure the accuracy of the results. Processing two different dilutions of the same extract will allow verification that the result is reproducible at both concentrations.

18. Several alternative methods are possible as described in the modified protocol to avoid interference caused by ribonucleotides (20). For example, Taq DNA polymerase can be used instead of Thermo Sequenase, and Klenow enzyme can be used, with some restrictions, for dATP and dTTP determinations. Here, we detail the protocol with the particular conditions that we use for dNTP determinations, but modifications of this protocol according to ref. 20 are also possible.

19. The master mixes should be prepared immediately before use and kept on ice before starting the reaction. The simultaneous presence of primed oligonucleotides, radiolabeled dNTP and Thermo Sequenase, in the absence of the sample, may promote nonspecific incorporation of radiolabel into the primed oligonucleotides, thus increasing the background signal and affecting the detection limit of the assay.

20. Label with pencil. Do not use ink because a further washing step with ethanol would erase the label.

21. It is advisable to include the following controls for every assay: Measure the specific activity of the [³H]-dATP and [³H]-dTTP reagents in the beta counter. In addition, process DE81 discs with master mixes spotted, omitting the washing steps, to measure the total radioactivity loaded for every master mix.

22. The discs can be washed in four pools, one for each dNTP to be measured. Each wash can be done in 200–300 ml of washing solution in beakers or other recipients of the appropriate volume, using a horizontal shaker.

23. Since readings below 100 cpm are not expected (not even for the zero standard, which will have the lowest radiolabel), a reading time of 1 min in the beta counter is sufficient to obtain reliable counts.

24. In our experience, the calibration curves are acceptably linear within the range of 5–200 nmol/L. The slopes and offset of the curves depend on the specific activity of the $^3$H-dATP and $^3$H-dTTP used. The offset values should be similar to the dpm obtained for the standard zero.

## Acknowledgment

Dr. Martí is supported by a grant from the Spanish Instituto de Salud Carlos III (PS09/01591). Dr. Hirano is supported by NIH grants R01 HD056103 (cofunded by NICHD and the NIH Office of Dietary Supplements), R01 HD057543, and RC1 NS070232; MDA grant 115567; and the Marriott Mitochondrial Disorder Clinical Research Fund.

## References

1. Bourdon A, Minai L, Serre V, Jais JP, Sarzi E, Aubert S, et al. (2007). Mutation of RRM2B, encoding p53-controlled ribonucleotide reductase (p53R2), causes severe mitochondrial DNA depletion. *Nat Genet* 39:776–780.

2. Mandel H, Szargel R, Labay V, Elpeleg O, Saada A, Shalata A, et al. (2001). The deoxyguanosine kinase gene is mutated in individuals with depleted hepatocerebral mitochondrial DNA. *Nat Genet* 29:337–341.

3. Nishino I, Spinazzola A, Hirano M (1999). Thymidine phosphorylase gene mutations in MNGIE, a human mitochondrial disorder. *Science* 283:689–692.

4. Saada A, Shaag A, Mandel H, Nevo Y, Eriksson S, Elpeleg O (2001). Mutant mitochondrial thymidine kinase in mitochondrial DNA depletion myopathy. *Nat Genet* 29:342–344.

5. Pohjoismaki JL, Holmes JB, Wood SR, Yang MY, Yasukawa T, Reyes A, et al. (2010). Mammalian mitochondrial DNA replication intermediates are essentially duplex but contain extensive tracts of RNA/DNA hybrid. *J Mol Biol* 397:1144–1155.

6. Yasukawa T, Reyes A, Cluett TJ, Yang MY, Bowmaker M, Jacobs HT, et al. (2006). Replication of vertebrate mitochondrial DNA entails transient ribonucleotide incorporation throughout the lagging strand. *Embo J* 25:5358–5371.

7. Kaukonen J, Juselius JK, Tiranti V, Kyttala A, Zeviani M, Comi GP, et al. (2000). Role of adenine nucleotide translocator 1 in mtDNA maintenance. *Science* 289:782–785.

8. Chen P, Liu Z, Liu S, Xie Z, Aimiuwu J, Pang J, et al. (2009). A LC-MS/MS method for the analysis of intracellular nucleoside triphosphate levels. *Pharm Res* 26:1504–1515.

9. Decosterd LA, Cottin E, Chen X, Lejeune F, Mirimanoff RO, Biollaz J, et al. (1999). Simultaneous determination of deoxyribonucleoside in the presence of ribonucleoside triphosphates in human carcinoma cells by high-performance liquid chromatography. *Anal Biochem* 270:59–68.

10. Di Pierro D, Tavazzi B, Perno CF, Bartolini M, Balestra E, Calio R, et al. (1995). An ion-pairing high-performance liquid chromatographic method for the direct simultaneous determination of nucleotides, deoxynucleotides, nicotinic coenzymes, oxypurines, nucleosides, and bases in perchloric acid cell extracts. *Anal Biochem* 231:407–412.

11. Hennere G, Becher F, Pruvost A, Goujard C, Grassi J, Benech H (2003). Liquid chromatography-tandem mass spectrometry assays for

intracellular deoxyribonucleotide triphosphate competitors of nucleoside antiretrovirals. *J Chromatogr B Analyt Technol Biomed Life Sci* 789:273–281.

12. Shewach DS (1992). Quantitation of deoxyribonucleoside 5′-triphosphates by a sequential boronate and anion-exchange high-pressure liquid chromatographic procedure. *Anal Biochem* 206:178–182.

13. Solter AW, Handschumacher RE (1969). A rapid quantitative determination of deoxyribonucleoside triphosphates based on the enzymatic synthesis of DNA. *Biochim Biophys Acta* 174:585–590.

14. Sherman PA, Fyfe JA (1989). Enzymatic assay for deoxyribonucleoside triphosphates using synthetic oligonucleotides as template primers. *Anal Biochem* 180:222–226.

15. Bianchi V, Borella S, Rampazzo C, Ferraro P, Calderazzo F, Bianchi LC, et al. (1997). Cell cycle-dependent metabolism of pyrimidine deoxynucleoside triphosphates in CEM cells. *J Biol Chem* 272:16118–16124.

16. Dorado B, Area E, Akman HO, Hirano M Onset and organ specificity of Tk2 deficiency depends on Tk1 down-regulation and transcriptional compensation. *Hum Mol Genet* 20:155–164.

17. Lopez LC, Akman HO, Garcia-Cazorla A, Dorado B, Marti R, Nishino I, et al. (2009). Unbalanced deoxynucleotide pools cause mitochondrial DNA instability in thymidine phosphorylase-deficient mice. *Hum Mol Genet* 18:714–722.

18. Saada A, Ben-Shalom E, Zyslin R, Miller C, Mandel H, Elpeleg O (2003). Mitochondrial deoxyribonucleoside triphosphate pools in thymidine kinase 2 deficiency. *Biochem Biophys Res Commun* 310:963–966.

19. Song S, Wheeler LJ, Mathews CK (2003). Deoxyribonucleotide pool imbalance stimulates deletions in HeLa cell mitochondrial DNA. *J Biol Chem* 278:43893–43896.

20. Ferraro P, Franzolin E, Pontarin G, Reichard P, Bianchi V (2010). Quantitation of cellular deoxynucleoside triphosphates. *Nucleic Acids Res* 38:e85.

21. Fernandez-Vizarra E, Ferrin G, Perez-Martos A, Fernandez-Silva P, Zeviani M, Enriquez JA Isolation of mitochondria for biogenetical studies: An update. *Mitochondrion* 10:253–262.

22. Frezza C, Cipolat S, Scorrano L (2007). Organelle isolation: functional mitochondria from mouse liver, muscle and cultured fibroblasts. *Nat Protoc* 2:287–295.

23. Pontarin G, Gallinaro L, Ferraro P, Reichard P, Bianchi V (2003). Origins of mitochondrial thymidine triphosphate: dynamic relations to cytosolic pools. *Proc Natl Acad Sci USA* 100:12159–12164.

24. Ferraro P, Pontarin G, Crocco L, Fabris S, Reichard P, Bianchi V (2005). Mitochondrial deoxynucleotide pools in quiescent fibroblasts: a possible model for mitochondrial neurogastrointestinal encephalomyopathy (MNGIE). *J Biol Chem* 280:24472–24480.

25. Ferraro P, Nicolosi L, Bernardi P, Reichard P, Bianchi V (2006). Mitochondrial deoxynucleotide pool sizes in mouse liver and evidence for a transport mechanism for thymidine monophosphate. *Proc Natl Acad Sci USA* 103:18586–18591.

26. Bradford MM (1976). A rapid and sensitive method for the quantitation of microgram quantities of protein utilizing the principle of protein-dye binding. *Anal Biochem* 72:248–254.

# Chapter 10

# Measurement of Oxidized and Reduced Coenzyme Q in Biological Fluids, Cells, and Tissues: An HPLC-EC Method

## Peter H. Tang and Michael V. Miles

## Abstract

Direct measure of coenzyme Q (CoQ) in biological specimens may provide important advantages. Precise and selective high-performance liquid chromatography (HPLC) methods with electrochemical (EC) detection have been developed for the measurement of reduced (ubiquinol) and oxidized (ubiquinone) CoQ in biological fluids, cells, and tissues. EC detection is preferred for measurement of CoQ because of its high sensitivity. Reduced and oxidized CoQ are first extracted from biological specimens using 1-propanol. After centrifugation, the 1-propanol supernatant is directly injected into HPLC and monitored at a dual-electrode. The EC reactions occur at the electrode surface. The first electrode transforms ubiquinone into ubiquinol, and the second electrode measures the current produced by the oxidation of the hydroquinone group of ubiquinol. The methods described provide rapid, precise, and simple procedures for determination of reduced and oxidized CoQ in biological fluids, cells, and tissues. The methods have been successfully adapted to meet regulatory requirements for clinical laboratories, and have been proven reliable for analysis of clinical and research samples for clinical trials and animal studies involving large numbers of specimens.

**Key words:** Coenzyme Q, Ubiquinol, Ubiquinone, Redox, HPLC, Coulometric, Electrochemical, Analysis, Plasma, Milk, Platelet, Muscle, Liver, Brain

## 1. Introduction

The measurement of coenzyme Q (CoQ) has become important because of growing evidence that it may be a useful biomarker for evaluation of mitochondrial density and redox state in disorders of energy metabolism. The structure of CoQ is composed of a hydrophilic quinone ring and a hydrophobic isoprenoid side chain. The nonpolar side chain provides a lipid-soluble characteristic which allows association of CoQ with lipid bilayers, including the inner membrane of mitochondria (IMM). The IMM is an important example because it is the site where CoQ is integrated in the

Lee-Jun C. Wong (ed.), *Mitochondrial Disorders: Biochemical and Molecular Analysis*, Methods in Molecular Biology, vol. 837,
DOI 10.1007/978-1-61779-504-6_10, © Springer Science+Business Media, LLC 2012

mitochondrial respiratory chain. A significant CoQ deficiency in biological cells and tissues will contribute to respiratory chain dysfunction and promote the production of damaging free radicals.

CoQ is well known for its antioxidant properties, but has other important functions as well, including regulation of mitochondrial permeability pores, activation of mitochondrial uncoupling proteins, multiple anti-inflammatory effects, and protection from lipid peroxidation (1, 2). The quinone ring can undergo a two-step reversible reduction, which results in transformation into its quinol or antioxidant form. The lipid-soluble side chain enables CoQ to function as an electron shuttle in the mitochondrial respiratory chain. The proportion of reduced (ubiquinol) and oxidized (ubiquinone) formed in fluids and tissues is dependent upon several factors, including tissue type, efficiency of mitochondrial function, energy requirement, nutritional status, and oxidative stress (1–3). It should be noted that CoQ exists as several homologues in living organisms, which vary in relation to the length of the isoprenoid side chain. For example, in fungi, CoQ exists in forms ranging from CoQ6 to CoQ10. In rodents, it exists as two homologues, mainly CoQ9 with lesser amounts of CoQ10. In humans, CoQ10 is the predominate form with only trace quantities of CoQ9. The unique biochemistry of CoQ presents a significant challenge for the measurement of CoQ content and redox state in tissues and fluids (3).

## 1.1. CoQ Function in the Electron Transport Chain

Maintenance of adequate quantities of CoQ is essential for the normal function of the mitochondrial electron transport chain (ETC) (2). ETC complexes I and II receive electrons from NADH and succinate, respectively. CoQ is an electron acceptor from complexes I and II, and electron donor to complex III. Complex III oxidizes ubiquinol, which in turn reduces cytochrome $c$ oxidase (COX), another compound that moves freely through the IMM. Complex IV couples the oxidation of COX to the reduction of $O_2$. As a result of the electron transport shuttle, protons are pumped into the intermembrane space. In the final step, complex V releases these protons back into the mitochondrial matrix, which drives the formation of adenosine triphosphate (ATP).

It is important to understand the potential significance of CoQ10 deficiency in humans because it may contribute to the increased production of free radicals and oxidative stress, and even trigger mitochondrial degradation (2, 4, 5). In certain human tissues, CoQ content has been correlated with mitochondrial proliferation. Total CoQ10 (ubiquinol-10 plus ubiquinone-10) content was correlated with the percentage of subsarcolemmal mitochondrial aggregates (SSMA) in skeletal muscle of 47 patients (6). Increased SSMA, which is a pathological measure of mitochondrial proliferation, is common in muscles of younger patients with mitochondrial myopathies (7, 8). Another report noted correlations between total CoQ10 (TCoQ10) and activities of ETC complexes I, II, IV, I + III, and II + III in muscles of 82 children with suspected mitochondrial myopathy (9). Interestingly, a very recent study found *plasma* TCoQ10 was correlated with ETC complexes II and I + III in muscles of children requiring prolonged elemental nutrition because of severe food intolerance and allergies (10). It has been recommended that patients with significant deficiency of either ETC complex I + III or II + III have direct quantitation of CoQ10 to confirm deficiency (8, 11).

**1.2. CoQ Redox State and Oxidative Stress**

An important function of CoQ is protection of cells and membranes against lipid peroxidation and oxidative stress (2). In the circulation, CoQ is carried by lipoproteins, and is predominantly in the reduced form ($CoQH_2$). However, $CoQH_2$ in low-density lipoprotein (LDL) is easily oxidized to ubiquinone. $CoQH_2$ is thought to be the first antioxidant depleted when LDL is subjected to oxidative stress in vivo. It has been postulated that $CoQH_2$ prevents the initiation and/or the propagation of lipid peroxidation in plasma lipoproteins and biological membranes. The antioxidant activity of $CoQH_2$ depends not only on its concentration but also on mitochondrial redox status. In certain conditions, the redox status of CoQ in target tissues may be a unique biomarker for oxidative stress and mitochondrial content (3, 6).

**1.3. Clinical Considerations**

A wide range of applications has been developed for the quantitation of CoQ10 content and redox status in biological fluids, cells, and tissues (Table 1). However, only a few of these methods have been validated for analytical reliability. The methods outlined in this paper have been evaluated extensively and shown to provide excellent analytical precision and sensitivity. It is important to emphasize that the reliability of CoQ testing results is largely dependent upon careful specimen collection and storage procedures. In addition, meticulous attention is required during CoQ extraction and analysis to prevent the oxidation of ubiquinol.

The clinical importance of CoQ10 deficiency was underscored recently when gene mutations associated with CoQ biosynthesis were identified (12, 31, 32). As a result, both primary and secondary causes of CoQ10 deficiency have been reported (Table 2). It is expected that as new studies are conducted evaluating the function of CoQ10 in mitochondrial disorders, the importance of CoQ monitoring will continue to increase.

CoQ10 is also important therapeutically because it is one of only a few agents which can provide benefits to patients with mitochondrial disorders (1, 2, 11, 20, 31). Monitoring plasma CoQ10 during supplementation has been widely utilized to evaluate the uptake and distribution of this agent (3, 46). Well-defined reference ranges have been established for CoQ10, which enable clinicians to readily identify patients with CoQ10 deficiency. Patients with

**Table 1**
**Experience in the measurement and evaluation of CoQ10 content and redox status in various biological media and clinical applications**

| Clinical considerations and applications | Specimen type *(References)* | | |
|---|---|---|---|
| | **Fluid** | **Cellular** | **Tissue** |
| | Plasma/serum (14–16) | Platelets (17, 18) | Muscle (6, 9, 19, 20) |
| | Breast milk (21) | Lymphocytes (22) | Liver (23) |
| | Cerebrospinal fluid (24) | Erythrocytes (25) | Brain (26) |
| | Seminal fluid (27) | Spermatozoa (27) | Kidney (28) |
| | | Fibroblasts (29, 30) | |
| Invasiveness of specimen collection procedure | +–++ | +–++ | +++ |
| Availability of established reference ranges | +++ | + | +–++ |
| Evaluation of CoQ content and redox status | +++ | ++ | +–++ |
| Evaluation of effects of CoQ10 supplementation | +++ | ++ | + |
| Evaluation of effects of disease/aging | +++ | ++ | + |
| Evaluation of nutritional effects | ++ | ++ | + |
| Evaluation of environmental effects | ++ | + | + |

Clinical application/experience: +, limited; ++, moderate; +++, extensive

CoQ10 deficiency are most likely to benefit from supplementation (2, 8, 13, 20, 32, 46). Measurement of plasma ubiquinol-10 and ubiquinone-10 is also important for the development and evaluation of the absorption characteristics of new CoQ10-related compounds and formulations (13).

**1.4. Analytical Considerations**

Previous CoQ studies have encountered problems with instability of $CoQH_2$ during sample collection, handling, processing, and storage (47–53). These studies indicated that $CoQH_2$ is unstable in blood, plasma, and hexane extracts at room temperature. As a result, the $CoQH_2$:TCoQ ratio changes considerably within an hour after a blood or tissue sample is obtained. The lability of $CoQH_2$ is due to the structure of the hydroquinone moiety which makes $CoQH_2$ sensitive to oxygen, and at room temperature,

**Table 2**
**Evaluation of CoQ10 content and redox status in humans**

| Disease/disorder/condition | Principal effect | References |
|---|---|---|
| *Genetic-related* | | |
| Primary CoQ10 deficiency | Markedly decreased CoQ10 in target tissues | (12, 31) |
| Respiratory chain defect | Decreased CoQ10 in target tissues | (9, 20, 29, 32) |
| Trisomy 21 | Decreased reduced:total CoQ10 redox ratio | (33) |
| Mitochondrial DNA depletion syndrome | Decreased CoQ10 in muscle | (34) |
| Phenylketonuria | Decreased CoQ10 in lymphocytes | (22) |
| Smith-Lemli-Opitz syndrome | Decreased CoQ10 in platelets | (35) |
| *Disease-/disorder-related* | | |
| Parkinson's disease | Increased oxidized CoQ10 in plasma | (36) |
| Friedreich's ataxia | Decreased CoQ10 in serum | (37) |
| Postnatal stress changes | Decreased CoQ10 in erythrocyte membranes | (38) |
| Preeclampsia | Increased plasma oxidized:reduced CoQ10 redox ratio | (39) |
| Chronic headache disorders | Plasma CoQ10 deficiency in 33% of young patients | (40) |
| Cystic fibrosis | Low serum CoQ10 in cystic fibrosis patients with pancreatic insufficiency | (41) |
| Hypoadrenalism | Low plasma CoQ10 in patients with isolated hypoadrenalism | (42) |
| Severe food intolerance/allergy | Decreased muscle CoQ10 in children requiring prolonged elemental diet therapy | (10) |
| *Drug-treatment-/toxin-related* | | |
| Environmental toxins | Increased plasma oxidized CoQ10 associated with methylmercury and PCB exposure in Inuit people | (43) |
| Atorvastatin treatment effects in patients with coronary artery disease | Plasma CoQ10 was inversely related to brain natriuretic peptide with long-term atorvastatin treatment | (44) |
| Acute lymphoblastic leukemia | Increased reduced CoQ10 in children during chemotherapy-induction phase of treatment | (45) |

it is spontaneously oxidized at a rate of ~2 nM/min. This problem is very obvious in many studies which have reported wide variability in the $CoQH_2$:TCoQ ratio in biological specimens (47–53). Determination of CoQ redox status in biological specimens requires the utmost care to insure reliable estimates of the $CoQH_2$:TCoQ ratio. Because of this instability, other investigators recommended that plasma samples be individually thawed, extracted, and analyzed as a continuous process (51–53). This is obviously very impractical for analyzing significant numbers of clinical specimens. This paper describes methods which maintain CoQ redox state and provide for rapid and precise measurement.

An electrochemical (EC) detector is preferred for detection of $CoQH_2$ due to its high sensitivity. The electrochemical reactions

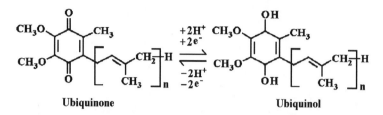

Fig. 1. Electrochemical reactions of coenzyme Q on the surface of electrodes.

are monitored at electrodes that measure the current produced by the reduction of the hydroquinone group of CoQ or by the oxidation of the hydroquinol group of $CoQH_2$ (Fig. 1).

## 2. Materials

### 2.1. General Guidelines

1. Prepare all solutions using HPLC grade solvents, analytical grade reagents, and ultrapure water (prepared by using Millipore water system to purify deionized water to attain a sensitivity of 18 MΩ cm at 25°C).

2. Prepare and store all reagents at room temperature (unless indicated otherwise).

3. Do not add sodium azide to the reagents.

4. Obtain the following HPLC grade solvents: methanol, ethanol, 1-propanol, 2-propanol, and hexane.

5. Obtain the following analytical grade chemicals: ubiquinone-9, ubiquinone-10, ubiquinol-10, ubiquinone-11, sodium acetate, and glacial acetic acid. All chemicals are used without further purification.

6. Diligently follow all waste disposal regulations when disposing waste materials.

### 2.2. Preparation of Standard Solutions

1. Ubiquinone-10 stock solution (100 μg/mL): Add 10 mL of hexane to a 100-mL volumetric flask. Weigh 10 mg of ubiquinone-10 and transfer to the flask. Make up to 100 mL with 1-propanol. Cap and mix to dissolve 10 mg of ubiquinone-10. Store at −20°C (see Note 1).

2. Ubiquinone-10 working solution (4 μg/mL): Pipette 4 mL of ubiquinone-10 stock solution into a 100-mL volumetric flask. Make up to 100 mL with 1-propanol. Calculate the exact concentration of ubiquinone-10 working solution by reading the absorbance at the spectrophotometer (275 nm, 1-cm light path quartz cuvette) using ε = 14,200. Store at −20°C (see Note 1).

3. Ubiquinol-10 stock solution (100 μg/mL): Weigh 10 mg of ubiquinol-10 to a 100-mL volumetric flask. Add 10 mL of hexane to the flask, mix, and make up to 100 mL with 1-propanol. Cap and mix to dissolve 10 mg of ubiquinol-10. Calculate the exact concentration of ubiquinol-10 stock solution by reading the absorbance at spectrophotometer (290 nm, 1-cm light path quartz cuvette) using $\varepsilon = 4,100$. Store at –20°C (see Note 2).

4. Ubiquinol-10 working solution (4 μg/mL): Pipette 4 mL of ubiquinol-10 stock solution into a 100-mL volumetric flask. Make up to 100 mL with 1-propanol. Store at –20°C (see Note 1).

5. Ubiquinone-9 stock solution (100 μg/mL): Add 10 mL of hexane to a 100-mL volumetric flask. Weigh 10 mg of ubiquinone-9 and transfer to the flask. Make up to 100 mL with 1-propanol. Cap and mix to dissolve 10 mg of ubiquinone-9. Store at –20°C (see Note 3).

6. Ubiquinone-9 working solution (2 μg/mL): Pipette 2 mL of ubiquinone-9 stock solution into a 100-mL volumetric flask. Make up to 100 mL with 1-propanol. Store at –20°C (see Note 1).

7. Ubiquinone-11 stock solution (100 μg/mL): Add 10 mL of hexane to a 100-mL volumetric flask. Weigh 10 mg of ubiquinone-11 and transfer to the flask. Make up to 100 mL with 1-propanol. Cap and mix to dissolve 10 mg of ubiquinone-11. Store at –20°C (see Note 4).

8. Ubiquinone-11 working solution (2 μg/mL): Pipette 2 mL of ubiquinone-11 stock solution into a 100-mL volumetric flask. Make up to 100 mL with 1-propanol. Store at –20°C (see Note 1).

*2.3. Instrumentation*

1. HPLC-EC system: ESA Model 582 Solvent Delivery Module equipped with a double plunger reciprocating pump, an autosampler, an analytical column, an ESA CouloChem II Model 5200A electrochemical detector, and a computer/controller with chromatography software. The system also comprises two guard cells (pre- and postcolumn) and an analytical cell. The first in-line filter is placed after the injector valve (see Note 5), the second in-line filter is placed after the guard column, and the third in-line filter is placed right after the analytical column (Fig. 2).

2. Autosampler: An autosampler equipped with a 100-μL injection loop. Set up needle height at 1.5 mm. Set up the autosampler temperature at ~0°C.

3. Analytical column: A reverse-phase Microsorb-MV column (5 μm, 4.6 mm × 15 cm). A reverse-phase C18 guard column

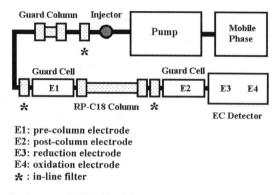

E1: pre-column electrode
E2: post-column electrode
E3: reduction electrode
E4: oxidation electrode
✶ : in-line filter

Fig. 2. Schematic diagram of HPLC-EC system.

(5 μm, 10×4.6 mm) is used to protect the analytical column. Guard cell is a single coulometric electrode (ESA Model 5020). Guard cells are installed before and after the analytical column.

4. Analytical cell: An ESA Model 5010 consists of a series of two coulometric electrodes. Connect analytical cell in series to postcolumn guard cell. Operate the analytical cell operated in the reduction mode with the first electrode for reduction of ubiquinone and in the oxidation mode with the second electrode for detection of ubiquinol (see Note 6).

## 3. Methods

### 3.1. Measurement of Ubiquinol-10 and Ubiquinone-10 in Human Plasma

*3.1.1. Calibration of Ubiquinone-10*

1. Prepare six calibration solutions of ubiquinone-10 containing 0.01, 0.1, 0.5, 1.0, 2.0, and 4.0 μg/mL by pipetting 0.025, 0.25, 1.25, 2.5, 5.0, and 10.0 mL of ubiquinone-10 solution (4 μg/mL) into 10-mL volumetric flasks and make up to 10 mL with 1-propanol.

2. Prepare and number six polypropylene tubes from 1 to 6.

3. Pipette 100 μL of ubiquinone-10 calibration solution, 100 μL of 2 μg/mL ubiquinone-9 solution, 500 μL of cold 1-propanol, and 100 μL of water into each numbered tube.

4. Vortex-mix all six tubes for 20 s at room temperature.

5. Transfer each mixture into a numbered autosampler vial.

6. Place all sample vials to the autosampler tray at a temperature of ~0°C.

7. Inject 50 μL of the mixture onto the HPLC system.

8. Obtain the peak heights of ubiquinone-9 and ubiquinone-10 for each injection.

9. Measure the peak height ratio of ubiquinone-10 versus ubiquinone-9 for each ubiquinone-10 concentration.

10. Establish a calibration curve for ubiquinone-10 by using a least squares linear regression from the peak height ratios of ubiquinone-10 versus ubiquinone-9 as functions of ubiquinone-10 concentrations.

*3.1.2. Calibration*
*of Ubiquinol-10*
*and Ubiquinone-10*

1. Prepare six calibration solutions of ubiquinol-10 containing 0.01, 0.1, 0.5, 1.0, 2.0, and 4.0 µg/mL by pipetting 0.025, 0.25, 1.25, 2.5, 5.0, and 10.0 mL of ubiquinol-10 solution (4 µg/mL) into 10-mL volumetric flasks and make up to 10 mL with 1-propanol.

2. Pipette 100 µL of ubiquinol-10 calibration solution, 100 µL of 2 µg/mL ubiquinone-9 solution, 500 µL of cold 1-propanol, and 100 µL of water into each numbered tube.

3. Vortex-mix all six tubes for 20 s at room temperature.

4. Transfer each mixture into a numbered autosampler vial.

5. Place all sample vials to the autosampler tray at a temperature of ~0°C.

6. Inject 50 µL of the mixture onto the HPLC system.

7. Obtain the peak heights of ubiquinol-10, ubiquinone-9, and ubiquinone-10 for each injection (see Note 7).

8. Measure the concentration of ubiquinone-10 in each injection by using the calibration curve of ubiquinone-10 in Subheading 3.1.1.

9. Subtract the ubiquinone-10 concentration from the initial concentration of ubiquinol-10 in each injection to obtain the true concentration of ubiquinol-10.

10. Establish two calibration curves for simultaneous determination of ubiquinol-10 and ubiquinone-10 concentrations by using least squares linear regression from the peak height ratios of ubiquinol-10 and ubiquinone-10 versus ubiquinone-9 as functions of ubiquinol-10 and ubiquinone-10 concentrations, respectively.

*3.1.3. Specimen Collection,*
*Processing, and Storage*

1. Draw venous blood into a Vacutainer® tube containing heparin as anticoagulant.

2. Do not open the Vacutainer® tube to ambient air (see Note 8).

3. Do not freeze the Vacutainer® tube (see Note 9).

4. Keep the Vacutainer® tube in ice bath or refrigerator before processing.

5. Process the Vacutainer® tube within 4 h.

6. Centrifuge the Vacutainer® tube at $2,000 \times g$ for 10 min at 4°C.

7. Transfer the plasma to a polypropylene tube and cap it.

8. Store immediately at or below −80°C until analysis.

### 3.1.4. CoQ10 Extraction

1. Process plasma samples in batches of 20 or less.

2. Thaw plasma samples at room temperature.

3. Pipette 100 μL of plasma, 100 μL of 2 μg/mL ubiquinone-9 solution, and 600 μL of cold 1-propanol into a numbered polypropylene tube.

4. Vortex-mix all tubes for 2 min at room temperature.

5. Centrifuge all tubes for 10 min at $21,000 \times g$ and ~0°C.

6. Transfer the supernatant to glass autosampler vials.

7. Place all sample vials to the autosampler tray at a temperature of ~0°C.

### 3.1.5. HPLC Conditions and Sample Analysis

1. Prepare mobile phase by mixing sodium acetate (4.2 g), 15 mL of glacial acetic acid, and 15 mL of 2-propanol to 695 mL of methanol and 275 mL of hexane in a 1-L flask.

2. Filter mobile phase by using a 0.2-μm-pore-sized, 47-mm nylon filter.

3. Set up the HPLC run at room temperature and a flow rate of 1.0 mL/min.

4. Place all sample vials to the autosampler tray at a temperature of ~0°C.

5. Inject immediately 50 μL of the extract onto the HPLC system.

6. Obtain three peaks of ubiquinol-10, ubiquinone-9, and ubiquinone-10 for each injection (Fig. 3).

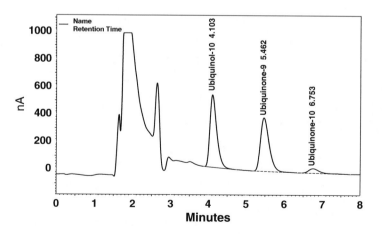

Fig. 3. Typical HPLC-EC chromatogram of a plasma sample.

7. Measure the concentrations of ubiquinol-10 and ubiquinone-10 in each injection by using the calibration curves of ubiquinol-10 and ubiquinone-10 in Subheading 3.1.2.

8. Sum up the concentrations of ubiquinol-10 and ubiquinone-10 to obtain TCoQ10 concentration.

9. Calculate ubiquinol-10 in TCoQ10 concentration to obtain the redox status.

### 3.2. Measurement of CoQ10 in Breast Milk

#### 3.2.1. Calibration of Ubiquinone-10

1. Prepare a series of calibration solutions containing 0.05, 0.25, 0.5, 1.0, and 2.0 μg/mL of ubiquinone-10 by pipetting 0.125, 0.625, 1.25, 2.5, and 5.0 mL of ubiquinone-10 solution (4 μg/mL) into 10-mL volumetric flasks and make up to 10 mL with 1-propanol (see Note 10).

2. Pipette 200 μL of ubiquinone-10 calibration solution, 100 μL of 2 μg/mL ubiquinone-9 solution, 100 μL of hexane, 600 μL of cold 1-propanol, and 200 μL of water into each numbered tube.

3. Vortex-mix all tubes for 20 s room temperature.

4. Transfer each mixture into a numbered autosampler vial.

5. Place all sample vials to the autosampler tray at a temperature of ~0°C.

6. Inject 50 μL of the mixture onto the HPLC system.

7. Obtain the peak heights of ubiquinone-9 and ubiquinone-10 for each injection.

8. Measure the peak height ratio of ubiquinone-10 versus ubiquinone-9 for each ubiquinone-10 concentration.

9. Establish a calibration curve for ubiquinone-10 by using a least squares linear regression from the peak height ratios of ubiquinone-10 versus ubiquinone-9 as functions of ubiquinone-10 concentrations.

#### 3.2.2. Specimen Collection, Processing, and Storage

1. Collect milk sample into a polypropylene tube and cap it.

2. Keep the sample tube in ice bath or refrigerator before processing.

3. Store all tubes at or below −80°C until analysis.

#### 3.2.3. CoQ10 Extraction

1. Process milk samples in batches of 20 or less.

2. Thaw milk samples at room temperature.

3. Pipette 200 μL of sample, 100 μL of ubiquinone-9, 800 μL of cold 1-propanol, and 100 μL of hexane into a numbered polypropylene tube.

4. Vortex-mix all tubes for 1 min at room temperature.

5. Centrifuge all tubes for 10 min at $21,000 \times g$ and ~0°C.

6. Transfer the supernatant to glass autosampler vials.

7. Place all sample vials to the autosampler tray at a temperature of ~0°C.

*3.2.4. HPLC Conditions*
*and Sample Analysis*

1. Prepare mobile phase by mixing sodium acetate (4.2 g), 15 mL of glacial acetic acid, and 15 mL of 2-propanol to 695 mL of methanol and 275 mL of hexane in a 1-L flask.

2. Filter mobile phase by using a 0.2-μm-pore-sized, 47-mm nylon filter.

3. Set up the HPLC run at room temperature and a flow rate of 1.0 mL/min.

4. Set up the precolumn guard cell at +500 mV or higher (see Note 11).

5. Inject immediately 50 μL of the extract onto the HPLC system.

6. Obtain the peak heights of ubiquinone-9 and ubiquinone-10 for each injection (Fig. 4).

7. Measure the concentration of ubiquinone-10 in each injection by using the calibration curves of ubiquinone-10 in Subheading 3.2.1.

### 3.3. Measurement of CoQ10 in Blood Platelets

*3.3.1. Calibration of Ubiquinone-10*

1. Prepare five concentrations of ubiquinone-10 solutions containing 0.1, 0.5, 1.0, 1.5, and 2.0 μg/mL (see Note 12).

2. Pipette 100 μL of ubiquinone-10 solution, 50 μL of 2 μg/mL ubiquinone-9 internal standard, 200 μL of 1-propanol, and 50 μL of water into a numbered tube for each concentration.

3. Vortex-mix for 20 s and then transfer each mixture into a numbered autosampler vial.

4. Place all sample vials to the autosampler tray at a temperature of ~0°C.

5. Inject 50 μL of the mixture onto the HPLC system.

6. Obtain the peak heights of ubiquinone-9 and ubiquinone-10 for each injection.

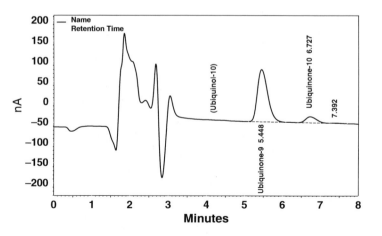

Fig. 4. Typical HPLC-EC chromatogram of a breast milk sample.

7. Establish a calibration curve for ubiquinone-10 by a least squares linear regression from the peak height ratios of ubiquinone-10 versus ubiquinol-9 as functions of ubiquinone-10 concentrations.

*3.3.2. Calibration of Ubiquinol-10 and Ubiquinone-10*

1. Prepare five concentrations of ubiquinol-10 solutions containing 0.1, 0.5, 1.0, 1.5, and 2.0 μg/mL (see Note 13).

2. Pipette 100 μL of ubiquinol-10 solution, 50 μL of 2 μg/mL ubiquinone-9 internal standard, 200 μL of 1-propanol, and 50 μL of water into a numbered tube for each concentration.

3. Vortex-mix for 20 s and then transfer each mixture into a numbered autosampler vial.

4. Inject 50 μL of the above mixture onto the HPLC system.

5. Obtain the peak heights of ubiquinol-10, ubiquinone-9, and ubiquinone-10 for each injection.

6. Measure the concentration of ubiquinone-10 in each injection by using the calibration curve of ubiquinone-10 in Subheading 3.3.1.

7. Subtract the ubiquinone-10 concentration from the initial concentration of ubiquinol-10 in each injection to obtain the true concentration of ubiquinol-10.

8. Establish two calibration curves for simultaneous determination of ubiquinol-10 and ubiquinone-10 concentrations by using least squares linear regression from the peak height ratios of ubiquinol-10 and ubiquinone-10 versus ubiquinone-9 as functions of ubiquinol-10 and ubiquinone-10 concentrations, respectively.

*3.3.3. Blood Platelet Preparation*

1. Draw 3 mL of blood into a trisodium citrate (13.2 g/L), citric acid (4.8 g/L), and dextrose (14.7 g/L) (ACD) anticoagulated Vacutainer® tube.

2. Centrifuge the blood at $250 \times g$ for 20 min at 22°C.

3. Pipette approximately 1.5 mL of platelet-rich plasma (PRP) from the upper portion of the plasma layer of ACD tube. Mix *gently* the PRP specimen.

4. Pipette 1.0 mL of PRP into a microtube. Determine a complete blood count (CBC) on the remaining PRP.

5. Centrifuge the microtube containing 1.0 mL of PRP at $3,500 \times g$ for 10 min at 22°C to pellet the platelet.

6. Wash the platelet pellet with 1.0 mL of sterile saline solution using low speed on a vortex mixer for 2 min.

7. Centrifuge again the isolated platelets at $3,500 \times g$ for 10 min at 22°C to reform the pellet.

8. Remove the saline from the specimen and store the remaining platelets at –70°C until analysis.

*3.3.4. CoQ10 Extraction*

1. Vortex-mix frozen platelet sample with 50 μL of 2 μg/mL ubiquinone-9 solution and 300 μL of 1-propanol for 1 min.

2. Centrifuge the above mixture at $21,000 \times g$ for 5 min to spin down the platelet.

3. Add 50 μL of water to the tube and transfer the supernatant to the autosampler vial.

4. Place all sample vials to the autosampler tray at a temperature of ~0°C.

*3.3.5. HPLC Conditions and Sample Analysis*

1. Prepare mobile phase by mixing sodium acetate (4.2 g), 15 mL of glacial acetic acid, and 15 mL of 2-propanol to 695 mL of methanol and 275 mL of hexane in a 1-L flask.

2. Filter mobile phase by using a 0.2-μm-pore-sized, 47-mm nylon filter.

3. Set up the HPLC run at room temperature and a flow rate of 1.0 mL/min.

4. Inject immediately 50 μL of the extract onto the HPLC system.

5. Obtain the peak heights of ubiquinol-10, ubiquinone-9, and ubiquinone-10 for each injection (Fig. 5).

6. Measure the concentrations of ubiquinol-10 and ubiquinone-10 in each injection by using the calibration curves of ubiquinol-10 and ubiquinone-10 in Subheading 3.3.2.

7. Sum up the concentrations of ubiquinol-10 and ubiquinone-10 to obtain TCoQ10 concentration.

8. Calculate ubiquinol-10 in TCoQ to obtain the redox status.

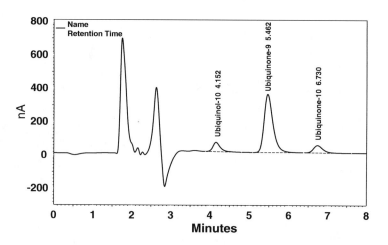

Fig. 5. Typical HPLC-EC chromatogram of a blood platelets sample.

**3.4. Measurement of Ubiquinol-10 and Ubiquinone-10 in Muscle, Liver, and Brain**

*3.4.1. Calibration of Ubiquinone-10*

1. Prepare six calibration solutions of ubiquinone-10 containing 0.01, 0.1, 0.5, 1.0, 2.0, and 4.0 μg/mL.

2. Pipette 200 μL of ubiquinone-10 calibration solution, 200 μL of 2 μg/mL ubiquinone-11, 400 μL of cold 1-propanol, and 200 μL of water into a numbered tube.

3. Vortex-mix all six tubes for 20 s at room temperature.

4. Transfer each mixture into a numbered autosampler vial.

5. Place all vials to the autosampler tray at a temperature of ~0°C.

6. Inject 50 μL of the mixture onto the HPLC system.

7. Obtain the peak heights of ubiquinone-10 and ubiquinone-11 for each injection.

8. Measure the peak height ratio of ubiquinone-10 versus ubiquinone-11 for each ubiquinone-10 concentration (see Note 14).

9. Establish a calibration curve for ubiquinone-10 by using a least squares linear regression from the peak height ratios of ubiquinone-10 versus ubiquinone-11 as functions of ubiquinone-10 concentrations.

*3.4.2. Calibration of Ubiquinol-10 and Ubiquinone-10*

1. Prepare six calibration solutions of ubiquinol-10 containing 0.01, 0.1, 0.5, 1.0, 2.0, and 4.0 μg/mL.

2. Pipette 200 μL of ubiquinol-10 calibration solution, 200 μL of 2 μg/mL ubiquinone-11, 400 μL of cold 1-propanol, and 200 μL of water into a numbered tube.

3. Vortex-mix all six tubes for 20 s at room temperature.

4. Transfer each mixture into a numbered autosampler vial.

5. Place all sample vials to the autosampler tray at a temperature of ~0°C.

6. Inject 50 μL of the mixture onto the HPLC system.

7. Obtain the peak heights of ubiquinol-10, ubiquinone-10, and ubiquinone-11 for each injection.

8. Measure the concentration of ubiquinone-10 in each injection by using the calibration curve of ubiquinone-10 in Subheading 3.4.1.

9. Subtract the ubiquinone-10 concentration from the initial concentration of ubiquinol-10 in each injection to obtain the true concentration of ubiquinol-10.

10. Establish two calibration curves for simultaneous determination of ubiquinol-10 and ubiquinone-10 concentrations by using least squares linear regression from the peak height ratios of ubiquinol-10 and ubiquinone-10 versus ubiquinone-11 as functions of ubiquinol-10 and ubiquinone-10 concentrations, respectively.

*3.4.3. Specimen Collection, Processing, and Storage*

1. Resect tissue from the subject and maintain specimen on ice until it can be flash-frozen (preferably within 5–10 min) to minimize the oxidation of ubiquinol-10.

2. Store sample at −80°C until analysis.

*3.4.4. CoQ Extraction*

1. Weigh a piece of frozen tissue with a size between 5 and 50 mg.

2. Homogenize the weighed tissue with 600 µL of cold 1-propanol and 200 µL of ubiquinone-11 in a tissue homogenizer on ice bath.

3. Add 200 µL of cold water to the above mixture and vortex-mix for 20 s.

4. Transfer the above mixture to a capped polypropylene tube and centrifuge at $21,000 \times g$ for 5 min.

5. Transfer the supernatant to glass autosampler vials.

6. Place all sample vials to the autosampler tray at a temperature of ~0°C.

*3.4.5. HPLC Conditions and Sample Analysis*

1. Prepare mobile phase by mixing sodium acetate (4.2 g), 20 mL of glacial acetic acid, and 20 mL of 2-propanol to 760 mL of methanol and 200 mL of hexane in a 1-L flask.

2. Filter mobile phase by using a 0.2-µm-pore-sized, 47-mm nylon filter.

3. Set up the HPLC run at 40°C and a flow rate of 1.2 mL/min.

4. Inject immediately 50 µL of the extract onto the HPLC system.

5. Obtain the peak heights of ubiquinol-10, ubiquinone-10, and ubiquinone-11 for each injection (Figs. 6–8).

6. Measure the concentrations of ubiquinol-10 and ubiquinone-10 in each injection by using the calibration curves of ubiquinol-10 and ubiquinone-10 in Subheading 3.4.2.

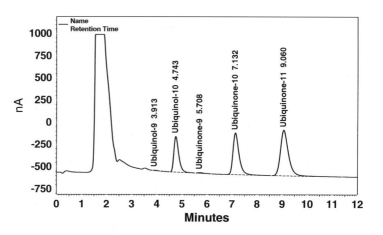

Fig. 6. Typical HPLC-EC chromatogram of a muscle sample.

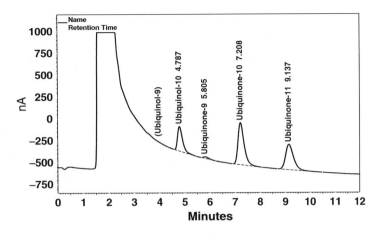

Fig. 7. Typical HPLC-EC chromatogram of a liver sample.

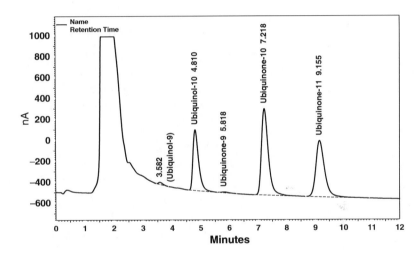

Fig. 8. Typical HPLC-EC chromatogram of a brain sample.

7. Sum up the concentrations of ubiquinol-10 and ubiquinone-10 to obtain TCoQ10 concentration.

8. Calculate ubiquinol-10 in TCoQ10 to obtain the redox status.

# 4. Notes

1. The preparation work is carried out under a dim light to avoid photochemical decomposition of ubiquinone and ubiquinol. The solution is thoroughly vortex-mixed until complete dissolution of ubiquinone and ubiquinol.

2. Ubiquinol-10 solution is prepared freshly due to its instability in air.

3. Use the concentration of ubiquinone-9 as labeled in the certificate of analysis from the manufacturer.

4. Use the concentration of ubiquinone-11 as labeled in the certificate of analysis from the manufacturer.

5. The first in-line filter collects particles from injections of real samples. Replace the in-line filter as needed when the cumulus of particles creates higher pressure to the system.

6. The reduction mode can be set up at an applied potential between −600 and −1,000 mV. The oxidation mode can be set up at an applied potential between +400 and +800 mV. However, a higher potential generates a higher background due to overcharge.

7. Ubiquinol is prone to air oxidation to ubiquinone during handling, storage, and processing. Therefore, a ubiquinol solution contains both ubiquinol and ubiquinone.

8. Ubiquinol is stable in the unopened Vacutainer® tube up to 4 h.

9. Blood cells disrupt after freezing so that no plasma can be obtained.

10. The concentrations of ubiquinone-10 listed in the text are equivalent to 0.06, 0.29, 0.579, 1.158, and 2.317 μmol/L, respectively. Different calibration ranges of ubiquinone-10 are used according to its physiological concentrations.

11. Interference to the ubiquinol-10 peak is removed by applying oxidation potential to the precolumn guard cell.

12. The concentrations of ubiquinone-10 listed in the text are equivalent to 116, 579, 579, 1,158, and 2,317 nmol/L, respectively. Different calibration ranges of ubiquinone-10 are used according to its physiological concentrations.

13. The concentrations of ubiquinol-10 listed in the text are equivalent to 116, 579, 579, 1,158, and 2,317 nmol/L, respectively. Different calibration ranges of ubiquinol-10 are used according to its physiological concentrations.

14. Human tissues contain both coenzyme Q10 and small quantities of coenzyme Q9; therefore, coenzyme Q11 is used as an internal standard.

## References

1. Turunen M., Olsson J., and Dallner G. (2004) Metabolism and function of coenzyme Q. *Biochim Biophys Acta* 1660, 171–199.

2. Bentinger M., Tekle M., and Dallner G. (2010) Coenzyme Q-biosynthesis and functions. *Biochem Biophys Res Commun.* 21, 396, 74–79.

3. Barshop B.A. and Gangoiti J.A. (2007) Analysis of coenzyme Q in human blood and tissues. *Mitochondrion* 7 Suppl, S89–93.

4. Quinzii C.M., López L.C., Gilkerson R.W., et al. (2010) Reactive oxygen species, oxidative stress, and cell death correlate with level of CoQ10 deficiency. *FASEB J.* 24, 3733–3743.

5. Rodríguez-Hernández A., Cordero M.D., Salviati L., et al. (2009) Coenzyme Q deficiency triggers mitochondria degradation by mitophagy. *Autophagy.* 5, 19–32.

6. Miles L., Miles M.V., Tang P.H., et al. (2005) Coenzyme Q: a potential marker for mitochondrial activity and redox status in skeletal muscle. *Pediatr Neurol* 32, 318–324.

7. Miles L., Wong B.L., Dinopoulos A., et al. (2006) Investigation of children for mitochondriopathy confirms need for strict patient selection, improved morphological criteria, and better laboratory methods. *Hum Pathol* 37, 173–184.

8. Haas R.H., Parikh S., Falk M.J., et al. (2008) The in-depth evaluation of suspected mitochondrial disease. *Mol Genet Metab* 94, 16–37.

9. Miles M.V., Miles L., Tang P.H., et al. (2008) Systematic evaluation of muscle coenzyme Q10 content in children with mitochondrial respiratory chain deficiencies. *Mitochondrion* 8, 170–180.

10. Miles M.V., Putnam P.E., Miles L., et al. (2011) Acquired coenzyme Q10 deficiency in children with recurrent food intolerance and allergies. *Mitochondrion* 11, 127–135.

11. Quinzii C.M. and Hirano M. (2010) Coenzyme Q and mitochondrial disease. *Dev Disabil Res Rev* 16, 183–188.

12. Quinzii C.M., López L.C., Von-Moltke J., et al. (2008) Respiratory chain dysfunction and oxidative stress correlate with severity of primary CoQ10 deficiency. *FASEB J* 22, 1874–1885.

13. Villalba J.M., Parrado C., Santos-Gonzalez M., et al. (2010) Therapeutic use of coenzyme Q10 and coenzyme Q10-related compounds and formulations. *Expert Opin Investig Drugs* 19, 535–554.

14. Tang P.H., Miles M.V., DeGrauw A., et al. (2001) HPLC analysis of reduced and oxidized coenzyme Q10 in human plasma. *Clin Chem* 47, 256–265.

15. Miles M.V., Horn P., Morrison J.A., et al. (2003) Plasma coenzyme Q10 reference intervals, but not redox status, are affected by gender and race in self-reported healthy adults. *Clin Chim Acta* 332, 123–132.

16. Miles M.V., Horn P.S., Tang P.H., et al. (2004) Age-related changes in plasma coenzyme Q10 concentrations and redox state in apparently healthy children and adults. *Clin Chim Acta* 347, 139–144.

17. Niklowitz P., Andler W., and Menke T. (2006) Coenzyme Q10 concentration in plasma and blood cells: what about diurnal changes? *Biofactors* 28, 47–54.

18. Miles M.V., Tang P.H., Miles L., et al. (2008) Validation and application of an HPLC-EC method for analysis of coenzyme Q10 in blood platelets. *Biomed Chromatogr* 22, 1403–1408.

19. Tang P.H., Miles M.V., Miles L., et al. (2004) Measurement of reduced and oxidized coenzyme Q9 and coenzyme Q10 levels in mouse tissues by HPLC with coulometric detection. *Clin Chim Acta* 341, 173–184.

20. Sacconi S., Trevisson E., Salviati L., et al. (2010) Coenzyme Q10 is frequently reduced in muscle of patients with mitochondrial myopathy. *Neuromuscul Disord* 20, 44–48.

21. Tang P.H., Miles M.V., Steele P., et al. (2006) Determination of coenzyme Q10 in human breast milk by high-performance liquid chromatography. *Biomed Chromatogr* 20, 1336–1343.

22. Colomé C., Artuch R., Vilaseca M.A. et al. (2002) Ubiquinone-10 content in lymphocytes of phenylketonuric patients. *Clin Biochem* 35, 81–84.

23. Kohli R., Kirby M., Xanthakos S.A., et al. (2010) High-fructose, medium chain trans fat diet induces liver fibrosis and elevates plasma coenzyme Q9 in a novel murine model of obesity and nonalcoholic steatohepatitis. *Hepatology* 52, 934–944.

24. Isobe C., Abe T., and Terayama Y. (2010) Levels of reduced and oxidized coenzyme Q-10 and 8-hydroxy-2'-deoxyguanosine in the cerebrospinal fluid of patients with living Parkinson's disease demonstrate that mitochondrial oxidative damage and/or oxidative DNA damage contributes to the neurodegenerative process. *Neurosci Lett* 469, 159–163.

25. Niklowitz P., Menke T., Wiesel T., et al. (2002) Coenzyme Q10 in plasma and erythrocytes: comparison of antioxidant levels in healthy probands after oral supplementation and in patients suffering from sickle cell anemia. *Clin Chim Acta* 326, 155–161.

26. Albano C.B., Muralikrishnan D., and Ebadi M. (2002) Distribution of coenzyme Q homologues in brain. *Neurochem Res* 27, 359–368.

27. Balercia G., Buldreghini E., Vignini A., et al. (2009) Coenzyme Q10 treatment in infertile men with idiopathic asthenozoospermia: a placebo-controlled, double-blind randomized trial. *Fertil Steril* 91, 1785–1792.

28. Diomedi-Camassei F., Di Giandomenico S., Santorelli F.M., et al. (2007) COQ2 nephropathy: a newly described inherited mitochondriopathy with primary renal involvement. *J Am Soc Nephrol* 18, 2773–2780.

29. Montero R., Sánchez-Alcázar J.A., Briones P., et al. (2008) Analysis of coenzyme Q10 in muscle and fibroblasts for the diagnosis of CoQ10 deficiency syndromes. *Clin Biochem* 41, 697–700.

30. Haas D., Niklowitz P., Hörster F., et al. (2009) Coenzyme Q(10) is decreased in fibroblasts of

patients with methylmalonic aciduria but not in mevalonic aciduria. *J Inherit Metab Dis* 32, 570–575.

31. DiMauro S., Quinzii C.M., and Hirano M. (2007) Mutations in coenzyme Q10 biosynthetic genes. *J Clin Invest* 117, 587–589.

32. Debray F.G., Lambert M., and Mitchell G.A. (2008) Disorders of mitochondrial function. *Curr Opin Pediatr* 20, 471–482.

33. Miles M.V., Patterson B.J., Chalfonte–Evans M., et al. (2007) Ubiquinol-10 improves oxidative imbalance in children with trisomy 21. *Pediatr Neurol* 37, 398–403.

34. Montero R., Sánchez-Alcázar J.A., Briones P., et al. (2009) Coenzyme Q10 deficiency associated with a mitochondrial DNA depletion syndrome: a case report. *Clin Biochem* 42, 742–745.

35. Haas D., Niklowitz P., Hoffmann G.F., et al. (2008) Plasma and thrombocyte levels of coenzyme Q10 in children with Smith-Lemli-Opitz syndrome (SLOS) and the influence of HMG-CoA reductase inhibitors. *Biofactors* 32, 191–197.

36. Sohmiya M., Tanaka M., Tak N.W., et al. (2004) Redox status of plasma coenzyme Q10 indicates elevated systemic oxidative stress in Parkinson's disease. *J Neurol Sci* 223, 161–166.

37. Cooper J.M., Korlipara L.V., Hart P.E., et al. (2008) Coenzyme Q10 and vitamin E deficiency in Friedreich's ataxia: predictor of efficacy of vitamin E and coenzyme Q10 therapy. *Eur J Neurol* 15, 1371–1379.

38. Ochoa J.J., Ramirez-Tortosa M.C., Quiles J.L., et al. (2003) Oxidative stress in erythrocytes from premature and full-term infants during their first 72 h of life. *Free Radic Res* 3, 317–322.

39. Roland L., Gagné A., Bélanger M.C., et al. (2010) Existence of compensatory defense mechanisms against oxidative stress and hypertension in preeclampsia. *Hypertens Pregnancy* 29, 21–37.

40. Hershey A.D., Powers S.W., Vockell A.L., et al. (2007) Coenzyme Q10 deficiency and response to supplementation in pediatric and adolescent migraine. *Headache* 47, 73–80.

41. Laguna T.A., Sontag M.K., Osberg I., et al. (2008) Decreased total serum coenzyme-Q10 concentrations: a longitudinal study in children with cystic fibrosis. *J Pediatr* 153, 402–407.

42. Mancini A., Leone E., Silvestrini A., et al. (2010) Evaluation of antioxidant systems in pituitary-adrenal axis diseases. *Pituitary* 13, 138–145.

43. Bélanger M.C., Mirault M.E., Dewailly E., et al. (2008) Environmental contaminants and redox status of coenzyme Q10 and vitamin E in Inuit from Nunavik. *Metabolism* 57, 927–933.

44. Suzuki T., Nozawa T., Sobajima M., et al. (2008) Atorvastatin-induced changes in plasma coenzyme Q10 and brain natriuretic peptide in patients with coronary artery disease. *Int Heart J* 49, 423–433.

45. Niklowitz P., Wiesel T., Andler W., and Menke T. (2007) Coenzyme Q10 concentration in the plasma of children suffering from acute lymphoblastic leukemia before and during induction treatment. *Biofactors* 29, 83–89.

46. Steele P.E., Tang P.H., DeGrauw T.J., and Miles M.V. (2004) Clinical laboratory monitoring of coenzyme Q10 use in neurologic and muscular diseases. *Am J Clin Pathol* 121 (Suppl 1), S113–S120.

47. Takada M., Ikenoya S., Yuzuriha T., and Katayama K. (1984) Simultaneous determination of reduced and oxidized ubiquinones. *Methods Enzymol* 105, 147–155.

48. Lang J.K. and Packer L. (1987) Quantitative determination of vitamin E and oxidized and reduced coenzyme Q by high-performance liquid chromatography with in-line ultraviolet and electrochemical detection. *J Chromatogr* 385, 109–117.

49. Edlund P.O. (1988) Determination of coenzyme Q10, alpha-tocopherol and cholesterol in biological samples by coupled-column liquid chromatography with coulometric and ultraviolet detection. *J Chromatogr* 425, 87–97.

50. Okamoto T., Fukunaga Y., Ida Y., and Kishi T. (1988) Determination of reduced and total ubiquinones in biological materials by liquid chromatography with electrochemical detection. *J Chromatogr* 430, 11–19.

51. Grossi G., Bargossi A.M., Fiorella P.L., and Piazzi S. (1992) Improved high-performance liquid chromatographic method for the determination of coenzyme Q10 in plasma. *J Chromatogr* 593, 217–226.

52. Wakabayashi H., Yamato S., Nakajima M., and Shimada K. (1994) Simultaneous determination of oxidized and reduced coenzyme Q and alpha-tocopherol in biological samples by high performance liquid chromatography with platinum catalyst reduction and electrochemical detection. *Biol Pharm Bull* 17, 997–1002.

53. Finckh B., Kontush A., Commentz J., et al. (1995) Monitoring of ubiquinol-10, ubiquinone-10, carotenoids, and tocopherols in neonatal plasma microsamples using high-performance liquid chromatography with coulometric electrochemical detection. *Anal Biochem* 232, 210–216.

# Chapter 11

## Assay to Measure Oxidized and Reduced Forms of CoQ by LC–MS/MS

### Si Houn Hahn, Sandra Kerfoot, and Valeria Vasta

### Abstract

The redox status of mitochondrial coenzyme Q (CoQ) is an important marker for oxidative stress associated with several disorders such as Parkinson disease and Alzheimer disease. Altered redox status may be present in mitochondrial electron transport complex disorders. Intracellular CoQ levels reflect the functional status of the mitochondrial electron transport complex better than plasma levels. Here, we describe the method to determine the reduced and oxidized form of CoQ in white blood cells using LC–MS/MS.

**Key words:** CoQ, Altered redox status, Reduced form, Oxidized form, Tandem mass spectrometry, Mitochondrial disorders, CoQ deficiency

### 1. Introduction

Mitochondrial coenzyme Q (CoQ) in its reduced form serves as an antioxidant to protect the cell from ROS damages and also an electron shuttle molecule in the mitochondrial electron transport chain, from both complex I and II to complex III (1). It is generally accepted that intracellular CoQ levels reflect the functional status of the mitochondrial respiratory chain complex better than plasma levels (2). The redox status of CoQ has been shown to be an important marker for metabolic and oxidative stress associated with diseases (3, 4), and altered redox status may be present in respiratory complex deficiencies (5). Measuring CoQ levels in muscle, plasma, or blood cells is a standard practice to diagnose primary or secondary CoQ deficiency in patients suspected of mitochondrial disorders (6, 7). Primary deficiency of CoQ10 is caused by heterogeneous autosomal recessive diseases. Mutations

Lee-Jun C. Wong (ed.), *Mitochondrial Disorders: Biochemical and Molecular Analysis*, Methods in Molecular Biology, vol. 837, DOI 10.1007/978-1-61779-504-6_11, © Springer Science+Business Media, LLC 2012

in OH-benzoate polyprenyltransferase (COQ2), prenyldiphosphate synthase subunit 1 (PDSS1), and decaprenyl diphosphate synthase subunit 2 (PDSS2) have been identified as CoQ10 biosynthesis defects (8–10). The ratio between the reduced (ubiquinol) and oxidized (ubiquinone) forms of CoQ10 can provide significant information in patients with mitochondrial ETC defects. In addition, markedly decreased CoQ10/CoQ9 ratio has been reported in patients with PDSS1 and CoQ2 mutations (8).

This method simultaneously quantifies CoQ10 and CoQ9 in both reduced and oxidized forms in white blood cells (WBC) by liquid chromatography–tandem mass spectrometry as previously described (11) with minor modification (5) using di-propoxy-CoQ10 as the internal standard (6). The coenzymes are extracted from the mononuclear white blood cells (WBC) pellet with methanol containing an internal standard (IS), vortex-mixed to precipitate proteins, and then centrifuged. The resulting supernatant is transferred to a LCMS vial, and 5 µl of sample is injected onto a Waters Acquity C18 (2.1 × 50 mm; 1.7 µm) UPLC column. All analytes are separated by isocratic UPLC method using mobile phase consisting of 2 mM ammonium formate in methanol. Waters Acquity UPLC system coupled to Quattro Premier tandem mass spectrometer is used for analyses. Multiple reaction monitoring mass spectrometry in the positive electrospray ionization mode is performed. The ratios of the extracted peak areas of the reduced and oxidized forms of CoQ9 and CoQ10 to the internal standard (IS) di-propoxy-CoQ10 (DP-Q10) are used to calculate the concentration of the analytes present. Results are normalized to total proteins in sample as determined by Bradford assay.

## 2. Materials

### 2.1. Chemicals and Reagents

1. 40 mg/ml KOH in 1-propanol: Accurately weigh one KOH pellet and dissolve in 1-propanol to obtain 40 mg/ml. Prepare fresh for the day.

2. 4 mg/ml CoQ9 in hexanes: Dissolve 2 mg vial of CoQ9 in 0.5 ml hexanes, store at –80°C, stable for 1 year.

3. 10 mg/ml CoQ10 in hexanes: Dissolve 10 mg CoQ10 in 1 ml hexanes, store at –80°C, stable for 1 year.

4. 1 mg/ml DP-Q10 internal standard stock solution:

   (a) Weight out approximately 10 mg CoQ10, use glass tubes (Kimble 73790 and caps Caliper 63911/0) (see Note 1).

   (b) Add 500 µl hexane, keep covered with foil.

   (c) Add 2 ml 1-propanol.

(d) Add 50 μl of 40 mg/ml KOH in 1-propanol.

(e) Vortex and incubate at room temperature for 25 min.

(f) Add 50 μl glacial acetic acid.

(g) Add 2.5 ml hexane.

(h) Add 5 ml HPLC grade $H_2O$ and vortex at full speed for 30 s. Centrifuge at $1,000 \times g$ for 1 min. Use a Pasteur pipette to carefully remove the bottom aqueous layer, reserving the upper phase. Repeat this step two times.

(i) Dry the upper phase under nitrogen gas and reconstitute in 500 μl hexane by vortexing.

(j) Determine concentration of DP-Q10 by UV absorbance (see Note 2).

(k) Dilute DP-Q10 to 1 mg/ml stock solution with methanol.

(l) Store at −80°C, stable for 1 year.

5. 0.1 μg/ml DP-Q10 working internal standard (IS) solution: Add 50 ml methanol to a 50 ml Falcon tube. Add 5 μl 1 mg/ml stock DP-Q10 to the methanol, cap, and vortex 10 s on full speed.

6. 50 μg/ml CoQ9 in working IS: Confirm concentration of stock CoQ9 by measuring absorbance (see Note 2). Dilute 5 μl stock CoQ9 to 50 μg/ml with working IS.

7. 50 μg/ml CoQ10 in working IS: Confirm concentration of stock CoQ10 by measuring absorbance (see Note 2). Dilute 5 μl stock CoQ10 to 50 μg/ml with working IS.

8. 10 M ammonium formate solution: Add 63 g ammonium formate to 50 ml of deionized water. Dissolve completely then add deionized water to make the final volume of 100 ml. Store the solution at room temperature.

9. HPLC mobile phase (2 mM ammonium formate in methanol): Add 200 μl of 10 M ammonium formate to 1 L HPLC grade methanol.

10. 100 mM Sodium hydroxide (NaOH) in water solution: Dissolve 400 mg NaOH in 100 ml deionized $H_2O$; store at room temperature.

11. GE Healthcare Ficoll-Paque.

12. Phosphate Buffered Saline (PBS): 137 mM NaCl, 2.7 mM KCl, 8.1 mM sodium phosphate, and 1.76 mM potassium phosphate. Commercially prepared PBS is acceptable.

13. 1.8% NaCl: Dissolve 180 g NaCl in 80 ml deionized $H_2O$. Add deionized $H_2O$ to make 100 ml solution.

14. Thermo Scientific Pierce Coomassie Plus (Bradford) Protein Assay Reagent.

15. 10 mg/ml Bovine Serum Albumin in deionized $H_2O$.

16. Nitrogen as API gas for LC/MS instrument.

17. Argon as collision gas for LC/MS instrument. 99.999% Ultra High Purity.

### 2.2. Equipment

1. Personal protective equipment: gown and gloves.

2. Repeating dispenser to dispense 200 µl.

3. Adjustable pipettes with disposable tips to pipette 100 µl.

4. 1.5 ml Polypropylene microcentrifuge tubes.

5. BD Falcon 15 ml conical centrifuge tubes.

6. BD Falcon 50 ml conical centrifuge tubes.

7. Glass transfer pipettes, five 3/4 in.

8. Glass autosampler vials: Target DP vials.

9. Glass insert for autosampler vials.

10. Caps in blue, green, red, and yellow with Teflon/silicone septa.

11. Leucosep tubes (Greiner Bio-One Inc.).

12. Vortex Mixer, Genie 2 Digital.

13. MixMate mixer.

14. Microcentrifuge.

15. Acquity UPLC® BEH C18, 1.7 µm particle size, 2.1×50 mm LC column.

16. Waters Quattro Premier™ tandem mass spectrometer with electrospray probe, coupled to Waters Acquity UPLC system.

17. MassLynx software.

## 3. Methods

### 3.1. Calibration

1. Starting standard solution (1 µg/ml CoQ9 and 1 µg/ml CoQ10): Add 10 µl 50 µg/ml CoQ9 in working IS and 10 µl 50 µg/ml CoQ10 in working IS to 480 µl working IS solution in a 1.5-ml Eppendorf tube. Cap and vortex at full speed for 5 s.

2. The highest standard, standard 6, is made by adding 120 µl starting solution to 680 µl working IS and mixing well. Follow the chart below to prepare remaining standards. Mix each standard well before proceeding to make the next standard.

| Standard | Dilution | Final concentration ng/ml CoQ9 | Final concentration ng/ml CoQ10 |
|---|---|---|---|
| 6 | | 150.0 | 150.0 |
| 5 | 2 (0.4 ml Std 6):1 (0.2 ml IS) | 100.0 | 100.0 |
| 4 | 1 (0.2 ml Std 5):1 (0.2 ml IS) | 50.0 | 50.0 |
| 3 | 1 (0.1 ml Std 4):5 (0.4 ml IS) | 10.0 | 10.0 |
| 2 | 1 (0.2 ml Std 3):1 (0.2 ml IS) | 5.0 | 5.0 |
| 1 | 1 (0.2 ml Std 2):1 (0.2 ml IS) | 2.5 | 2.5 |

3. Run all six standards, standards 1–6 plus working IS for standard 0 with every sample set.

4. MassLynx software will use a previously programmed quantification method to calculate and create a calibration curve and quantitate all controls and analytes based on this curve (see Subheading 3.7 later in this procedure).

5. Run new lot of standards (controls) against standards (controls) currently in use.

*3.2. Quality Control*

1. Use separate stock solutions for preparing standards and controls.

2. Three levels of controls (low, mid, and high) are run with each standard curve and sample set. Results must fall within ±2 SD of the mean values established for each control.

3. The mean and standard deviation are calculated for the new preparations with a minimum of 20 sets of control values.

4. For controls preparation, use 50 µg/ml CoQ9 in working IS and 50 µg/ml CoQ10 in working IS. Details for each control are listed in the table below.

   (a) Starting control solution (1 µg/ml CoQ9 and 1 µg/ml CoQ10): Add 10 µl 50 µg/ml CoQ9 in working IS and 10 µl 50 µg/ml CoQ10 in working IS to 480 µl working IS solution in a 1.5 ml Eppendorf tube. Cap and vortex at full speed for 5 s.

   (b) High control: Add 125 µl starting solution to 875 µl working IS and mix well. Follow the chart below to prepare remaining standards. Mix each standard well before proceeding to the next standard.

| Control | Dilution | Final concentration ng/ml CoQ9 | Final concentration ng/ml CoQ10 |
|---|---|---|---|
| High | | 125.0 | 125.0 |
| Mid | 1 (0.1 ml high control):5 (0.4 ml IS) | 25.0 | 25.0 |
| Low | 3 (0.06 ml mid control):10 (0.14 ml IS) | 7.5 | 7.5 |

**3.3. Mononuclear WBC Pellet Preparation**

1. Add 3 ml of room temperature Ficoll (GE Healthcare) to Leucosep tubes (Greiner Bio-One Inc.), and spin briefly to pass through filter.

2. Dispense 4 ml of ACD treated blood into the Leucosep tube.

3. Centrifuge at $1,000 \times g$ for 10 min at room temperature (see Note 3).

4. Transfer the WBC phase to a 15 ml Falcon tube.

5. Add 10 ml PBS at room temperature, invert tube, and centrifuge at $1,000 \times g$ for 10 min.

6. Remove supernatant with a pipette; some liquid will stay with the pellet – that is expected.

7. Add 2 ml cold water to pellet, resuspend by pipetting, check that pellet is resuspended without clumps, then set in ice, and after 30 s add 2 ml cold 1.8% NaCl (see Note 4).

8. Add 9 ml PBS.

9. Centrifuge at $1,000 \times g$ for 10 min, remove supernatant with a pipette.

10. Resuspend in 0.8 ml PBS and divide in eight Eppendorf tubes (see Note 4).

11. Centrifuge for 1 min at max velocity in microfuge, remove all supernatant, and store dry pellet at −80°C until use.

**3.4. Extraction of CoQ9 and CoQ10**

1. Add 200 μl working IS solution to each WBC pellet (see Note 5) and vortex for 30 s at full speed. Once the pellet is dispersed, vortex 15 s on max speed and keep in ice for 15 min.

2. Centrifuge for 5 min at max velocity in microcentrifuge at room temperature.

3. Pipette supernatant into LC/MS autosampler vials with glass inserts (see Note 6).

4. Store protein pellet at −20°C for Bradford analysis (see Note 7).

**3.5. LC–MS/MS Conditions**

1. Autosampler temperature should be 7°C.

2. Keep the UPLC column at 45°C for the experiment.

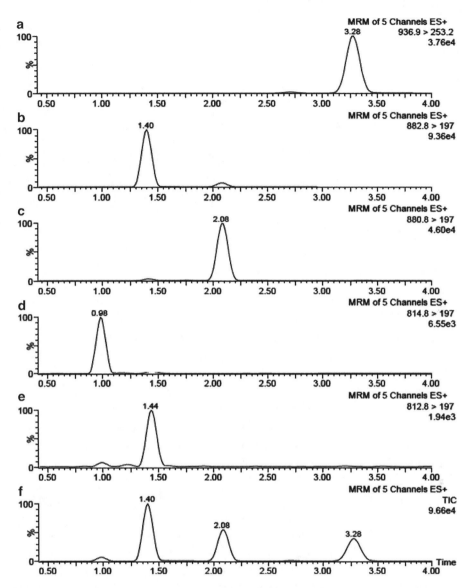

Fig. 1. LC–MS/MS of reduced and oxidized form of Q10 and Q9 in a WBC extract. DP-Q10 is used as IS. (**a**) DP-Q10 transition; (**b**) reduced CoQ10 transition; (**c**) oxidized CoQ10 transition; (**d**) reduced CoQ9 transition; (**e**) oxidized CoQ9 transition; and (**f**) total ion scan.

3. Analytes are eluted at 0.8 ml/min flow rate with an isocratic separation using 2 mM ammonium formate in 100% methanol for a total cycle time of 3.5 min (Fig. 1).

4. Directly couple the effluent from LC column to the electrospray source of a Waters Quattro Premier triple-quadrupole mass spectrometer.

5. Perform MS/MS analysis in positive ion mode with the capillary voltage set to 1.7 kV.

6. Set source temperature to 100°C and desolvation temperature to 400°C.

7. Set the desolvation gas to 800 L/h and the cone gas to 0 L/h.

8. The collision energy and cone voltage is described below (the precursor ions monitored are ammoniated adducts).

9. Monitored transitions (Fig. 1):

| Compound | MRM transition (*m/z*) | Cone voltage (V) | Collision energy (eV) |
|---|---|---|---|
| CoQ9 | 812.9 → 197 | 30 | 22 |
| Reduced CoQ9 | 814.8 → 197 | 30 | 22 |
| CoQ10 | 880.9 → 197 | 30 | 22 |
| Reduced CoQ10 | 882.8 → 197 | 30 | 22 |
| DP-Q10 | 936.9 → 253.2 | 30 | 20 |

### 3.6. Protein Quantification

1. To quantify total proteins, dissolve residual pellet after methanol extraction in 200 µl 100 mM NaOH. Vortex well and let sit at room temperature for 45 min (see Note 8).

2. Assay proteins with 5–10 µl of the suspension using the Bradford Assay and BSA as standard (see Note 9).

### 3.7. Quantification of CoQ9 and CoQ10

1. Check that retention times and peak areas are in correct range. Pay attention to peak shapes (see Note 10).

2. There is no significant sample or system carryover by injection of methanol blank after highest standard and control. If any carryover is observed in the blank, results should be discussed with the technical specialist or supervisor.

3. Quantification of analytes is carried out by monitoring the transition from the precursors to the product ions (see above transition table).

4. Integrate the area under the chromatographic peak using QuanLynx. Smooth peaks using 1 iteration of Savitzky-Golay smoothing, width set to 1.

5. Generate standard curve in QuanLynx normalizing to the internal standard DP-Q10 using a linear calibration function with $1/x$ weighting function. The concentration of both reduced and oxidized CoQ9 and CoQ10 is calculated based on the area of the standards which are in oxidized form.

6. Normalize values by dividing the concentrations of analytes by the protein concentration.

7. Results are reported as follows:

Total CoQ10 (ng/mg protein).

% Reduced CoQ10 (see Note 11).

Total CoQ10/total CoQ9 ratio (only reported if the total CoQ10 is out of the normal range).

8. Results are reported to one decimal point. If there is no detectable peak at all, the result is reported as ND.

9. Results with value greater than the highest standard are repeated after making a dilution using working IS. Subsequent result will be multiplied by the dilution factor and reported.

10. The normal range is listed in the table below.

| | Normal range in WBC |
|---|---|
| Total [CoQ10] ng/mg protein | 140–325 |
| % Reduced | 60–80 |
| [CoQ10]/[CoQ9] | 8–20 |

## 4. Notes

1. Any solvent-resistant tubes should be appropriate. Oxidation is not an issue since the standards are fully oxidized.

2. Measure concentration of stock solutions using UV absorbance. Dilute aliquots in triplicate (DP-Q10 and CoQ10: dilute 5 μl aliquot with 1,995 μl ethanol; CoQ9: dilute 10 μl aliquot with 790 μl). Measure UV absorbance at 275 nm. Average the results. Multiply the average absorbance by the constant listed in the table below; this will give you the concentration in mg/ml. Each constant is obtained from the molar extinction coefficient, unit conversion (from molar to mg/ml), and dilution factors. This step should be done each time the stock solution is used because hexanes are very volatile and the concentration of the stock solution may increase with time.

| Compound | Constant | Molar extinction coefficient |
|---|---|---|
| DP-Q10 | 25.19 | 14,600 |
| CoQ9 | 4.33 | 14,700 |
| CoQ10 | 23.65 | 14,600 |

3. Maximum deceleration of centrifuge is okay.

4. Process at most two pellets at once. If you choose to start with a volume different than 4 ml, it is important that in this step, each pellet is prepared from 0.5 ml of blood. For example, if you start with only 3 ml, you will have six pellets at this point.

5. If a pellet is not resuspended, use pipette tip to break up the pellet by pipetting up and down repeatedly. Once the pellet is dissolved, vortex 15 s on max speed and keep in ice for 15 min.

6. Transfer as much supernatant as possible without disturbing pellet. Check for bubbles in the tip of the insert and flick vial to remove them.

7. Protein pellet can be stored at −20°C until prepared for protein assay.

8. Check that all pellets are resuspended well before Bradford assay. If the sample is viscous, due to DNA, pay attention in pipetting the correct aliquot.

9. The Bradford protein assay is a spectroscopic analytical procedure that measures the total protein concentration in a solution. The assay can be performed using a commercially available kit (Coomassie Plus, Pierce) or by using Coomassie stain and a protein standard. Protocols are typically included in kits; however, for more information, see Bradford, M.M. (1976), "Rapid and sensitive method for the quantitation of microgram quantities of protein utilizing the principle of protein-dye binding," *Anal. Biochem.* 72:248–254.

10. Samples can be re-injected on the LC/MS/MS if necessary; however, the percentage reduced is not accurate after 24 h. There is a sufficient volume for multiple injections per sample, as long as more than 100 µl of supernatant was transferred to the glass insert.

11. Percent reduced is calculated: [Reduced CoQ10]/[Reduced CoQ10 + Oxidized CoQ10].

## References

1. Bentinger, M., Brismar, K., and Dallner, G. (2007) The antioxidant role of coenzyme Q, *Mitochondrion 7 Suppl,* S41–50.

2. Haas, R. H., Parikh, S., Falk, M. J., Saneto, R. P., Wolf, N. I., Darin, N., Wong, L. J., Cohen, B. H., and Naviaux, R. K. (2008) The in-depth evaluation of suspected mitochondrial disease, *Mol Genet Metab 94,* 16–37.

3. Gotz, M. E., Gerstner, A., Harth, R., Dirr, A., Janetzky, B., Kuhn, W., Riederer, P., and Gerlach, M. (2000) Altered redox state of platelet coenzyme Q10 in Parkinson's disease, *J Neural Transm 107,* 41–48.

4. Tang, P. H., Miles, M. V., Miles, L., Quinlan, J., Wong, B., Wenisch, A., and Bove, K. (2004) Measurement of reduced and oxidized coenzyme Q9 and coenzyme Q10 levels in mouse tissues by HPLC with coulometric detection, *Clin Chim Acta 341,* 173–184.

5. Vasta, V., Sedensky, M., Morgan, P., and Hahn, S. H. (2011) Altered redox status of coenzyme Q9 reflects mitochondrial electron transport chain deficiencies in Caenorhabditis elegans, *Mitochondrion 11,* 136–138.

6. Duncan, A. J., Heales, S. J., Mills, K., Eaton, S., Land, J. M., and Hargreaves, I. P. (2005) Determination of coenzyme Q10 status in blood mononuclear cells, skeletal muscle, and plasma by HPLC with di-propoxy-coenzyme Q10 as an internal standard, *Clin Chem 51,* 2380–2382.

7. Sacconi, S., Trevisson, E., Salviati, L., Ayme, S., Rigal, O., Redondo, A. G., Mancuso, M., Siciliano, G., Tonin, P., Angelini, C., Aure, K.,

Lombes, A., and Desnuelle, C. (2010) Coenzyme Q10 is frequently reduced in muscle of patients with mitochondrial myopathy, *Neuromuscul Disord 20*, 44–48.

8. Mollet, J., Giurgea, I., Schlemmer, D., Dallner, G., Chretien, D., Delahodde, A., Bacq, D., de Lonlay, P., Munnich, A., and Rotig, A. (2007) Prenyldiphosphate synthase, subunit 1 (PDSS1) and OH-benzoate polyprenyltransferase (COQ2) mutations in ubiquinone deficiency and oxidative phosphorylation disorders, J. Clin. Invest. 117:765–772.

9. Lopez, L. C., Schuelke, M., Quinzii, C. M., Kanki, T., Rodenburg, R.J., Naini, A., Dimauro, S., and Hirano, M. (2006) Leigh syndrome with nephropathy and CoQ10 deficiency due to decaprenyl diphosphate synthase subunit 2 (PDSS2) mutations. Am. J. Hum. Genet 79, 1125–1129.

10. Quizii, C., Niani, A., Salviati, L., Trevission, E., Navas, P., DiMauro, S., and Hirano, M (2006) A mutation in parahydroxybenoate-polyprenyltransferase (COQ2) mutations in ubiquinone deficiency and oxidative phosphorylation disorders. J. Clin. Invest. 117, 765–772.

11. Ruiz-Jimenez, J., Priego-Capote, F., Mata-Granados, J. M., Quesada, J. M., and Luque de Castro, M. D. (2007) Determination of the ubiquinol-10 and ubiquinone-10 (coenzyme Q10) in human serum by liquid chromatography tandem mass spectrometry to evaluate the oxidative stress, *J Chromatogr A 1175*, 242–248.

# Chapter 12

## Morphological Assessment of Mitochondrial Respiratory Chain Function on Tissue Sections

### Kurenai Tanji

### Abstract

In recent decades, genetic, biochemical, immunological, and cell biological techniques have been applied not only for better understanding of pathogenesis of known mitochondrial encephalomyopathies but also for exploring the possibility of mitochondrial involvement in other neurological, cardiac, gastrointestinal, and urological diseases. Techniques applied in a coordinated fashion have made it clear that mitochondrial dysfunction plays an important role in a wide range of human diseases including degenerative, toxic, metabolic, and neoplastic disorders. In this chapter, we provide updated protocols of essential histochemical and immunohistochemical methods that, in our opinion, are the most reliable morphological tool to visualize respiratory chain abnormality on tissue sections.

**Key words:** Histochemistry, Immunohistochemistry, Paraffin sections, Cryosections

## 1. Introduction

The human mitochondrial genome is a 16,569-bp double-stranded DNA. It is highly compact and contains only 37 genes: 2 genes encode ribosomal RNAs (rRNAs), 22 encode transfer RNAs (tRNAs), and 13 encode polypeptides. All 13 polypeptides are components of the respiratory chain, including seven subunits of complex I or NADH dehydrogenase–ubiquinone oxidoreductase, one subunit of complex III or ubiquinone–cytochrome $c$ oxidoreductase, three subunits of complex IV or cytochrome $c$ oxidase (COX), and two subunits of complex V or ATP synthase (1). The respiratory complexes also contain nuclear DNA (nDNA)-encoded subunits, which are imported into the organelle from the cytosol and assembled together with the mitochondrial DNA (mtDNA)-encoded subunits, into the respective holoenzymes in

Lee-Jun C. Wong (ed.), *Mitochondrial Disorders: Biochemical and Molecular Analysis*, Methods in Molecular Biology, vol. 837, DOI 10.1007/978-1-61779-504-6_12, © Springer Science+Business Media, LLC 2012

the mitochondrial inner membrane (2). Complex II or succinate dehydrogenase (SDH)–ubiquinone oxidoreductase contains only nDNA-encoded subunits. The research on mitochondrial disease has uncovered an increasing number of disorders that are caused by not only mutations in mtDNA but also in nuclear genes encoding the subunits of the respiratory chain or other proteins that are essential for the biosynthesis of specific cofactors or assembly of the complexes (3). Brain, nerve, and muscle, whose functions are highly dependent on oxidative metabolism, are most significantly affected tissues in the mitochondrial disorders (4, 5), and muscle biopsy remains a most useful diagnostic tool for diagnosis (6). Furthermore, genetic, biochemical, and morphological studies of muscle and brain from patients with mitochondrial disease have been proven fundamental to understand the pathogenesis of mitochondrial dysfunction at the level of individual muscle fibers and in different brain regions (7–10). The aim of this chapter is to present the histochemical and immunohistochemical methods that appear to be most reliable for correct identification of mitochondria on frozen or paraffin tissue sections. While the described protocols refer to skeletal muscle and brain tissue, the methods can be applied to other cell types (11–14). It is beyond the scope of this chapter to cover every study or method related to morphological analysis of mitochondria, but rather to provide sufficient information to allow investigators to apply these useful tools to answer some scientific or diagnostic questions.

## 2. Materials

Prepare all necessary solutions fresh before each staining procedure. It is also important to use freshly cut cryosections to prevent the occurrence of enzymatic degradation in sections.

### 2.1. Succinate Dehydrogenase Histochemistry

1. Cryosections: 0.1% poly-L-lysine-coated coverslips (or commercially available precoated slides), glass slides.

2. Incubation medium: 5 mM ethylenediaminetetraacetic acid (EDTA), 1 mM potassium cyanide (KCN), 0.2 mM phenazine methosulfate (PMS), 50 mM succinic acid, 1.5 mM nitroblue tetrazolium (NBT).

3. Dissolve the agents described in Subheading 2.1, step 2 in 5 mM Na phosphate buffer (pH 7.4), adjust pH to 7.6 with HCl (1N and 0.1N), and filter the solution with Whitman No. 1 filter paper.

4. For negative control sections, 0.01 M sodium malonate is added to the incubation medium.

**2.2. Cytochrome c Oxidase Histochemistry**

1. Cryosections: 0.1% poly-L-lysine-coated coverslips (or commercially available precoated slides), glass slides.

2. Incubation medium: 0.1% 3,3'-diaminobenzidine (DAB), 0.1% cytochrome $c$ (from horse heart), 0.02% catalase.

3. Dissolve the agents described in Subheading 2.2, step 2 in 5 mM Na phosphate buffer, adjust pH to 7.4 with NaOH (1N and 0.1N), and filter the solution with filter paper No. 1 (see Note 1).

4. For negative control sections, 0.01 M potassium cyanide is added to the incubation medium.

**2.3. Immunohisto-chemical Preparations**

1. Deparaffinization of paraffin-embedded sections: xylene, ethanol (EtOH) series (100, 95, 80, and 75%).

2. Avidin–biotin–peroxidase complex (ABC) methods: primary antibody for mitochondrial protein, phosphate buffer saline (PBS) (137 mM NaCl, 2.7 mM KCl, 8.1 mM $Na_2HPO_4 \cdot 2H_2O$, 1.76 mM $KH_2PO_4$, pH 7.4), methanol containing 0.3% $H_2O_2$, biotinylated second antibody, ABC mixed solution (use a commercially available ABC kit), normal serum (ideally from the same species as the host of the second antibody), bovine serum albumin (BSA), $DAB–H_2O_2$ solution (40 mg of 3,3'-diaminobenzidine tetrahydrochloride dissolved in 10 ml of PBS or 0.05 M Tris–HCl buffer (pH 7.6) containing 0.01% $H_2O_2$), hematoxylin (for counter staining), Permount (synthetic resin).

3. Immunofluorescent double staining for mtDNA-encoded and nDNA-encoded respiratory chain subunits: primary antibodies (see Subheading 3), PBS, BSA, biotinylated anti-mouse IgG, streptavidin-Texas red, anti-rabbit IgG FITC, 50% glycerol (or fluorescein-preserving aqueous mounting medium).

4. Immunofluorescent staining for anti-DNA: monoclonal anti-DNA antibody, PBS, BSA, 4% formaldehyde in 0.1 M CaCl2 (pH 7.0), biotinylated anti-mouse IgG, Streptavidin-FITC or streptavidin-Texas red.

## 3. Methods

Carry out all staining procedures at room temperature (RT) unless otherwise specified.

**3.1. Histochemical Staining (see Note 2)**

*3.1.1. SDH Histochemical Staining (see Note 3)*

1. Collect 8-μm-thick cryostat sections on 0.1% poly-L-lysine-coated coverslips (see Note 4).

2. Incubate sections in SDH solution (see Subheading 2.1, step 3) for 20 min at 37°C (see Note 5).

3. Rinse three times with distilled water, each time at RT for 5 min.

4. Mount on glass slides with warm glycerin gel.

*3.1.2. COX Histochemical Staining (see Note 6)*

1. Collect 8-μm-thick cryostat sections on 0.1% poly-L-lysine-coated coverslips.

2. Incubate sections in COX solution (see Subheading 2.2, step 3) for 1 h at 37°C (see Note 5).

3. Rinse three times with distilled water, each time at RT for 5 min.

4. Mount on glass slides with warm glycerin gel.

**3.2. Immunohisto-chemical Staining (see Note 7)**

*3.2.1. Immunolocalization of Mitochondrial Proteins in Paraffin-Embedded Sections Using ABC Method (see Note 8)*

1. Collect 4-μm paraffin sections on coated slides. Deparaffinize the sections through xylene and descending ethanol series (100, 95, 80, and 75%).

2. Incubate the sections in methanol containing 0.3% $H_2O_2$ for 30 min at RT.

3. Place the slides in PBS and then incubate the samples with 5% normal serum for 1 h at RT.

4. Incubate the slides with the primary antibody (1:1,000–1:2,000) diluted in PBS containing 1% bovine serum albumin (1% BSA/PBS) overnight at 4°C.

5. Rinse the slides with PBS three times, each time at RT for 5 min.

6. Incubate the slides with the biotinylated second antibody at the optimal concentrations (1:100–1:300) diluted in 1% BSA/PBS for 1 h at RT.

7. Rinse the slides with PBS three times, each time at RT for 5 min.

8. Incubate the slides with ABC solution (Prepare 30 min prior to use).

9. Rinse the slides with PBS three times, each time at RT for 5 min.

10. Incubate the slides with DAB–$H_2O_2$ solution for 1–3 min at RT.

11. Rinse the slides with distilled water ($dH_2O$) several times.

12. Counterstain the slides briefly with hematoxylin.

13. Rinse the slides with $dH_2O$, dehydrate through ascending ethanol series (75, 80, 95, and 100%), and clear in xylene.

14. Mount the slides with synthetic resin (Permount).

*3.2.2. Simultaneous Visualization F (Double Labeling) of mtDNA- and nDNA-Encoded Subunits of the Respiratory Chain Using Different Fluorochromes (see Note 9)*

1. Collect 4-µm-thick cryostat sections on coated coverslip or glass slides.

2. Incubate the sections for 2 h at RT in a humid chamber with a polyclonal antibody against mtDNA-encoded subunit (e.g., COX II) and with a monoclonal antibody against nDNA-encoded subunit (e.g., COX IV) at optimal dilutions (1:100 to 1:500) in 1% BSA/PBS. Control sections are incubated without the primary antibodies (see Note 10).

3. Rinse the slides with PBS three times, each time at RT for 5 min.

4. Incubate the sections for 1 h at RT (in a humid chamber) with biotinylated anti-mouse IgG (1:100) in 1% BSA/PBS.

5. Rinse the slides with PBS three times, each time at RT for 5 min.

6. Incubate the sections for 30 min. at RT (in a humid chamber) with streptavidin-Texas red (1:250) and with anti-rabbit IgG FITC (1:100) in 1% BSA/PBS to visualize the mtDNA probe in green and the nDNA probe in red.

7. Rinse the samples with PBS three times, each time at RT for 5 min.

8. Mount on slides with 50% glycerol in PBS (or fluorescein-preserving aqueous mounting medium).

*3.2.3. Localization of Mitochondrial DNA Using Immunofluorescent Probe (see Note 11)*

1. Collect 4-µm-thick cryostat sections on coated coverslip or glass slides.

2. Fix the sections in 4% formaldehyde in 0.1 M $CaCl_2$ pH 7, for 1 h at RT.

3. Dehydrate the sections in 70, 80, and 90% EtOH for 5 min each, and in 100% EtOH for 15 min.

4. Rinse the slides with PBS three times, each time at RT for 5 min.

5. Incubate the sections for 2 h at RT (in a humid chamber) with anti-DNA monoclonal antibody (1:100–1:500) in 1% BSA/PBS. Control sections are incubated without the primary antibodies.

6. Rinse the slides with PBS three times, each time at RT for 5 min.

7. Incubate the sections for 30 min at RT (in a humid chamber) with biotinylated anti-mouse IgG (1:100) in 1% BSA/PBS.

8. Rinse the slides with PBS three times, each time at RT for 5 min.

9. Incubate the sections for 30 min at RT (in a humid chamber) with streptavidin-Texas red or streptavidin-FITC (1:250) in 1% BSA/PBS.

10. Rinse the slides with PBS three times, each time at RT for 5 min.

11. Mount slides with 50% glycerol in PBS (or fluorescein-preserving aqueous mounting medium).

## 4. Notes

1. DAB is light sensitive. The COX incubation medium is, therefore, to be made in a beaker sealed with the aluminum foil to protect the agent from light exposure.

2. The most informative histochemical alteration of mitochondria in skeletal muscle is ragged red fiber (RRF), observed on frozen sections traditionally with the modified Gomori trichrome method (15). Since accumulation of material other than mitochondria sometimes simulates an RRF appearance, the identification of deposits suspected of being mitochondrial proliferation should be confirmed by histochemical staining of oxidative enzymes. In our experience, enzyme histochemistry for the activity of SDH and COX has proven to be the most reliable methods for the correct visualization of normal mitochondria, and for the diagnosis of some of mitochondrial disorders affecting skeletal muscle (6, 16).

3. The histochemical method for the microscopic demonstration of SDH activity on frozen sections is based upon the use of a tetrazolium salt (nitroblue tetrazolium, NBT) as an electron acceptor with phenazine methosulfate (PMS) serving as intermediate electron donor to NBT (17, 18). The specificity of the method may be tested by performing negative control test in which an SDH inhibitor, sodium malonate (0.01 M), is added to the incubation medium.

   SDH is the enzyme that catalyzes the conversion of succinate to fumarate in the tricarboxylic acid cycle, consisting of four subunits. Because complex II is the only component of the respiratory chain whose subunits are all encoded by the nDNA, SDH histochemistry is extremely useful for detecting variation in the sarcoplasmic distribution of mitochondria, being unaffected by any abnormality in mtDNA. In normal muscle, SDH detects two populations of fibers in a checkerboard pattern. Type II fibers, which rely upon glycolytic metabolism, show a light blue network-like stain. Type I fibers, whose metabolism is highly oxidative, show a darker and elaborate mitochondrial network (Fig. 1a). In samples with pathological proliferation of mitochondria, RRFs are highlighted with an intense blue SDH reaction corresponding to the distribution of the mitochondria within the fiber (Fig. 2), namely,

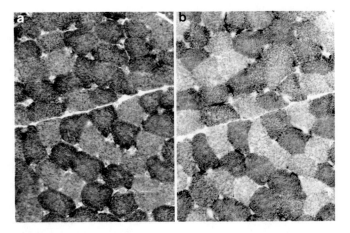

Fig. 1. (**a** and **b**) Normal muscle stained by SDH and COX. A checkerboard staining pattern is demonstrated by SDH (**a**) and COX (**b**) in a normal muscle biopsy: Mitochondria-rich type I fibers are stained darker than type II fibers (×200).

Fig. 2. Ragged "blue" fibers. RRFs highlighted by SDH that show an intense *blue* staining, indicating accumulation of mitochondria in the subsarcolemma as well as in the interfibrillar spaces (×250).

exhibiting "ragged blue" appearances. RRFs or "ragged blue" fibers" can be seen in most mtDNA defects (deletions and tRNA point mutations), but can also appear in other disorders such as myopathic form of mtDNA depletion syndrome, and fatal and benign COX-deficient myopathies of infancy (6, 16). SDH histochemistry is certainly useful for the diagnosis of complex II deficiency, as several patients with myopathy and isolated complex II deficiency have been reported (19).

4. Commercially available precoated glass slides can be alternatively used. Mount sections then with glycerin gel and coverslips.

5. When SDH–COX double staining (7) is desired, slides are first incubated with COX medium for 30 min, briefly washed, and are incubated with SDH medium for additional 30 min.

6. The histochemical method to visualize COX activity is based upon the use of 3,3'-diaminobenzidine (DAB) as electron donor for COX (20). The reaction product on oxidation of DAB occurs in the form of a brown pigmentation corresponding to the distribution of mitochondria in the tissue. The specificity of the method may also be tested by performing control experiments in which the COX inhibitor, potassium cyanide (0.01 M), is added to the incubation medium.

COX, the last component of the respiratory chain, catalyzes the transfer of reducing equivalents from cytochrome $c$ to molecular oxygen. In mammals, the apoenzyme is composed of 13 subunits. The three largest subunits (I, II, and III), which are encoded by mtDNA and synthesized within the mitochondria, confer the catalytic and proton-pumping activities to the enzyme. The ten small subunits are synthesized in the cytoplasm under the control of nuclear genes, which are presumed to modulate tissue specificity, thus adjusting the enzymatic activity to the metabolic demands of different tissues (21). Additional nDNA-encoded factors are required for the assembly of COX, and several COX assembly genes have been identified in yeast (22, 23). In human, pathogenic mutations in the human homologues of two of these genes have been identified in patients with COX-deficient Leigh syndrome (24, 25) and in patients with COX-deficient cardioencephalomyopathy (26, 27). Comprehending this complex genetic makeup of the enzyme, the availability of reliable histochemical method to visualize its activity has made COX one of the ideal tools for investigations of mitochondrial biogenesis, nDNA–mtDNA interactions, and for human mitochondrial disorders at both light and electron microscopic levels (28–31).

As in the case of SDH, staining of normal muscle for COX activity also shows a checkerboard pattern. Type I fibers stain darker due to their dependence upon oxidative metabolism and more abundant mitochondria content while type II fibers show a finer and less intensely stained mitochondria network (Fig. 1b). Muscle from mitochondrial myopathy associated with mtDNA mutations tends to show a mosaic expression of COX consisting of a variable number of COX-deficient and COX-positive fibers, and RRFs can be COX negative or COX positive (Fig. 3). At the advent of molecular genetics, it has become evident that the mosaic was an indicator of the heteroplasmic nature of the genetic defects. In another word, the mosaic pattern of COX expression is now considered as the "histochemical signature" of a heteroplasmic mtDNA mutation affecting the expression of mtDNA-encoded genes in skeletal muscle (7, 32). Muscle biopsies from patient with mitochondrial encephalomyopathy, lactic acidosis, and stroke-like episodes (MELAS) show COX-positive RRFs, but also

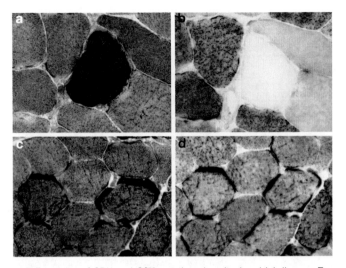

Fig. 3. (**a–d**) Examples of SDH and COX reactions in mitochondrial disease. Two sets of serial sections (**a** and **b**: biopsy from a KSS patient who has common deletion of mtDNA/**c** and **d**: biopsy from a MLEAS patient who has mtDNA A3243G tRNA^Leu(UUR) gene mutation) demonstrate RRFs stained by SDH (**a**, **c**) and COX (**b**, **d**). RRFs are COX deficient in **b**, whereas COX positive in **d** (×400).

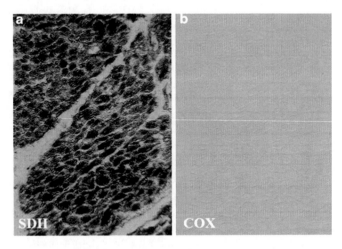

Fig. 4. (**a** and **b**) An example of global COX deficiency in mitochondrial encephalomyopathy. Muscle biopsy from a patient with SCO2 mutation stained by SDH (**a**) and COX (**b**) shows a global pattern of profound COX deficiency (×150).

reveal COX-deficient fibers. The enzymatic reactivity is decreased in the center of the fibers and largely preserved in the subsarcolemmal regions. A more diffuse pattern of COX deficiency is observed in infants with mtDNA depletion, and with fatal or benign COX-deficient myopathy of infancy (16). Children with COX-deficient Leigh syndrome resulting from mutations in either SURF1 or SCO2 genes show on biopsy a global pattern of COX deficiency including extra- and intrafusal muscle fibers as well as vascular smooth muscle (33) (Fig. 4).

7. There are several immunohistochemical methods for study of mitochondria on tissue sections. These include enzyme-linked methods (peroxidase, alkaline phosphatase, and glucose oxidase) and methods based upon the application of fluorochromes. For studies of mitochondria on formalin-fixed and paraffin-embedded samples, we routinely employ ABC method (34, 35).

8. We have often used ABC method for studies of the mitochondrial respiratory chain on paraffin-embedded samples of brain, in order to obtain information regarding the pathogenesis of neuronal dysfunction in mitochondrial disorders (10, 36, 37). It is also potentially useful to gain an insight into increasingly recognized roles of mitochondria in neurodegenerative disorders (38).

9. For studies on frozen tissue sections such as muscle biopsy, we favor the use of fluorochromes because they allow for the direct visualization of the antigen–antibody binding sites, and because they are more flexible for double-labeling experiments. In our laboratory, we routinely use a monoclonal antibody against COX IV as a probe for nDNA-encoded mitochondrial protein and a polyclonal antibody against COX II as a probe for mtDNA-encoded protein. However, any other combination of antibodies can be alternatively used since the entire mitochondrial genome has been sequenced, and any mtDNA-encoded respiratory chain subunit is potentially available for immunohistochemical studies. It is also anticipated that the same will soon be true for all the nDNA-encoded subunits of the respiratory chain.

    For example, in muscle from patients with Kearns–Sayre syndrome (KSS) harboring a documented deletion of mitochondrial DNA, COX-deficient RRFs typically show lack or marked reduction of COX II, whereas the COX IV immunostaining is typically normal in both COX-positive and COX-deficient fibers (Fig. 5). Presumably, this is due to the fact that even the smallest deletion eliminates essential tRNAs that are required for translation of the mitochondrial genome (39), and cybrid studies have confirmed this hypothesis (40).

10. In our laboratory, we carry out the study with unfixed frozen sections, but with some antibodies, permeabilization of the mitochondrial membranes may be required to uncover the antigenetic sequences or to facilitate the penetration of the probes into the inner mitochondrial compartment. In agreement with Johnson et al. (41), using 4% formaldehyde in 0.1 M $CaCl_2$, pH 7, followed by dehydration provides good results (see Subheading 3.2.3).

11. Immunohistochemistry using anti-DNA antibodies has been applied an alternative method to in situ hybridization for the

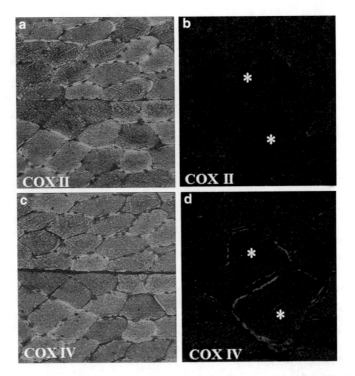

Fig. 5. (**a–d**) Immunofluorescent staining using antibodies against complex IV subunits. Serial sections of a normal muscle (**a**, **c**: ×200) and of muscle from a KSS patient (**b**, **d**: ×400) are immunolabeled with the mtDNA-encoded COX II (**a**, **b**) and the nDNA-encoded COX IV (**c**, **d**) antibodies using the immunofluorescent technique. In **b** (KSS), two RRFs (*asterisks*) lack COX II immunoreaction, as compared to the bright staining of COX IV.

studies of localization and distribution of mtDNA in normal and pathological conditions (42, 43). The advantages of this method are that both mtDNA and nDNA are detected simultaneously at the single cell level, and that the nuclear signal can be used as internal control. The method can be modified for double labeling using another antibody against mitochondrial protein. It is particularly useful and precise for the diagnosis of mtDNA depletion when it is confined to only a subpopulation of fibers (13, 44). In frozen muscle sections from normal controls and from patients without depletion of mtDNA, these antibodies show an intense staining of both the nuclei and a cytoplasmic network correlating with mitochondrial localization. Conversely, when muscle biopsies from patients with mtDNA depletion are analyzed, the cytoplasmic immunostaining of mtDNA is undetectable, or it is only present in a small number of muscle fibers (Fig. 6). The intensity of nDNA immunostaining is unchanged, compared to nondepleted controls.

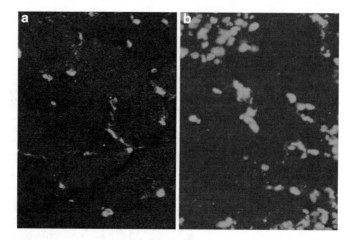

Fig. 6. (**a** and **b**) Immunofluorescent staining using anti-DNA antibody. Immunolocalization of mtDNA and nDNA can be done by using the anti-DNA antibody. In a normal muscle (**a**), fibers show both nuclear and cytoplasmic staining, reflecting nDNA and mtDNA, respectively. In a patient with mtDNA depletion (**b**), the cytoplasmic staining is invisible, while the nuclear staining remains intact (×400).

## Acknowledgment

This work was supported by grants from the National Institutes of Health 2P01HD32062-16.

## References

1. Attadi G, Schatz G. (1988) Biogenesis of mito-chondria. Annu Rev Cell Biol 4, 289–333.

2. Neupert W. (1997) Protein import into mito-chondria. Ann Rev Biochem 66, 863–917.

3. Fernandez-Vizarra E, Tiranti V, Zeviani M. (2009) Assembly of the oxidative phosphoryla-tion system in humans: What we have learned by studying its defects. Biochim Biophys Acta 1793, 200–211.

4. Zeviani M, Di Donato S. (2004) Mitochondrial disorders. Brain 124, 2153–2171.

5. DiMauro S, Schon EA. (2003) Mitochondrial respiratory-chain diseases. N Eng J Med 348, 2656–2668.

6. Hays AP, Oskoui M, Tanji K, et al. (2006) Mitochondrial neurology II: myopathies and peripheral neuropathies. In: DiMauro S, et al. (ed) Mitochondrial medicine, Informa Healthcare, Oxon, UK, pp. 45–74.

7. Bonilla E, Sciacco M, Tanji K, et al. (1992) New morphological approaches to the study of mitochondrial encephalomyopathies. Brain Pathol 2, 113–119.

8. Sparaco M, Schon EA, DiMauro S, et al. (1995) Myoclonic epilepsy with ragged red fibers: an immunohistochemical study of the brain. Brain Pathol 5, 125–133.

9. Tanji K, Vu TH, Schon EA, et al. (1999) Kearns-Sayre syndrome: unusual pattern of expression of subunits of the respiratory chain in the cerebellar system. Ann Neurol 45, 377–383.

10. Tanji K, Bonilla E. (2000) Neuropathologic aspects of cytochrome c oxidase deficiency. Brain Pathol 10, 422–430.

11. Tanji N, Tanji K, Kambham N, et al. (2001) Adefovir nephrotoxicity: Possible role of mito-chondrial DNA depletion. Hum. Pathol 32,734–740.

12. Tanji K, Gamez J, Cervera C, et al. (2003) The A8344G mutation in mitochondrial DNA asso-ciated with stroke-like episodes and gastroin-testinal dysfunction. Acta Neuropathol 105, 69–75.

13. Vu TH, Tanji K, Holve SA, et al. (2001) Navajo neurohepatopathy: A mitochondrial DNA

depletion syndrome? Hepatology 34(1), 116–120.

14. Tanji K, Bhagat G, Monzon BS, et al. (2003) Oncocytic hepatocytes; Are they a consequence of mitochondrial dysfunction. Liver Int 13, 1–7.

15. Engel WK, Cunningham GG. (1963) Rapid examination of muscle tissue. An improved trichrome method for fresh-frozen biopsy sections. Neurology 13, 919–923.

16. DiMauro S, Bonilla E. (2004) Mitochondrial encephalomyopathies. In: Engle EG, Franzini-Armstrong (ed) Myology, 3rd edition, McGraw-Hill, New York, pp. 1623–1662.

17. Seligman AM, Rutenburg RM. (1951) The histochemical demonstration of succinic dehydrogenase. Science 113, 317–320.

18. Pette D. (1981) Microphotometric measurement of initial maximum reaction rates in quantitative enzyme histochemistry in situ. Histochem J 13, 319–327.

19. Sugimoto J, Shimohira M, Osawa Y, et al. (2000) A patient with mitochondrial myopathy associated with isolated succinate dehydrogenase deficiency. Brain Dev 22(3), 158–162.

20. Seligman AM, Karnovsky MJ, Wasserkrug HL, et al. (1968) Nondroplet ultrastructural demonstration of cytochrome c oxidase activity with a polymerizing osmiophilic reagent, diaminobenzidine (DAB). J Cell Biol 38(1), 1–14.

21. Ludwig B, Bender E, Arnold S, et al. (2001) Cytochrome c oxidase and the regulation of oxidative phosphorylation. ChemBio Chem 2, 392–403.

22. Kloeckerner-Gruissem B, McEwen JE, Poyton RO. (1987) Nuclear functions required for cytochrome c oxidase biogenesis in Saccharomyces cerevisiae: multiple trans-acting nuclear genes exert specific effects on expression of each of the cytochrome c oxidase subunits encoded on mitochondrial DNA. Current Genet 12(5), 311–322.

23. Pel HJ, Tzagaloff A, Grivell LA. (1992) The identification of 18 nuclear genes required for the expression of the yeast mitochondrial gene encoding cytochrome c oxidase subunit 1. Curr Genet 21(2), 139–146.

24. Zhu Z, Yao J, Johns T, et al. (1998) SURF1, encoding a factor involved in the biosynthesis of cytochrome c oxidase, is mutated in Leigh syndrome. Nat gent 20, 337–343.

25. Tirani V, Hoertnagel K, Carrozzo R, et al. (1998) Mutations of SURF-1 in Leigh disease associated with cytochrome c oxidase deficiency. Am J Hum Genet 63, 1609–1621.

26. Papadopoulou LC, Sue CM, Davidson MM, et al. (1999) Fatal infantile cardioencephalomyopathy with COX deficiency and mutations in SCO2, a COX assembly gene. Nat Get 23, 333–337.

27. Jaksch M, Ogilvie I, Yao J, et al. (2000) Mutations in SCO2 are associated with a distinct form of hypertrophic cardiomyopathy and cytochrome c oxidase deficiency. Hum Mol Genet 9, 795–801.

28. Bonilla E, Shortland DL, DiMauro S, et al. (1975) Electron cytochemistry of crystalline inclusions in human skeletal muscle mitochondria. J Ultrastruct Res 51, 404–408.

29. Johnson MA, Turnbull DM, Dick HS, et al. (1983) A partial deficiency of cytochrome c oxidase in chronic progressive external ophthalmoplegia. J Neurol Sci 60(1), 75–90.

30. Wong-Riley MT. (1989) Cytochrome c oxidase: an endogenous metabolic marker for neuronal activity. Trends Neurosci 12, 94–101.

31. Tanji K, Bonilla E. (2006) Optical imaging techniques (Histochemical, immunohistochemical and *in situ* hybridization staining methods) to visualize mitochondria. Methods Cell Biol 80, 135–154.

32. Attardi G, Yoneda M, Chomyn A. (1995) Complementation and segregation behavior of disease-causing mitochondrial DNA mutations in cellular model system. Biochim Biophys Acta 1271(1), 241–248.

33. Sue CM, Karadimas C, Checcarelli N, et al. (2000) Differential features of patients with mutations in two COX assembly genes, SURF-1 and SCO2. Ann Neurol 47(5), 589–595.

34. Hsu SM, Raine L. (1981) Protein A, avidin, and biotin in immunohistochemistry. J Histochem Cytochem 29(11), 1349–1353.

35. Bedetti CD. (1985) Immunocytochemical demonstration of cytochrome c oxidase with an immunoperoxidase method: a specific stain for mitochondria in formalin-fixed and paraffin embedded human tissue. J Histochem Cytochem 33(5), 446–452.

36. Hirano M, Kaufmann P, De Vivo D, et al. Mitochondrial neurology I: Encephalopathies. (2006) In: DiMauro S, et al (ed) Mitochondrial medicine. Informa Healthcare, Oxon UK, pp. 27–44.

37. Tanji K, Kunimatsu T, Vu TH, et al. (2001) Neurological features of mitochondrial disorders. Semin Cell Dev Biol 12(6), 429–39.

38. Bonilla E, Tanji K, Vu TH, et al. (1999) Mitochondrial involvement in Alzheimer's disease. Biochem. Biophys Acta 1410, 171–182.

39. Nakase H, Moraes CT, Rizzuto R, et al. (1990) Transcription and translation of deleted mitochondrial genomes in Kearns-Sayre syndromes: implications for pathogenesis. Am J Hum Genet 46(3), 418–427.

40. Hayashi J, Ohta S, Kikuchi A, et al. (1991) Introduction of disease-related mitochondrial DNA deletions into HeLa cells lacking mitochondrial DNA results in mitochondrial dysfunction. Proc Natl Acd USA 88, 10614–10618.

41. Johnson MA, Turnbull DM, Dick HS, et al. (1988) Immunocytochemical studies of cytochrome *c* oxidase subunits in skeletal muscle of patients with partial cytochrome *c* oxidase deficiencies. Neurol Sci 87, 75–90.

42. Andreatta F, Tritschler HJ, Schon EA, et al. (1991) J Neurol Sci 205, 88–92.

43. Tritschler HJ, Andreetta F, Moraes CT, et al. (1992) Mitochondrial myopathy of childhood associated with depletion of mitochondrial DNA. Neurology 42, 209–217.

44. Vu TH, Tanji K, Valsamis H, et al. (1998) Mitochondrial DNA depletion in a patient with long survival. Neurology 42, 209–217.

# Chapter 13

# Blue Native Polyacrylamide Gel Electrophoresis: A Powerful Diagnostic Tool for the Detection of Assembly Defects in the Enzyme Complexes of Oxidative Phosphorylation

## Scot C. Leary

## Abstract

The bulk of ATP consumed by various cellular processes is normally produced by five multimeric protein complexes embedded within the inner mitochondrial membrane in a process known as oxidative phosphorylation (OXPHOS). Mutations that impair the assembly, and therefore the function, of one or more of these enzyme complexes severely compromise energy homeostasis and are a frequent cause of human disease. Because mitochondrial diseases are a clinically heterogeneous group of genetic disorders, biochemical and molecular diagnostic analyses are often an essential first step in confirming suspected cases and ultimately aid in identifying the genetic basis of disease in affected individuals. Blue native polyacrylamide gel electrophoresis has proven to be particularly invaluable in this regard, providing researchers with a facile approach for analyzing the assembly, total abundance, and residual enzymatic activity of individual OXPHOS complexes. As such, this technique has greatly facilitated the more thorough molecular genetic investigation of diseases that are caused by isolated and combined deficiencies in the enzymes that comprise the OXPHOS system.

**Key words:** Human disease, Mitochondria, Oxidative phosphorylation, Holoenzyme assembly, Blue native polyacrylamide gel electrophoresis

## 1. Introduction

Mitochondria are organelles that fulfill important roles in several cellular processes, including apoptosis and the metabolism and storage of essential micronutrients such as iron and copper (1, 2). Their most widely recognized role, however, is in the maintenance of energy homeostasis. In fact, mitochondria are generally responsible for producing the bulk of the ATP that is consumed within the cell.

Lee-Jun C. Wong (ed.), *Mitochondrial Disorders: Biochemical and Molecular Analysis*, Methods in Molecular Biology, vol. 837, DOI 10.1007/978-1-61779-504-6_13, © Springer Science+Business Media, LLC 2012

Aerobic ATP production is dependent upon the coordinate activity of five multimeric protein complexes (I–V) localized to the inner mitochondrial membrane. Complexes I–IV form the respiratory chain, which functions as a unit to transfer reducing equivalents obtained from NADH and $FADH_2$ to electron acceptors of increasing affinity while simultaneously pumping protons from the matrix into the intermembrane space to establish an electrochemical gradient across the inner mitochondrial membrane. The resultant chemiosmotic potential is then exploited by complex V, the ATP synthase, to synthesize ATP in a process known as oxidative phosphorylation (OXPHOS).

Because of their aggregate turnover, maintaining basal levels of a given OXPHOS complex requires the de novo synthesis of its individual structural subunits and their subsequent assembly into a functional holoenzyme complex. Thirteen proteins critical to the biogenesis of complexes I, III, IV, and V are encoded by mitochondrial DNA, and a large number of nuclear-encoded accessory factors regulate their expression at both the transcriptional and translational levels (3). The stability of these newly synthesized mitochondrial proteins is dependent upon their insertion into the inner mitochondrial membrane, and subsequent assembly with relevant nuclear-encoded structural partners to form functional holoenzyme complexes, a process that is facilitated by an expanding "family" of nuclear-encoded accessory proteins known as assembly factors (4). The apparent interdependence of the biogenesis of some OXPHOS complexes may be explained by their eventual organization into higher order structures termed supercomplexes (5). Genetic lesions that impair the assembly of one or more of these enzyme complexes severely perturb cellular energy homeostasis and are a frequent cause of human disease (6–8).

Traditionally, biochemical defects in OXPHOS function have been identified by spectrophotometric, polarographic, or histochemical analysis of patient cells or tissues; however, each of these methods has its limitations. Histochemical staining is at best semiquantitative, while polarography and enzymology both require a large amount of starting material (9). Importantly, none of these techniques directly addresses the process by which individual structural subunits are assembled into a mature holoenzyme. In contrast, differences in both the assembly of OXPHOS complexes and their absolute content can be readily assessed by blue native polyacrylamide gel electrophoresis (BN-PAGE), followed by conventional immunoblotting with commercially available antibodies (10). Residual enzymatic activity of most OXPHOS complexes also can be measured semiquantitatively by BN-PAGE using established in-gel activity stains (11, 12). Therefore, with very little starting material, this straightforward technique may provide considerable mechanistic insight into the molecular genetic basis of human diseases caused by mitochondrial dysfunction.

## 2. Materials

### 2.1. Cell Culture and Sample Preparation

1. Dulbecco's modified Eagle's medium (DMEM) supplemented with 10% FBS. Store at 4°C.

2. Phosphate buffered saline (PBS): In double distilled water, prepare a solution containing 2.7 mM potassium chloride, 137 mM sodium chloride, and 10 mM potassium phosphate (pH 7.4). Sterilize by autoclaving and store at room temperature.

3. 0.05% Trypsin solution in PBS. Store at 4°C.

4. 1 mg/mL Bovine serum albumin (BSA) in double distilled water and 1× Bradford reagent.

5. 4 mg/mL Digitonin in PBS (see Note 1).

6. 3× Gel buffer: To 100 mL double distilled water, add 19.68 g of aminocaproic acid and 3.14 g of Bis-tris. Adjust to pH 7.0 and store at 4°C.

7. Blue native (BN) sample buffer: Combine 0.5 mL 3× Gel buffer, 0.5 mL 2 M aminocaproic acid, and 4 μL 500 mM EDTA (see Note 2). Stable for 6–12 months when stored at 4°C.

8. 10% Lauryl maltoside solution in double distilled water (see Note 3).

9. 5% Coomassie brilliant blue G-250 (SBG) in 0.75 mM aminocaproic acid solution (see Note 4). Store indefinitely at 4°C.

### 2.2. BN-PAGE

1. Acrylamide/bisacrylamide (AB) mix: To double distilled water, add 24.0 g of acrylamide and 0.75 g of bisacrylamide (see Note 5). Adjust to a final volume of 50 mL and store for 6–12 months at 4°C.

2. 3× Gel buffer (see Subheading 2.1).

3. 87% Glycerol stock solution (in water). Store indefinitely at room temperature.

4. Colorless cathode buffer: To 1 L double distilled water, add 3.14 g of Bis-tris and 8.96 g of Tricine. Adjust to pH 7.0.

5. Blue cathode buffer: Colorless cathode buffer containing 0.02% SBG.

6. Anode buffer: To 2 L double distilled water, add 20.93 g of Bis-tris, and adjust to pH 7.0 (see Note 6).

7. $N,N,N',N'$-Tetramethyl-ethylenediamine (TEMED). Store at 4°C.

8. 10% Ammonium persulfate (APS) solution: Prepare in double distilled water. Make fresh as required.

9. Protein standards: High molecular weight native marker kit (Pharmacia) (see Note 7).

**2.3. Western Blotting and Immunodetection**

1. Semidry transfer apparatus.

2. Transfer buffer: To 1 L double distilled water, add 5.8 g Tris base, 2.93 g glycine, 0.75 g SDS, and 200 mL methanol. Do not pH. Store at room temperature.

3. Nitrocellulose and Whatman 3 M paper.

4. 10× Tris buffered saline (TBS): To 2 L of double distilled water, add 48.4 g Tris base and 160 g NaCl. Do not pH (see Note 8). Store indefinitely at room temperature.

5. Rotary shaker.

6. Blocking solution: 5% BSA dissolved in 1× TBS supplemented with 0.1% Tween-20 (TBS-T).

7. Primary and secondary antibody solution: TBS-T supplemented with 2% BSA.

8. Primary antibodies (Mitosciences, Eugene, OR) for complexes I (anti-39 kDa), II (anti-SDHA), III (anti-core 1), IV (anti-COX I or anti-COX IV), and V (anti-ATPase α).

9. Secondary antibody: Anti-mouse IgG conjugated to horseradish peroxidase. Enhanced chemiluminescent (ECL) reagents and autoradiography film.

10. Stripping solution: To 0.5 L of double distilled water, add 31.5 mL Tris–HCl (pH 7.5), 10 g SDS, and 3.9 mL β-ME (see Note 9).

11. Temperature-controlled, shaking water bath.

# 3. Methods

Pioneered in the early 1990s by Schagger and colleagues (13, 14), BN-PAGE has emerged as the technique of choice with which to examine the assembly and abundance of OXPHOS complexes within the inner mitochondrial membrane. In its original form, Coomassie dyes were used to impart the charge shift necessary for detergent-solubilized proteins to be fractionated by size in a non-ionic gel and buffer system. The technique has since been refined to permit studies of the organization of OXPHOS complexes into higher order structures known as supercomplexes (15); however, it is described here in its simplest form, which results in the release of OXPHOS complexes from the inner mitochondrial membrane in either their monomeric (complexes I, II, IV, and V) or dimeric forms (complex III) (see Fig. 1). Detailed instructions are provided for using digitonin to prepare an enriched mitoplast fraction starting from whole cells (16), an approach we favor because much smaller amounts of starting material are required for downstream analyses. If starting material is either abundant or prevents the use

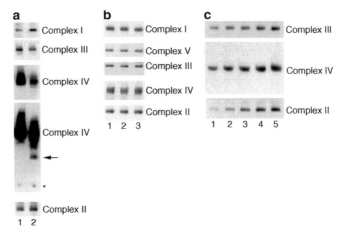

Fig. 1. Representative BN-PAGE gels of samples prepared by differential digitonin permeabilization. (**a**) Control human fibroblasts either without (1) or with (2) stable expression of a short-hairpin RNA that results in the knockdown of COX 11, an assembly factor critical to complex IV biogenesis. While the assembly of complexes I, II, and III is unaffected, there is a significant accumulation of both monomeric COX IV (*asterisk*) and the S2 assembly intermediate (*arrow*). (**b**) The abundance of OXPHOS complexes in mouse spinal cord extracts (1–3). (**c**) The levels of complexes II, III, and IV in HEK293 cells exposed to an increasing digitonin to protein ratio (1–5: 0.2–1.6 mg digitonin: mg protein). In all *panels*, a total of 10 µg of protein was loaded per lane.

of such an approach (e.g., autopsy/biopsy material), mitochondria or a mitoplast fraction may be isolated from homogenates by differential centrifugation prior to solubilization in lauryl maltoside.

***3.1. Cell Culture and Sample Preparation***

1. Harvest a confluent 100-mm plate of cells by washing once in PBS, incubating in 4 mL of 0.05% trypsin solution for 10 min, and neutralizing with an equal volume of culture medium (see Note 10).

2. Pellet the cells in a 15-mL Falcon tube, resuspend the pellet in 1 mL ice-cold PBS, transfer to an Eppendorf tube, and spin at $14,000 \times g$ for 2 min at 4°C.

3. Resuspend the pellet in 500–1,000 µL of ice-cold PBS depending on the size of the pellet and quantify the protein concentration using the Bradford assay according to manufacturer's instructions.

4. Repellet the cells and resuspend in ice-cold PBS to a final concentration of 5 mg/mL for human fibroblasts (see Note 11).

5. Add an equal volume of a 4 mg/mL digitonin solution, mix the tube twice by inversion, and incubate on ice for 10 min.

6. Dilute with ice-cold PBS to a final volume of 1.5 mL and spin for 10 min at $10,000 \times g$ at 4°C.

7. Remove supernatant without disturbing the pellet and wash it gently with 1 mL ice-cold PBS to completely remove residual digitonin.

8. Add BN sample buffer at half the volume that was initially required to resuspend the cell pellet at 5 mg/mL and add lauryl maltoside at 1/10th of the BN sample buffer volume (see Note 12).

9. Resuspend pellet carefully by pipetting up and down 10–20 times, taking care not to foam the detergent (see Note 13).

10. Following a 20-min extraction on ice, spin at $20,000 \times g$ for 20 min at 4°C.

11. Carefully remove the supernatant, transfer to a new Eppendorf tube, and quantify the protein concentration using the Bradford assay (see Note 14).

12. Add a volume of SBG that corresponds to half the volume of lauryl maltoside used in step 8 and store at –20°C (see Note 15).

### 3.2. BN-PAGE

1. Rinse the glass plates, spacers, combs, and casting stand gaskets for either the Bio-Rad Mini Protean II or Mini Protean 3 gel system several times with deionized, then double distilled water, followed by a final rinse in 70–95% ethanol. Air-dry.

2. Prepare 10 mL of a 6% gel mixture in a 15-mL Falcon tube by combining 3.3 mL of 3× Gel buffer, 0.6 mL AB mix, and 5.44 mL double distilled water. Prepare the same volume of a 15% gel mixture in another 15-mL Falcon tube by combining 3.3 mL of 3× Gel buffer, 3.0 mL AB mix, 1.68 mL double distilled water, and 2 mL 87% glycerol. Chill both solutions on ice for at least half an hour.

3. Add 60 μL 10% APS and 4 μL TEMED to the 6% gel mixture, and 10 μL 10% APS and 2 μL TEMED to the 15% gel mixture. Mix both solutions by inversion several times and place on ice.

4. Set the WIZ Peristaltic Pump to a flow rate of 75 and pour a 1.0-mm thick, 6–15% gradient gel using 2.8 and 2.3 mL of the 6% and 15% gel stock solutions, respectively (see Fig. 2 for detailed instructions). Gently overlay the gradient gel with double distilled water by gravity flow from 1-mL syringes with 24-gage needles and allow 1 h for polymerization (see Note 16).

5. While the gradient gel is polymerizing, prepare and chill 5 mL of stacking gel solution by combining 1.64 mL 3× Gel buffer, 0.4 mL AB mix, and 2.87 mL double distilled water in a 15-mL Falcon tube.

6. Once the gradient gel has polymerized, pour off the overlay and use a Kimwipe to ensure complete removal of all residual water. Add 60 μL 10% APS and 6 μL TEMED to the stacking

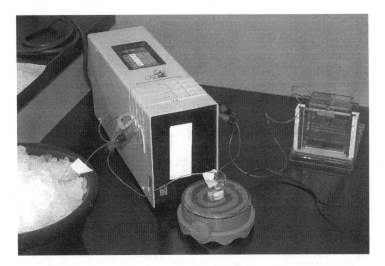

Fig. 2. Organization, assembly, and casting of BN-PAGE gels. Clean and assemble the components required to cast a BN-PAGE gel (see relevant Subheading 3 for detailed instructions). Run double distilled water through the Wiz peristaltic pump for roughly 5 min at a flow rate of 99 to ensure that the tubing is clean. Completely empty the tubing of all double distilled water and secure the relevant lines of tubing to either the beakers or the gel using tape. Once the tubing is secured to both beakers, add the lowest percentage acrylamide solution to the left most beaker, which is kept on ice and is not stirred while the gel is poured. Turn on the pump set to a flow rate of 75 and let the low percentage solution flow through the tubing until it almost reaches the beaker that will contain the highest percentage solution. Immediately add the high percentage solution to the right most beaker and exhaust the entire volume of both solutions prior to stopping the pump. Overlay gently with water and allow at least 1 h for polymerization.

gel solution, mix by inversion several times, and pour it on top of the separating gel until it begins to overflow. Insert a 15-well comb.

7. Upon polymerization of the stacking gel, gently remove the comb by slowly pulling it straight up. Rinse the wells three times with colorless cathode buffer.

8. Add blue cathode buffer to the wells and assemble the portion of the electrophoresis unit that will form the cathode (i.e., inner) chamber. Load wells with equal amounts of protein while reserving at least one well for the molecular weight markers (see Note 17).

9. Insert the cathode chamber into the gel tank and fill it with blue cathode buffer. Completely fill the gel tank (i.e., outer chamber) with anode buffer (see Note 18), fully assemble the electrophoresis unit, and connect to a power supply. Run the gel for 30–45 min at 35 V. Increase to 75 V for another half an hour, and to 90–100 V for the remainder of the run.

10. Once the dye front is one-third of the way through the gradient gel, stop the run and replace the blue cathode buffer with colorless cathode buffer (see Note 19). Continue electrophoresis until the dye front reaches the bottom of the gradient gel.

**3.3. Western Blotting and Immunodetection**

1. While the gel is running, cut six pieces of Whatman paper and nitrocellulose membrane to the exact dimensions of the gel(s) to be transferred.

2. Equilibrate the membrane in transfer buffer shortly before the end of the run. At the end of the run, disconnect the electrophoresis unit from the power supply and disassemble it. Separate the glass plates sandwiching the gel by twisting one of the spacers. Remove and discard the stacking gel and cut the bottom right-hand corner of the gradient gel to mark its orientation.

3. Leaving the gel on the glass plate to which it has adhered, fully immerse it in a vessel containing transfer buffer and gently rock it back and forth until it physically separates from the plate.

4. Prepare the apparatus for transfer by wetting a piece of Whatman paper and placing it on the cathode plate of the transfer apparatus. Carefully remove all bubbles by rolling a borosilicate tube over the Whatman paper. Dab away excess transfer buffer with Kimwipes as required (see Note 20).

5. Repeat step 4 with two more pieces of Whatman paper and finally with the nitrocellulose membrane.

6. Handling the gel by its bottom end, carefully place it on the membrane. Remove any bubbles between it and the membrane by very gently rolling the borosilicate tube over the gel as many times as necessary.

7. Repeat step 4 three more times.

8. Wet the anode plate with double distilled water, fully assemble the apparatus, and transfer for 1 h at a constant milliamperage of $0.8$ mA/cm$^2$ of nitrocellulose membrane.

9. At the end of the transfer, disassemble the apparatus, discard the gel(s), and place the nitrocellulose membrane in a vessel containing an ample volume of TBST.

10. Shake the membrane for 5 min, replace TBST with blocking solution, and continue shaking at room temperature for at least 1 h.

11. Replace the blocking solution with that containing the primary antibody solution and rock overnight at 4°C (see Note 21).

12. The following day, remove the primary antibody and wash the membrane 6 times for 5 min per wash with shaking at room temperature using ample volumes of TBST.

13. Incubate the membrane for 1 h in the secondary antibody solution and repeat washes as outlined in step 12.

14. Combine both ECL reagents, diluting with double distilled water if necessary, to obtain a final volume of 5 mL per membrane to be exposed. Remove TBST from the vessel as completely as possible and replace with the ECL solution. Rock for 1 min. Using a pair of forceps, dab the membrane against the side of the vessel to remove excess ECL solution, and place it in a transparent acetate leaflet. Remove all bubbles with the help of a borosilicate tube.

15. Once in the darkroom, take multiple exposures of the membrane and develop the film.

16. Return the membrane to a vessel containing TBST. If stripping is not required, repeat steps 10–15 with a different primary antibody. Should stripping be necessary, incubate the membrane in stripping buffer for 30 min while gently rocking in a water bath set at 50°C. Promptly remove stripping solution and wash the membrane extensively with TBST (4× 15 min) while shaking at room temperature. Repeat steps 10–15 (see Note 22).

## 4. Notes

1. Commercially available sources of digitonin contain impurities that affect its solubility in aqueous solutions. Full dissolution of digitonin in PBS requires boiling for 5 min. The solution is then cooled on ice and is stable for approximately 4 h. It should not be reused for preparation of samples on subsequent days.

2. The addition of protease inhibitors is not required when preparing native extracts from cultured cells.

3. Lauryl maltoside requires minimal heating or vortexing to go into solution. While it is stable indefinitely at 4°C, its efficacy declines with time. As a result, it should be made fresh the day that samples are prepared to ensure the most consistent results.

4. The use of SBG from other commercial sources results in a considerable increase in the nonspecific background during Western blotting and therefore should be avoided.

5. In the event that the AB mix precipitates during storage, heat at 42°C and vortex periodically until it goes back into solution.

6. Cathode and anode buffers can be stored indefinitely at room temperature once they have been prepared.

7. High molecular weight markers are resuspended in 100-μL BN sample buffer, followed by the addition of 10 μL of 5% SBG. Loading 5–10 μL per lane allows for visualization of all five markers by Coomassie staining; however, only a subset of these are visible by Ponceau staining if they are transferred onto a nitrocellulose membrane.

8. We have found that this unbuffered form of TBS helps with consistent immunodetection of OXPHOS complexes. This may be attributable to the exposure by high pH of epitopes that are otherwise either partially or fully masked in this native gel system.

9. The stripping solution is supplemented with β-mercaptoethanol immediately prior to its use.

10. One confluent 100-mm plate of human fibroblasts or myoblasts will yield sufficient material to prepare and analyze a sample at least twice, assuming that 10–20 μg of total protein are loaded per lane.

11. If a different cell type or the same cell type from another species is being prepared for BN-PAGE, it is important to ensure that the digitonin to protein ratio that is used achieves maximal enrichment for OXPHOS complexes without promoting either their dissociation or degradation (see Fig. 1). In our experience, a digitonin to protein ratio of 0.8 and 1.2 is appropriate for generating enriched mitoplasts from human fibroblasts and myoblasts, respectively.

12. In our experience, roughly half of the total cellular protein is depleted upon treatment of human fibroblasts (and myoblasts) with digitonin. The purpose of reducing the volume of BN buffer used in the solubilization of the enriched mitoplast pellet reflects a desire to maintain a protein concentration of roughly 2–4 mg/mL at this stage of the isolation. If a different digitonin to protein ratio is used in the preparation of enriched mitoplasts from other cell lines, the volume of BN buffer to be used should therefore be adjusted accordingly.

13. It is impossible to generate a homogeneous solution when resuspending the pellet. Once the solution has been repeatedly pipetted up and down as described in step 9, small, tight fragments of "insoluble" material should be visible. Do not use more vigorous means of solubilizing the pellet (e.g., homogenizing) since this may promote the dissociation of OXPHOS complexes.

14. The final protein concentration should be between 1 and 3 mg/mL. Lower protein concentrations may result in the dissociation of OXPHOS complexes, while higher protein concentrations may lead to their anomalous or inconsistent

migration due to partial release from either the inner mito-chondrial membrane or higher order complexes (i.e., supercomplexes).

15. Freezing at −20°C only preserves the integrity of OXPHOS complexes for 1–2 months. For long-term storage, samples may be kept indefinitely at −80°C.

16. The volumes specified for pouring the gradient and stacking gels are sufficient to cast up to three gels. These may be scaled either up or down based on individual needs; however, no more than four gradient gels should be prepared from the same solutions as there is an increased risk of polymerization within the pump tubing.

17. Loading 10–20 μg of total protein per lane will permit for the immunodetection of all five OXPHOS complexes. Samples are loaded prior to filling the cathode chamber with blue cathode buffer due to the extreme difficulty one otherwise has visualizing the wells.

18. It is very difficult to generate a perfectly sealed cathode chamber. Completely filling the gel tank with anode buffer therefore serves to minimize the potential leaking of cathode buffer into the anode chamber.

19. The cathode buffer is changed from blue to colorless at this stage of electrophoresis to minimize the amount of Coomassie that remains in the gel at the end of the run. This will not affect the electrophoretic mobility of the proteins and greatly enhances the signal to noise ratio upon Western blotting.

20. The presence of excess transfer buffer will affect the quality of protein transfer to the membrane. Transferring more than two gels at a time using this particular system will also adversely affect the quality of transfer.

21. The monoclonal antisera raised against structural subunits of OXPHOS can be repeatedly frozen and thawed, thus extending their lifespan if their stock solution is diluted with TBST supplemented with BSA as opposed to milk. The only exception is anti-ATPase α, which even in BSA, retains its immuno-reactivity for a maximum of 2–3 freeze/thaw cycles.

22. Stripping is only necessary if the complexes to be detected are of similar sizes (e.g., complexes III and V), and repeated stripping should be avoided since it will result in an unwanted loss of protein from the membrane. Sequential blotting for complexes IV, V, and II, followed by stripping of the membrane, and subsequent blotting for complexes I then III is therefore preferable.

# References

1. McBride, H.M., Neuspiel, M. and Wasiak, S. (2006) Mitochondria: more than just a power-house. *Curr. Biol.* **16**, R551–R560.

2. Pierrel, F., Cobine, P.A. and Winge, D.R. (2007). Metal ion availability in mitochondria. *Biometals* **20**, 675–682.

3. Shoubridge, E.A. and Sasarman, F. Mitochondrial translation and human disease, in *Translational control in Biology and Medicine* (Mathews, M.B., Sonenberg, N. and Hershey, J.W.B., Ed.), CSHL Press, Cold Spring Harbor, NY, pp. 775–801.

4. Shoubridge, E.A. (2001) Nuclear genetic defects of oxidative phosphorylation. *Hum. Mol. Genet.* **10**, 2277–2284.

5. Acín-Pérez, R., Bayona-Bafaluy, M.P., Fernández-Silva, P., Moreno-Loshuertos, R., Pérez-Martos, A., Bruno, C., Moraes, C.T. and Enríquez, J.A. (2004) Respiratory complex III is required to maintain complex I in mammalian mitochondria. *Mol. Cell.* **13**, 805–815.

6. Smeitink, J.A., Elpeleg, O., Antonicka, H., Diepstra, H., Saada, A., Smits, P., Sasarman, F., Vriend, G., Jacob-Hirsch, J., Shaag, A., Rechavi, G., Welling, B., Horst, J., Rodenburg, R.J., van den Heuvel, B. and Shoubridge, E.A. (2006). Distinct clinical phenotypes associated with a mutation in the mitochondrial translation elongation factor EFTs. *Am. J. Hum. Genet.* **79**, 869–877.

7. Weraarpachai, W., Antonicka, H., Sasarman, F., Seeger, J., Schrank, B., Kolesar, J.E., Lochmuller, H., Chevrette, M., Kaufman, B.A., Horvath, R. and Shoubridge, E.A. (2009) Mutation in TACO1, encoding a translational activator of COX I, results in cytochrome c oxidase deficiency and late-onset Leigh syndrome. *Nat. Genet.* **41**, 833–837.

8. Ghezzi, D., Goffrini, P., Uziel, G., Horvath, R., Klopstock, T., Lochmuller, H., D'Adamo, P., Gasparini, P., Strom, T.M., Prokisch, H., Invernizzi, F., Ferrero, I. and Zeviani, M. (2009) SDAHF1, encoding a LYR complex-II specific assembly factor, is mutated in SDH-defective infantile leukoencephalopathy. *Nat. Genet.* **41**, 654–656.

9. Williams, S.L., Scholte, H.R., Gray, G.F., Leonard, J.V., Schapira, A.H.V. and Taanman, J.-W. (2001) Immunological phenotyping of fibroblast cultures from patients with a mitochondrial respiratory chain deficit. *Lab. Invest.* **81**, 1069–1077.

10. Leary, S.C., Kaufman, B.A., Pellecchia, G., Guercin, G.H., Mattman, A., Jaksch, M. and Shoubridge, E.A. (2004) Human SCO1 and SCO2 have independent, cooperative functions in copper delivery to cytochrome c oxidase. *Hum. Mol. Genet.* **13**, 1839–1848.

11. Zerbetto, E., Vergani, L. and Dabbeni-Sala, F. (1997) Quantification of muscle mitochondrial oxidative phosphorylation enzymes via histochemical staining of blue native polyacrylamide gels. *Electrophoresis* **18**, 2059–2064.

12. Diaz, F., Barrientos, A. and Fontanesi, F. (2009) Evaluation of the mitochondrial respiratory chain and oxidative phosphorylation system using blue native gel electrophoresis. *Curr. Protoc. Hum. Genet.* **63**, 19.4.1–19.4.12.

13. Schagger, H. and von Jagow G. (1991) Blue native electrophoresis for isolation of membrane protein complexes in enzymatically active form. *Anal. Biochem.* **199**, 223–231.

14. Schagger, H., Bentlage, H., Ruitenbeek, W., Pfeiffer, K., Rotter, S., Rother, C., Bottcher-Purkl, A. and Lodemann, E. (1996) Electrophoretic separation of multiprotein complexes from blood platelets and cell lines: technique for the analysis of diseases with defects in oxidative phosphorylation. *Electrophoresis* **17**, 709–714.

15. Pfeiffer, K., Gohil, V., Stuart, R.A., Hunte, C., Brandt, U., Greenberg, M.L. and Schagger, H. (2003) Cardiolipin stabilizes respiratory chain supercomplexes. *J. Biol. Chem.* **278**, 52873–52880.

16. Klement, P., Nijtmans, L.G., Van den Bogert, C. and Houstek, J. (1995) Analysis of oxidative phosphorylation complexes in cultured human fibroblasts and amniocytes by blue-native-electrophoresis using mitoplasts isolated with the help of digitonin. *Anal. Biochem.* **231**, 218–224.

# Chapter 14

## Radioactive Labeling of Mitochondrial Translation Products in Cultured Cells

### Florin Sasarman and Eric A. Shoubridge

### Abstract

The mammalian mitochondrial genome contains 37 genes, 13 of which encode polypeptide subunits in the enzyme complexes of the oxidative phosphorylation system. The other genes encode the rRNAs and tRNAs necessary for their translation. The mitochondrial translation machinery is located in the mitochondrial matrix, and is exclusively dedicated to the synthesis of these 13 enzyme subunits. Mitochondrial disease in humans is often associated with defects in mitochondrial translation. This can manifest as a global decrease in the rate of mitochondrial protein synthesis, a decrease in the synthesis of specific polypeptides, the synthesis of abnormal polypeptides, or in altered stability of specific translation products. All of these changes in the normal pattern of mitochondrial translation can be assessed by a straightforward technique that takes advantage of the insensitivity of the mitochondrial translation machinery to antibiotics that completely inhibit cytoplasmic translation. Thus, specific radioactive labeling of the mitochondrial translation products can be achieved in cultured cells, and the results can be visualized on gradient gels. The analysis of mitochondrial translation in cells cultured from patient biopsies is useful in the study of disease-causing mutations in both the mitochondrial and the nuclear genomes.

**Key words:** Mitochondria, Oxidative Phosphorylation (OXPHOS), Mitochondrial DNA (mtDNA), Pulse-chase labeling, Mitochondrial translation

## 1. Introduction

Pulse labeling of the mitochondrial translation products is useful for the assessment of both the individual and global rates of synthesis of the 13 proteins encoded by the mtDNA, and pulse-chase labeling permits the evaluation of the stability of the newly synthesized polypeptides. In both cases, cells are exposed to a mixture of radiolabeled methionine and cysteine in the presence of an inhibitor of cytoplasmic protein synthesis, which results in the specific radiolabeling of the mitochondrial translation products.

Lee-Jun C. Wong (ed.), *Mitochondrial Disorders: Biochemical and Molecular Analysis*, Methods in Molecular Biology, vol. 837, DOI 10.1007/978-1-61779-504-6_14, © Springer Science+Business Media, LLC 2012

Three main differences distinguish pulse, from pulse-chase labeling: first, the length of the chase, defined as the incubation time in regular, "cold" medium following removal of the radiolabel; second, the type of inhibition (i.e., irreversible versus reversible) of the cytoplasmic protein synthesis; and third, exposure to chloramphenicol, a reversible inhibitor of mitochondrial protein synthesis, which is used only in pulse-chase labeling. In pulse labeling, an irreversible inhibitor of cytoplasmic translation such as emetine can be used, given the short duration of the chase (10 min). The purpose of such short chases is to allow the ribosomes to finish translating any radiolabeled proteins; otherwise, any "hot" polypeptide shorter than the full-length species will be detected at a different size. By contrast, in pulse-chase labeling, when the rates of degradation of the radiolabeled proteins are assessed, cells need to be maintained in culture for longer periods of time (up to 17–18 h), which requires a reversible inhibitor of cytoplasmic translation. For this purpose, we use anisomycin, although other groups have used cycloheximide (1, 2). Finally, in pulse-chase labeling, cells are exposed to chloramphenicol prior to the incubation with radioisotope. This step allows the accumulation of a pool of nuclear-encoded structural subunits within mitochondria, which facilitates the assembly of radiolabeled mitochondrial subunits into nascent OXPHOS complexes following removal of the drug. It also results in increased labeling and preferential stabilization of the two subunits of Complex V that are encoded by the mtDNA (3, 4).

Changes in the normal pattern of mitochondrial protein synthesis, as revealed through radioactive labeling, have served to identify and confirm causal mutations, and to analyze the pathogenic mechanism of mitochondrial diseases caused by mutations in components of the mitochondrial translation system that are encoded by either the mitochondrial (5–10) or the nuclear genome (11–20).

## 2. Materials

### 2.1. Labeling of Mitochondrially Synthesized Proteins with Radioactive [$^{35}$S] Methionine and Cysteine

1. Labeling medium: Dulbecco's Modified Eagle's Medium (DMEM) without methionine and cysteine, supplemented with 10% dialyzed fetal bovine serum (FBS), 1× GlutaMax™-1, and 110 mg/L sodium pyruvate (see Note 1). Store at 4°C.

2. Regular DMEM supplemented with 10% FBS. Store at 4°C.

3. Phosphate buffered saline reconstituted from tablets according to supplier's instruction and sterilized by autoclaving. Store at room temperature.

4. Inhibitors of cytoplasmic translation: emetine for pulse labeling or anisomycin for pulse-chase labeling. In each case, prepare a

2 mg/ml solution in PBS, and sterilize by passing through a 0.2 μm syringe filter (Sarstedt, Nümbrect, Germany). Make fresh as required.

5. EasyTag Expre$^{35}$S$^{35}$S Protein Labeling Mix [$^{35}$S], >1,000 Ci/mmol (see Note 2). Store at 4°C. Observe handling and storage conditions required for this particular radioactive isotope.

6. Chloramphenicol (CAP) for pulse-chase labeling. Prepare a 1-mg/ml solution in regular DMEM without serum (see Note 3), then sterilize by passing through a 0.2 μm syringe filter. Stable at 4°C for up to 1 week.

7. Cell lifters (Corning, Inc. Life Sciences, Lowell, MA, USA).

### 2.2. Sample Preparation

1. Gel loading buffer (2×): 186-mM Tris–HCl, pH 6.7–6.8, 15% glycerol, 2% sodium dodecyl sulfate (SDS), 0.5-mg/ml bromophenol blue, 6% β-mercaptoethanol (β-ME). Store at room temperature. Add β-ME just before use.

2. Micro-BCA™ Protein Assay Kit (Thermo Scientific Pierce Protein Research Products, Rockford, IL, USA).

3. High Intensity Ultrasonic Processor (Sonics & Materials, Inc., Danbury, CT, USA).

### 2.3. SDS–PAGE

1. Separating buffer (4×): 1.5-M Tris–HCl (pH 8.8), 8-mM EDTA-Na$_2$, 0.4% SDS. Store at room temperature.

2. Stacking buffer (4×): 0.5-M Tris–HCl (pH 6.8), 8-mM EDTA-Na$_2$, 0.4% SDS. Store at room temperature.

3. Thirty percent acrylamide/bisacrylamide solution (37.5:1). Avoid exposure to unpolymerized solution as it is a neurotoxin. Store at 4°C.

4. N,N,N,N'-Tetramethylethylenediamine (TEMED). Store at 4°C.

5. Ammonium persulfate (APS): prepare a 10% solution in double-distilled water. Make fresh as required.

6. Running buffer (1×): to 3 L of double-distilled water (total volume required for one run), add 9.08-g Tris base, 43.25-g glycine, and 3.0-g SDS. Do not pH. Store at room temperature.

7. Molecular weight markers: Page Ruler™ Prestained Protein Ladder (Fermentas Canada, Burlington, ON, USA).

8. WIZ Peristaltic Pump (Teledyne Isco, Lincoln, NE, USA).

### 2.4. Generation and Analysis of the Data

1. SGD2000 Digital Slab Gel Dryer (Thermo Fisher Scientific, Waltham, MA, USA).

2. Storm 840 Gel and Blot Imaging System (GE Healthcare).

## 3. Methods

As outlined in this chapter, the pulse and pulse-chase labeling procedures can be applied to all types of adherent cells, independent of either the species of origin or their proliferative state (i.e., dividing versus terminally differentiated), regardless of whether they are transformed, primary or immortalized. We have also used this technique in our laboratory equally successfully for cells growing in suspension: in this case, the only change to the method presented hereafter is that at each wash, cells have to be spun down, then resuspended in the next wash solution or growing medium, as specified in the protocol.

It is important to recognize that the characteristic pattern of mitochondrial translation is unique to each individual species, even when the identical cell type is being considered. Variation across both individuals and tissues within a single species is also possible. This variation can be qualitative, with differences in the electrophoretic mobility of a specific protein, or quantitative, with differences in the overall abundance of mitochondrial translation products (see Fig. 1). Qualitative differences can be due to neutral polymorphisms or to pathogenic mutations, while quantitative differences may reflect different energetic requirements across cell types or a pathogenic event. It is therefore essential that all appropriate controls be included in each experiment.

### 3.1. Labeling of Mitochondrially Synthesized Proteins with Radioactive [$^{35}$S] Methionine and Cysteine

1. One 60 mm tissue culture plate between 75 and 90% confluent on the day of the experiment should provide sufficient material for analysis (see Note 4).

2. If cells will be pulse-chase labeled, prepare the CAP solution.

3. For pulse-chase labeling only, aspirate growth medium from each plate 22–24 h prior to the start of the labeling procedure and add 4.8 ml fresh growth medium and 200 µl CAP solution (total volume of 5 ml/plate, final CAP concentration of 40 µg/ml).

4. At least 30 min before the start of the labeling procedure, pipette the total volume of labeling medium (2 ml/plate) and of DMEM + 10% FBS (5 ml/plate) that are required for the entire experiment (see Note 5) into two separate tissue culture plates, and place the plates in the incubator. This step will allow the media to equilibrate to 5% $CO_2$ and 37°C.

5. For each plate to be labeled, aspirate the growth medium and wash twice with ~3-ml PBS each time.

6. Add 2-ml equilibrated labeling medium/plate, and incubate for 30 min (see Note 6). During this time, prepare and sterilize a 2-mg/ml solution of either emetine (pulse labeling) or anisomycin (pulse-chase labeling).

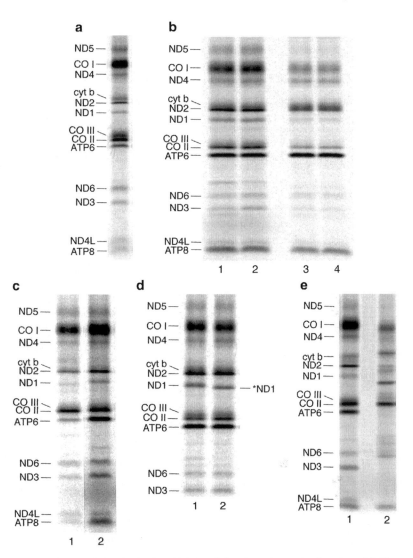

Fig. 1. In vivo analysis of mitochondrial translation by pulse and pulse-chase labeling. *Panel a*: Typical pattern of pulse-labeled mitochondrial translation products in human cultured cells, shown here for immortalized myoblasts. The 13 mitochondrially synthe-sized proteins are indicated at the *left* of the panel: *ND* subunits of Complex I, *CO* subunits of Complex IV, *ATP* subunits of Complex V, *cyt b* subunit of Complex III. *Panel b*: Pulse (1, 2) and pulse-chase (3, 4) labeling of two lines of immortalized human fibroblasts. Note that in pulse-chase labeling, the two Complex V subunits are preferentially stabilized, a characteristic event resulting from the addition of CAP. Each lane contains 50 μg of total cellular protein. *Panel c*: Increased levels of mitochondrial translation products in the transformed cell line HEK293 (2), when compared to a line of immortalized myotubes (1). Each lane contains 50 μg of total cellular protein. The two lanes are part of the same gel and have been placed side by side to facilitate comparison. *Panel d*: Different migration of the ND1 subunit (most likely due to a neutral polymorphism) in two different lines of immortalized human fibroblasts analyzed by pulse labeling. *Panel e*: Difference between human- and mouse-cultured cells in the overall pattern of pulse-labeled mitochondrial translation products, shown here for the human osteosarcoma line 143B (1) and the mouse myeloma line A9 (2).

7. Add 100 μl of the appropriate inhibitor of cytoplasmic translation (final concentration of 100 μg/ml) to each plate and incubate for 5 min.

8. Add 400 μCi of EasyTag labeling mixture to each plate (final concentration of 200 μCi/ml) and incubate for 60 min (see Note 7).

9. Remove labeling mixture from cells and dispose of it according to university guidelines for the handling of radioisotopes. For pulse labeling, add 5 ml of equilibrated DMEM + 10% FBS/plate and return plate(s) to the incubator for 10 min. For pulse-chase labeling, wash cells once with either DMEM + 10% FBS or with PBS, and chase in DMEM + 10% FBS (5 ml/plate) for up to 17–18 h (see Note 8).

10. Wash cells three times with PBS (see Note 9).

11. Using the cell lifter, scrape cells in 0.7- to 0.8-ml ice-cold PBS and then use a pipette to transfer the entire volume to an Eppendorf tube. Repeat with an additional 0.7- to 0.8-ml ice-cold PBS to collect cells remaining on the plate, and transfer to the same Eppendorf tube (total volume of ~1.5 ml) (see Note 10).

12. Collect cells by centrifugation at $1,500 \times g$ for 10 min at 4°C.

13. Aspirate PBS and resuspend the pellet in 200 μl ice-cold PBS. From this point onward, keep cells on ice until they are resuspended in gel loading buffer. Samples may now be stored at −80°C for later use, or the procedure may be continued.

**3.2. Sample Preparation**

1. Use the Micro-BCA™ Protein Assay Kit to determine the protein concentration of each sample. Duplicates (5 and 10 or 3 and 6 μl) of each sample should be measured, and the calculated protein concentration of the duplicates must be within no more than 10–15% of each other, otherwise the measurement should be repeated.

2. For each sample, spin down the desired amount of protein (usually 50 μg) by centrifugation at $>20,000 \times g$ for 20 min at 4°C (see Note 11).

3. Resuspend each pellet in 10 μl of 2× gel loading buffer (room temperature) and then add 10 μl of double-distilled water.

4. Sonicate each sample for 3–8 s at an output control of 60.

5. Spin samples at room temperature for 10–15 min at $>20,000 \times g$, or until the bubbles resulting from sonication have disappeared.

**3.3. SDS–PAGE**

1. These instructions assume the use of a PROTEAN II xi gel system from Bio-Rad Laboratories. Rinse the glass plates, spacers, combs, and casting stand gaskets several times with

deionized, then double-distilled water, followed by a final rinse in 70–95% ethanol. Air-dry.

2. Prepare 12 ml of a 15% gel mixture by combining 6 ml of acrylamide/bisacrylamide solution, 3 ml of 4× separating buffer, and 2.9 ml of double-distilled water, and then prepare 12 ml of a 20% gel mixture by combining 8 ml of acrylamide/bisacrylamide solution, 3 ml of 4× separating buffer, and 0.9 ml of double-distilled water. Just before pouring the gradient gel (see next step), add 60 μl of 10% APS and 6 μl of TEMED to each gel mixture.

3. Using the WIZ Peristaltic Pump at its maximum flow rate, pour a 1.0-mm-thick, 15–20% gradient gel by using the entire volume (24 ml) of the 15% and 20% gel solutions (see Fig. 2, for detailed instructions). Overlay the gradient gel with double-distilled water.

4. Once the gradient gel has polymerized (see Note 12), pour off the water overlay and dry the area above the gel with Whatman paper. Prepare the stacking gel by mixing 1.04 ml of acrylamide/bisacrylamide solution, 2.5 ml of 4× stacking buffer, 6.5 ml of double-distilled water, 50 μl of 10% APS, and 10 μl of TEMED, then pipette the mixture on top of the separating

Fig. 2. Organization, assembly, and casting of gradient gels. Clean and assemble the components required to cast the gradient gel (see 3.3.1. for detailed instructions). Run double-distilled water through the *WIZ Peristaltic Pump* for roughly 5 min at maximum flow rate to ensure that the tubing is clean. Completely empty the tubing of all double-distilled water, and secure the relevant lines of tubing to the *beakers* containing the 15 and 20% acrylamide mixtures and to the gel casting setup using tape, as shown. Once the tubing is secured, add the TEMED and 10% APS solutions to the *two beakers* and turn on the *pump set* to a flow rate of maximum. Exhaust the entire volume of both solutions prior to stopping the pump. Overlay gently with water and allow 45 min to 1 h for polymerization.

gel until it begins to overflow. Insert the comb and allow the stacking gel to polymerize (see Note 13).

5. Once the stacking gel has set, remove the comb by pulling it straight up slowly and gently. Rinse the wells three times with double-distilled water.

6. Assemble the electrophoresis unit and add the running buffer to the inner and outer chambers of the unit. Load the whole 20 μl of each sample in an individual well. Reserve at least one well for the prestained molecular weight marker (15–20 μl/well).

7. Complete the assembly of the electrophoresis unit and connect to a power supply. Run the gel at constant current (8–10 mA) for 15–17 h or until the lowermost (usually 11 kDa) molecular weight marker is at ~1 cm from the bottom of the glass plates.

**3.4. Generation and Analysis of the Data**

1. At the end of the run, disconnect the electrophoresis unit from the power supply and disassemble it. Separate the glass plates sandwiching the gel by vigorously twisting one of the spacers. Remove and discard the stacking gel, and cut one corner of the separating gel for orientation.

2. Rinse the gel by submerging it in a vessel containing double-distilled water, and transfer it by hand to a piece of thick filter paper cut to the dimensions of the gel. Cover the gel with Saran wrap.

3. Dry gel under vacuum at 60°C for 1 h by using the SGD2000 Digital Slab Gel Dryer or equivalent.

4. Expose to a phosphorimager cassette for at least 3 days, then scan with the Storm 840 Gel and Blot Imaging System. Analyze the resultant image with the help of the ImageQuant TL Software. For characteristic patterns of mitochondrial translation analyzed by pulse and pulse-chase labeling and visualized by this method, see Fig. 1.

# 4. Notes

1. Certain formulations of DMEM without methionine and cysteine contain sodium pyruvate, while others do not. Check before adding.

2. Pure [$^{35}$S] methionine results in the strongest signal and the best signal/noise ratio. However, the EasyTag mixture of [$^{35}$S] methionine/cysteine is considerably less expensive and gives a comparable result (approximately 75% of the signal intensity compared with pure [$^{35}$S] methionine).

3. Add warm medium to the CAP powder and incubate in a water bath at 37°C with occasional vortexing to help dissolve the powder.

4. Depending on the cell type and cell size, starting with a plate less than ~75% confluent might result in insufficient protein for SDS–Polyacrylamide Gel Electrophoresis (PAGE) analysis. At the same time, cycling cells should be less than 100% confluent, as they should still be able to divide during radiolabeling.

5. Label a maximum of six plates at a time, otherwise it will be difficult to respect the required timing for several steps of the procedure.

6. Throughout the labeling procedure, it is important that individual plates be placed directly on the shelf of the incubator rather than stacked on top of each other. This ensures that during short incubation times, all the plates equilibrate in terms of temperature and $CO_2$ concentration.

7. Remember to dispose appropriately of all materials that come in contact with the radioisotope: pipettes, cell plates, Eppendorf tubes, pipette tips, cell lifters, etc.

8. While short chases can be universally done in DMEM + 10% FBS, the longer chases required in pulse-chase labeling should be done in cell-specific medium (e.g., chase myoblasts in myoblast-specific medium).

9. Be gentle when washing cells loosely attached to the plate, such as large myotubes or certain transformed cell lines.

10. Alternatively, in the case of myotubes, an enriched population of fused cells can be obtained by selective trypsinization: trypsinize cells for about 2 min or until fused cells start lifting (unfused myoblasts will take a minimum of 5 min to trypsinize). Dilute trypsin by adding 5-ml PBS to the plate and transfer trypsinized cells to a 15-ml Falcon tube. Rinse plate with another 5-ml PBS and add to the same 15-ml tube. Collect cells by centrifugation at $1,500 \times g$ for 5 min. Aspirate PBS, then resuspend pellet in ~1.5-ml cold PBS and transfer to an Eppendorf tube.

11. Sometimes the pellet is not easily visible; always be extra careful when removing the PBS to not disturb the pellet.

12. To save time, prepare the running buffer while the separating gel polymerizes. Likewise, start preparing the samples after pouring the stacking gel.

13. This gel system allows the use of 15- and 20-well combs. While the 20-well comb has the obvious advantage of a higher number of samples per run, the 15-well comb will result in better definition of the bands and a higher resolution between lanes, both of which allow easier quantification of the signal.

## References

1. Chomyn, A. (1996) In vivo labeling and analysis of human mitochondrial translation products. *Methods Enzymol*, **264**, 197–211.

2. Fernandez-Silva, P., Acin-Perez, R., Fernandez-Vizarra, E., Perez-Martos, A. and Enriquez, J.A. (2007) In vivo and in organello analyses of mitochondrial translation. *Methods Cell Biol*, **80**, 571–588.

3. Costantino, P. and Attardi, G. (1977) Metabolic properties of the products of mitochondrial protein synthesis in HeLa cells. *J Biol Chem*, **252**, 1702–1711.

4. Mariottini, P., Chomyn, A., Doolittle, R.F. and Attardi, G. (1986) Antibodies against the COOH-terminal undecapeptide of subunit II, but not those against the NH2-terminal deca-peptide, immunoprecipitate the whole human cytochrome c oxidase complex. *J Biol Chem*, **261**, 3355–3362.

5. Chomyn, A., Martinuzzi, A., Yoneda, M., Daga, A., Hurko, O., Johns, D., Lai, S.T., Nonaka, I., Angelini, C. and Attardi, G. (1992) MELAS mutation in mtDNA binding site for transcription termination factor causes defects in protein synthesis and in respiration but no change in levels of upstream and downstream mature transcripts. *Proc Natl Acad Sci USA*, **89**, 4221–4225.

6. Chomyn, A., Meola, G., Bresolin, N., Lai, S.T., Scarlato, G. and Attardi, G. (1991) In vitro genetic transfer of protein synthesis and respiration defects to mitochondrial DNA-less cells with myopathy-patient mitochondria. *Mol Cell Biol*, **11**, 2236–2244.

7. Enriquez, J.A., Chomyn, A. and Attardi, G. (1995) MtDNA mutation in MERRF syndrome causes defective aminoacylation of tRNA(Lys) and premature translation termination. *Nat Genet*, **10**, 47–55.

8. Hayashi, J., Ohta, S., Kikuchi, A., Takemitsu, M., Goto, Y. and Nonaka, I. (1991) Introduction of disease-related mitochondrial DNA deletions into HeLa cells lacking mitochondrial DNA results in mitochondrial dysfunction. *Proc Natl Acad Sci USA*, **88**, 10614–10618.

9. King, M.P., Koga, Y., Davidson, M. and Schon, E.A. (1992) Defects in mitochondrial protein synthesis and respiratory chain activity segregate with the tRNA(Leu(UUR)) mutation associated with mitochondrial myopathy, encephalopathy, lactic acidosis, and strokelike episodes. *Mol Cell Biol*, **12**, 480–490.

10. Sasarman, F., Antonicka, H. and Shoubridge, E.A. (2008) The A3243G tRNALeu(UUR) MELAS mutation causes amino acid misincor-poration and a combined respiratory chain assembly defect partially suppressed by overexpression of EFTu and EFG2. *Hum Mol Genet*, **17**, 3697–3707.

11. Antonicka, H., Ostergaard, E., Sasarman, F., Weraarpachai, W., Wibrand, F., Pedersen, A.M., Rodenburg, R.J., van der Knaap, M.S., Smeitink, J.A., Chrzanowska-Lightowlers, Z.M. et al. (2010) Mutations in C12orf65 in patients with encephalomyopathy and a mitochondrial translation defect. *Am J Hum Genet*, **87**, 115–122.

12. Antonicka, H., Sasarman, F., Kennaway, N.G. and Shoubridge, E.A. (2006) The molecular basis for tissue specificity of the oxidative phosphorylation deficiencies in patients with mutations in the mitochondrial translation factor EFG1. *Hum Mol Genet*, **15**, 1835–1846.

13. Coenen, M.J., Antonicka, H., Ugalde, C., Sasarman, F., Rossi, R., Heister, J.G., Newbold, R.F., Trijbels, F.J., van den Heuvel, L.P., Shoubridge, E.A. et al. (2004) Mutant mitochondrial elongation factor G1 and combined oxidative phosphorylation deficiency. *N Engl J Med*, **351**, 2080–2086.

14. Fernandez-Vizarra, E., Berardinelli, A., Valente, L., Tiranti, V. and Zeviani, M. (2007) Nonsense mutation in pseudouridylate synthase 1 (PUS1) in two brothers affected by myopathy, lactic acidosis and sideroblastic anaemia (MLASA). *J Med Genet*, **44**, 173–180.

15. Kemp, J.P., Smith, P.M., Pyle, A., Neeve, V.C., Tuppen, H.A., Schara, U., Talim, B., Topaloglu, H., Holinski-Feder, E., Abicht, A. et al. (2011) Nuclear factors involved in mitochondrial translation cause a subgroup of combined respiratory chain deficiency. *Brain*, **134**, 183–195.

16. Miller, C., Saada, A., Shaul, N., Shabtai, N., Ben-Shalom, E., Shaag, A., Hershkovitz, E. and Elpeleg, O. (2004) Defective mitochondrial translation caused by a ribosomal protein (MRPS16) mutation. *Ann Neurol*, **56**, 734–738.

17. Riley, L.G., Cooper, S., Hickey, P., Rudinger-Thirion, J., McKenzie, M., Compton, A., Lim, S.C., Thorburn, D., Ryan, M.T., Giege, R. et al. (2010) Mutation of the mitochondrial tyrosyl-tRNA synthetase gene, YARS2, causes myopathy, lactic acidosis, and sideroblastic anemia – MLASA syndrome. *Am J Hum Genet*, **87**, 52–59.

18. Smeitink, J.A., Elpeleg, O., Antonicka, H., Diepstra, H., Saada, A., Smits, P., Sasarman, F., Vriend, G., Jacob-Hirsch, J., Shaag, A. et al. (2006) Distinct clinical phenotypes associated

with a mutation in the mitochondrial translation elongation factor EFTs. *Am J Hum Genet*,
79, 869–877.

19. Valente, L., Tiranti, V., Marsano, R.M.,
Malfatti, E., Fernandez-Vizarra, E., Donnini,
C., Mereghetti, P., De Gioia, L., Burlina, A.,
Castellan, C. *et al.* (2007) Infantile encephal-
opathy and defective mitochondrial DNA

translation in patients with mutations of
mitochondrial elongation factors EFG1 and
EFTu. *Am J Hum Genet*, 80, 44–58.

20. Zeharia, A., Shaag, A., Pappo, O., Mager-
Heckel, A.M., Saada, A., Beinat, M., Karicheva,
O., Mandel, H., Ofek, N., Segel, R. *et al.* (2009)
Acute infantile liver failure due to mutations in
the TRMU gene. *Am J Hum Genet*, 85, 401–407.

# Chapter 15

# Transmitochondrial Cybrids: Tools for Functional Studies of Mutant Mitochondria

## Sajna Antony Vithayathil, Yewei Ma, and Benny Abraham Kaipparettu

### Abstract

Mitochondrial functions are controlled by both mitochondrial DNA (mtDNA) and nuclear DNA. Hence, it is difficult to identify whether mitochondrial or nuclear genome is responsible for a particular mitochondrial defect. Cybrid is a useful tool to overcome this difficulty, where we can compare mitochondria from different sources in a defined nuclear background. Cybrids are constructed by fusing enucleated cells harboring wild type or altered mtDNA of interest with $\rho^0$ cells (cells lacking mtDNA) in which the endogenous mtDNA has been depleted. Therefore, cybrids are very useful in studying consequences of mtDNA alterations or other mitochondrial defects at the cellular level by excluding the influence of nuclear DNA mutations.

**Key words:** Cybrids, Rho0 cells, Mitochondria, Mitochondrial depletion, Transmitochondrial cybrids, mtDNA, Mitochondrial DNA mutation, Mitochondrial retrograde regulation

## 1. Introduction

Cross talk between mitochondria and the nucleus is important for a variety of cellular processes such as metabolism, differentiation, signal transduction, growth, and apoptosis (1, 2). Although in general, nuclear genes control mitochondrial activities, there are reports which suggest that mitochondria may also regulate nuclear gene expression (3–5).

Transmitochondrial cybrid system is a great utility for the study of the functional effects of mitochondria in a defined nuclear background. Cybrids are constructed by fusing Rho-zero cells ($\rho^0$ cells) with enucleated cells, harboring mitochondrial DNA (mtDNA) of interest (see Fig. 1 for details). Cells lacking mtDNA ($\rho^0$ cells) can be generated by prolonged incubation with ethidium bromide

Lee-Jun C. Wong (ed.), *Mitochondrial Disorders: Biochemical and Molecular Analysis*, Methods in Molecular Biology, vol. 837, DOI 10.1007/978-1-61779-504-6_15, © Springer Science+Business Media, LLC 2012

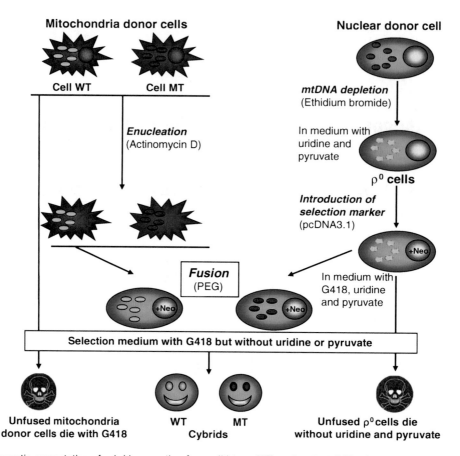

Fig. 1. Schematic presentation of cybrid generation from wild type (WT) and mutant (MT) mitochondrial donor cells for comparison. The ρ⁰ cells with a stable selection marker are made to fuse with enucleated WT and MT cells. Cybrids are selected in medium with the selection drug but without uridine or pyruvate.

(EtBr), a chemical which inhibits mtDNA replication. The $\rho^0$ cells do retain their own nuclear genome and can grow in rich culture medium supplemented with uridine and pyruvate. Prior to fusion, a drug selection marker needs to be introduced into the $\rho^0$ cells. However, it is not required to introduce the selection marker to one of the most commonly used $\rho^0$ cell models such as 143B TK⁻ derived from an osteosarcoma cell line (6). 143B TK⁻ is devoid of the cellular thymidine kinase (TK) enzyme and is thus resistant to thymidine analog 5-bromo-2-deoxy-uridine (BrdU)-induced cell death. After fusion with the enucleated mitochondrial donor cells, individual cybrid clones can be selected using medium containing the selection drug but without uridine or pyruvate. In case of 143B TK⁻ $\rho^0$ cells, BrdU can be used for cybrid selection. Unfused mitochondrial donor cells cannot survive after drug treatment, and unfused $\rho^0$ cells will stop growing without uridine and pyruvate. The mtDNA and selected nuclear genes should be genotyped to confirm nuclear and mitochondrial origins of cybrids.

The transmitochondrial cybrid is at present a widely used method to determine whether a nucleotide substitution in the mtDNA is pathogenic. Therefore, cybrids are very useful in studying consequences of mtDNA alterations and other mitochondrial defects at the cellular level by excluding the influence of nuclear variants.

## 2. Materials

All solutions need to be prepared in ultrapure water which can be made by purifying deionized water using Millipore™ water purifying system or other similar techniques to obtain a sensitivity of 18 MΩ cm at 25°C. Prepare all solutions using analytical grade reagents at room temperature unless otherwise indicated.

### 2.1. Cell Culture

1. Cells should always be grown in a humidified incubator with 5% carbon dioxide ($CO_2$) at 37°C (see Note 1).

2. Regular cell culture medium: DMEM supplemented with 10% heat-inactivated fetal bovine serum (FBS). Store at 4°C (see Note 2).

3. 100× Antibiotics stock solutions for cell culture: 10,000 U/ml penicillin and 10,000 μg/ml Streptomycin. Store at –20°C (see Note 3).

4. 1× Sterile phosphate buffered saline (PBS), pH 7.4. Store at 4°C (see Note 4).

5. 0.25% Trypsin–EDTA. Store at 4°C (see Note 5).

6. 2-ml and 10-ml Sterile, polystyrene pipettes.

7. 0–20, 200, and 1,000-μl Autoclaved pipette tips.

8. 100 and 3.5-mm Sterile cell culture dishes (see Note 6).

9. 15 and 50-ml Centrifuge tubes and 50-ml sterile glass beakers.

### 2.2. Generation of ρ⁰ Cells

1. 1.5-ml Eppendorf tubes.

2. 50 mg/ml Uridine: Measure 500 mg of uridine and dissolve it in 10 ml of ultrapure water in a clean 15-ml tube. Filter the solution using 0.2-μm nylon membrane Acrodisc syringe filters. Aliquot 1 ml each into sterile 1.5-ml Eppendorf tubes and store at –20°C.

3. 50 μg/ml Ethidium bromide: Dilute 5 μl of 1% (1 g/100 ml) EtBr solution (cat #BP1302-10, Fisher Biotech) to 1 ml using ultrapure water. Filter the solution using 0.2-μm nylon membrane filters. Store at 4°C (see Note 7).

**2.3. Transfection of Selection Marker**

1. 1 mg/ml Lipofectamine™ 2000 solution (Invitrogen). Store at 4°C (see Note 8).

2. OptiMEM™ reduced serum medium (Cat# 31985070, Invitrogen). Store at 4°C (see Note 9).

3. 500 ng/μl DNA from neo resistant plasmids, such as pcDNA3.1 (Invitrogen) in ultrapure water (see Note 10).

4. 50 mg/ml G418 (Invitrogen) for neo selection. Store 1-ml aliquots in 1.5-ml Eppendorf tubes and store at –20°C.

**2.4. Enucleation of Cells**

1 mg/ml Actinomycin D: Dissolve 1 mg of actinomycin D powder in 50 μl of DMSO and add 950 μl of 1× PBS into it. Store at –20°C and protect from light exposure.

**2.5. Generation of Cybrids**

1. Forty-five percent polyethylene glycol (PEG): Measure 22.5 g of PEG and transfer it into a 100-ml glass bottle. Bring the total volume to 50 ml using ultrapure water. Mix thoroughly, autoclave, and bring to room temperature before use (see Note 11).

2. Ten percent DMSO in PBS: Add 4 ml DMSO to 36 ml of sterile PBS in a 50-ml tube. Gently mix and use immediately.

# 3. Methods

EtBr has long time been used to inhibit transcription and replication of extrachromosomal DNA. Since mtDNA is an extrachromosomal genetic material, EtBr was used to deplete mtDNA without affecting chromosomal genes (7, 8). mtDNA-depleted $\rho^0$ cells are auxotrophic for pyruvate and uridine, and are incapable of aerobic respiration due to the lack of key respiratory chain components (9).

**3.1. Regular Cell Passages**

Cell culture protocols can be performed in room temperature unless otherwise indicated. Prior to use, cell culture medium and reagents need to be brought to room temperature or 37°C.

1. Warm the regular cell culture medium, PBS, and trypsin-EDTA in 37°C water bath for 10–15 min. Select the cell line of interest to be used as the nuclear background of cybrids (see Note 1).

2. Culture the cells in 100-mm culture plates in regular cell culture medium (see Note 2).

3. When the cells reach 80–100% confluency, remove the culture medium and wash the cells with 5 ml of PBS.

4. After removing PBS, add 1–2 ml of 0.25% trypsin–EDTA solution for 1–3 min (see Note 5).

5. Gently tap the plate and observe the cells under a microscope. Once the cells get detached from the plate, add 5 ml of culture medium and mix several times to suspend the cells.

6. Add 9 ml of culture medium to a new 100-mm plate.

7. Add 1 ml of cell suspension to the new plate and mix the cell suspension (see Note 12).

8. Examine the plates with cells in culture daily under a phase contrast microscope; observe the morphology of cells, the color of the medium, and the density of the cells. Repeat steps 3–8 each time when the cells reach about 80–100% confluence.

**3.2. EtBr Treatment and Uridine Supplementation**

1. Take two 100-mm cell culture plates and add 9 ml of regular cell culture medium. Label one plate as EB and the other one as control.

2. Add 1 ml each of cell suspension to both EB and control plates as described in Subheading 3.1, step 7.

3. Add 10 µl of 50 µg/ml EtBr only to the EB plate to make a final concentration of 50 ng/ml EtBr. Incubate the plates for 1 week and follow step 8 in Subheading 3.1 (see Note 7).

4. Add 50 ng/ml of EtBr to EB plate each time while changing the medium or passaging the cells. After 1 week of EtBr treatment, in addition to the EtBr, supplement the medium with 10 µl of 50 mg/ml uridine to make a final concentration of 50 µg/ml of uridine (see Note 13).

5. Continue the EtBr and uridine treatment in EB plate for 6–8 weeks. Control plate needs to be passaged in regular culture medium without EtBr and uridine. Frequently observe the cells under microscope for any changes in phenotype, growth rate, or other abnormalities in EB plate compared to the control plate (see Note 14).

6. After 6–8 weeks, extract DNA and analyze mtDNA content of cells in both EB and control plates as described in Chapter 22 (see Note 15)

7. EtBr treatment can be dismissed in EB plate after the mtDNA content reaches less than one copy number per diploid cell and continue to culture with 50 µg/ml uridine (see Note 16).

**3.3. Transfection of Selection Marker**

This can be done by stable transfection of pcDNA3.1 empty vector which contains neo resistant gene as a selection marker (see Note 10).

1. Plate $2 \times 10^5$ $\rho^0$ cells per well of a 35-mm plate in 1 ml of the appropriate regular cell culture medium.

2. After overnight culture, check the confluency of cells and make sure that the cells are about 40–60% confluent.

3. Take two sterile 1.5-ml Eppendorf tubes and label them A and B, respectively.

4. To tube A, add 1.5 μl (750 ng) of 500 ng/μl pcDNA 3.1 DNA into 98.5 μl of OptiMEM medium (see Note 9).

5. To tube B, add 5 μl of Lipofectamine 2000 into 95-μl OptiMEM and incubate for 5 min at room temperature (see Note 8).

6. Combine 100 μl of the diluted Lipofectamine 2000 from tube B with 100 μl of diluted DNA in tube A. Mix gently and incubate at room temperature for 20 min to allow DNA–Lipofectamine 2000 complexes to form.

7. Remove the medium from cells and replace with 800 μl of OptiMEM.

8. Add the 200-μl DNA–Lipofectamine 2000 complexes directly to the plate containing cells and mix gently.

9. Incubate the cells under normal cell culture conditions (37°C, 5% $CO_2$) for 6 h.

10. After 6 h, remove the transfection medium, wash cells once with 1× PBS, and add 2 ml of regular cell culture medium containing 50 μg/ml uridine and incubate for 36 h.

11. After 36 h of incubation, replace the medium with 10 ml of medium containing 400 μg/ml G418 (add 80 μl of 50 mg/ml G418 to 10-ml regular cell culture medium with 50 μg/ml uridine) (see Note 17).

12. Only cells transfected with the neo resistance gene will be able to survive the antibiotic treatment. Mark the areas in plate containing clones under a microscope and carefully pick individual clones using pipette tips.

13. Suspend each clone in 200 μl of medium containing 200 μg/ml G418 (Add 40 μl of 50 mg/ml G418 to 10-ml regular cell culture medium) and 50 μg/ml uridine. Transfer the individual clone suspension to each well of a 96-well cell culture plate (see Note 18).

14. As the cells grow, expand each clone to larger cell culture dishes (see Note 18).

15. Extract DNA from each clone and reconfirm the mtDNA depletion by qPCR as described in Chapter 22.

16. Select suitable $\rho^0$ cells with neo resistance for the generation of cybrids (see Note 18).

*3.4. Generation of Cybrids*

The protocol is based on the publication of King and Attardi (6), who demonstrated that, human $\rho^0$ cells can be repopulated with exogenous human mitochondria. The original study used 143B. TK⁻ cell line as the nuclear donor. Up-to-date, this is the most frequently used donor for $\rho^0$ cells. However, the protocols described in this article can be applied to generate any combination of cybrid, as long as the correct genotype of the nuclear and mitochondrial donors in the cybrids can be established. A cartoon explaining different steps for cybrid generation from cells with wild type (WT) and mutant (MT) mitochondria is shown in Fig. 1.

1. Culture $\rho^0$ cells in one 100-mm culture dish in regular cell culture medium containing 50 µg/ml uridine and 200 µg/ml G418 as described in Subheading 3.3 (see Notes 12, 16, and 17).

2. In parallel, culture mitochondrial donor cells containing WT or MT mtDNA in 100-mm culture dishes in regular cell culture medium (see Note 19).

3. When the mitochondrial donor cells reach 40–60% confluency, replace the medium in both WT and MT plates with 10-ml regular medium containing 5 µl of 1 mg/ml actinomycin D so as to obtain a final concentration of 0.5 µg/ml actinomycin D for enucleation (see Note 20).

4. Incubate the cells overnight. After incubation, cells with viable cytoplasts adhere to the culture plate. Make sure that at least 50–60% of cells adhere to the plate for cybrid fusion (see Notes 20 and 21).

5. Harvest the $\rho^0$ cells as described in Subheading 3.1 and make 20 ml of $1 \times 10^5$ cells/ml cell suspension in medium containing 50 µg/ml uridine (20 µl of 50 mg/ml uridine to 20 ml of regular cell culture medium).

6. Remove the medium from the enucleated WT and MT mitochondrial donor cells and gently wash once with 1× PBS (see Note 22).

7. Gently add the 10 ml of $\rho^0$ cell suspension on top of each enucleated mitochondrial donor cells (WT and MT).

8. Incubate the plates for 5–6 h so that the $\rho^0$ cells can adhere on top of the enucleated mitochondrial donor cells (see Note 23).

9. Remove the medium from the culture plates and add the 10 ml of 45% PEG on top of the cells using a 10-ml pipette. Wait for 60 s.

10. After 60 s, remove the PEG using a 10-ml pipette and discard it to a 50-ml beaker or similar container (see Note 24).

11. Wash the plates three times, each time with 10 ml of 10% DMSO in 1× PBS for 30 s (see Note 25).

12. Wash each plate one time with 10 ml 1× PBS without DMSO for 15 s.

13. Wash each plate one time with 10-ml culture medium for 15 s.

14. Add 10 ml of culture medium containing 50 µg/ml uridine to each plate and incubate under regular cell culture conditions (37°C, 5% $CO_2$) overnight.

15. After overnight culture, remove the culture medium from the plates and add new culture medium without uridine and with 400 μg/ml G418 (see Notes 17 and 26).

16. Unfused $\rho^0$ cells will die without uridine, and surviving mitochondrial donor cells will die because of G418.

17. After 1–2 weeks of culture in medium with G418, cybrid clones will appear in the plate.

18. Mark the areas of cybrid clones in the plate under a microscope and pick several healthy clones using pipette tips. Suspend each clone in 200 μl of medium with 200 μg/ml G418. Transfer the individual clone suspension to each well of a 96-well cell culture plate (see Note 17).

19. When individual clones reach confluence, trypsinize the clones and grow them in individual wells of a 24-well cell culture plate and further expand to larger dishes according to the experimental needs.

20. Extract DNA from each clone for mtDNA quantification (refer Chapter 22 for details).

21. Select healthy cybrids with matching mtDNA copy number (see Note 27).

22. Genotype both mtDNA and nuclear DNA from each clone to confirm the mitochondrial and nuclear origin. Also, verify the mtDNA mutation/variants in cybrids with MT cells (refer Chapter 19 for details) (see Note 28).

23. Select clones with confirmed correct genotype and matched mtDNA content for further mitochondrial functional studies and other experiments.

## 4. Notes

1. Selection of cell line depends on experimental objectives. The protocol described has been effectively working for established breast cancer cell lines and noncancerous breast epithelial cells. Cells can be obtained from American Type Culture Collection (ATCC) or any suppliers. Culture conditions vary widely for each cell type, and variation of conditions for a particular cell type can result in different phenotypes.

2. For most breast cancer cell lines, 10% FBS in DMEM high glucose 1× medium (Invitrogen) which contains 4.5 g/l D-glucose, 4 mM L-glutamine, and 110 mg/l sodium pyruvate is suitable. Use appropriate growth medium suggested for the cell lines of interest.

3. Cell culture medium usually includes antibiotics with 100 IU/ml penicillin and 100 μg/ml streptomycin for routine cell culture.

Antibiotics are not essential for cell culture. However, antibiotics can help to prevent bacterial contaminations. Thus, it is recommended to use antibiotics in the medium unless otherwise instructed.

4. This protocol usually uses sterile 1× PBS pH 7.4 without calcium and magnesium chloride from GIBCO, Invitrogen. However, sterile PBS can be easily prepared in the laboratory (refer Chapter 13).

5. Concentration of trypsin–EDTA and treatment time depends on the type of cell line. Primary cells and noncancerous cells may require more concentrated trypsin and/or longer duration of trypsinization.

6. The type of the culture plates or flasks to be used depends on the cell number required for particular experiments. For cybrid experiments, culture dishes are used instead of culture flasks for easy performance of cell fusion and selection of cybrid clones.

7. EtBr concentrations used may range from 50 to 500 ng/ml depending on the cell line. Different concentrations should be tested out for the maximum EtBr concentration which can be tolerated by the cell line of interest. EtBr is also available from other companies in various concentrations.

8. There are different transfection methods and transfection reagents available. It is important to determine the optimum amount of transfection reagent and duration of transfection for cells of interest before transfection.

9. It is not necessary to use OptiMEM for transfection. DMEM without serum and antibiotics will also work for most of the cells. However, in case of overnight transfection, it is better to use medium with low percentage of serum. Transfection efficacy may vary depending on the cell type. Primary cells may be difficult to transfect. In this case, viral transduction may be considered for difficult-to-transfect cells.

10. The selection marker can vary depending on the purpose of $\rho^0$ cells and mitochondrial donor cells. If mitochondrial donor cells already contain neo resistant gene, introduce other selection markers such as puromycin resistant gene in $\rho^0$ cells. When cybrids are generated using the most established 143B TK$^-$ $\rho^0$ cells, BrdU can be used for cybrid selection.

11. PEG can also be prepared by adding DMEM with 20% DMSO (10).

12. Passage ratio depends on the type of cells. Some cell lines may not survive under low density and require more cells during passage. However, some of the fast growing cells like cancer cells may be passaged in low density.

13. Most DMEM basal medium usually contains pyruvate in it. If the medium do not contain pyruvate, supplement the culture medium with 110 μg/ml of sodium pyruvate.

14. Parallel culture of a control plate is important especially if the $\rho^0$ cells from a particular cell line are being established for the first time. Changes in phenotype may be observed after long-term EtBr treatment, and a control plate can assure that such changes are not due to some contamination in the culture medium.

15. Concentration and duration of EtBr treatment depends on the cell type. Some cells require longer duration of treatment.

16. It is important to note that, most cell types cannot survive under $\rho^0$ condition for a long period of time. Thus, there is a possibility that mtDNA content may increase after culturing in the absence of EtBr for a long period of time. Check mtDNA content of $\rho^0$ cells from time to time if the culture continues and before performing important experiments. If there is increase in mtDNA copy number, continue EtBr treatment for a few more weeks to complete mtDNA depletion.

17. The drug used for selection depends on the transfected marker. Use about 400 μg/ml G418 for 2 weeks for clonal selection and 200 μg/ml for the maintenance of clones. However, optimal concentration for the selection of resistant clones in mammalian cells depends on the cell lines used as well as the plasmid carrying the resistance gene. Therefore, titration of antibiotic markers should be performed to find the best condition for each different experimental system. Resistant drug treatment of the resistant clones for selection may vary from 1 to 3 weeks depending on the cell line and the resistant marker.

18. Use of a single clone of $\rho^0$ cells for all cybrid experiments provides a uniform nuclear background. However, it is also possible to use pool of drug resistant $\rho^0$ cells as they are derived from the same cell line. Moreover, it is important to reconfirm the mtDNA depletion in clone(s) which will be used for cybrid generation.

19. The selection of mtDNA donor cells depends on the objective of each experiment. If the functional effect of an mtDNA mutation is to be studied, cells containing WT and MT mtDNA are used as mitochondrial donors for comparison in cybrids with the same nuclear background. However, if the effect of mitochondria from different disease stages is to be studied, then the mitochondria derived from each stage should be used.

20. Other agents including cytochalasin B may also be used for enucleation (11). The amount of actinomycin D used should be titrated for each cell type to determine the suitable amount at which enucleation will be optimal. After actinomycin D treatment, the cells will undergo irreversible death but with

viable cytoplasts (adhere to the culture plate) that could be used for fusion with $\rho^0$ cell (12). If high toxicity is observed, decrease the concentration of actinomycin D or the treatment time to 4–6 h.

21. Floating cells will be washed away while removing the medium. So there should be enough attached cells to generate cybrids.

22. Enucleated mitochondrial donor cells are easily detachable. So vigorous washing with PBS will flush out the adherent enucleated cells. Therefore, wash the cells with PBS gently.

23. Observe the cell plates under microscope to make sure that the $\rho^0$ cells have attached on to the enucleated mitochondrial donor-cell plates. The attachment time may vary from 2 to 8 h depending on the cell type and cell health.

24. Forty-five percent PEG in water is a very viscous solution. Removal of PEG using vacuum pump may clog the suction tubes. Hence, remove PEG using a 10-ml pipette into a 50-ml beaker or other container and discard it separately.

25. Wash gently. After the first wash, remove PBS-DMSO using a 10-ml pipette to the 50-ml container to avoid clogging. The rest of the wash solution can be removed by vacuum suction.

26. If possible, treat a plate of parental cells as control in parallel to the cybrids with same dose of G418, expecting complete death of control cells. This will make sure that the surviving cells will be cybrids with nucleus from $\rho^0$ cells with neo resistance.

27. It is important to select cybrids with matching mtDNA copy number for comparative mitochondrial functional studies. Thus, quantify mtDNA content for several clones and select matching cybrids from both WT and MT mitochondrial donor cells for comparative studies.

28. It is important to verify the mitochondrial and nuclear sources of cybrids by sequencing both mtDNA and selected nuclear genes to avoid parental cell contamination from mtDNA or nuclear donors. Sequence the parental cells before cybrid generation to identify mutational differences between the nuclear and mitochondrial donors. It is also important to quantify sequence variations if any. If there is minor contamination, such low population may overgrow during passage. *POLG* and *COX II* genes are good candidates to sequence for genotyping. However, the selection of nuclear gene depends on the cells used.

## Acknowledgements

This paper was partially supported by DOD W81XWH-11-1-0292 and NIH 1U54U54 CMCD grants to BAK.

## References

1. Chen, J. Q., Cammarata, P. R., Baines, C. P., and Yager, J. D. (2009) Regulation of mitochondrial respiratory chain biogenesis by estrogens/estrogen receptors and physiological, pathological and pharmacological implications. *Biochim Biophys Acta* 1793, 1540–1570.

2. Liu, Z., and Butow, R. A. (2006) Mitochondrial retrograde signaling. *Annu Rev Genet* 40, 159–185.

3. Ishikawa, K., Takenaga, K., Akimoto, M., Koshikawa, N., Yamaguchi, A., Imanishi, H., et al. (2008) ROS-generating mitochondrial DNA mutations can regulate tumor cell metastasis. *Science* 320, 661–664.

4. Petros, J. A., Baumann, A. K., Ruiz-Pesini, E., Amin, M. B., Sun, C. Q., Hall, J., et al. (2005) mtDNA mutations increase tumorigenicity in prostate cancer. *Proc Natl Acad Sci USA* 102, 719–724.

5. Ma, Y., Bai, R. K., Trieu, R., and Wong, L. J. Mitochondrial dysfunction in human breast cancer cells and their transmitochondrial cybrids. *Biochim Biophys Acta* 1797, 29–37.

6. King, M. P., and Attardi, G. (1989) Human cells lacking mtDNA: repopulation with exogenous mitochondria by complementation. *Science* 246, 500–503.

7. Hayakawa, T., Noda, M., Yasuda, K., Yorifuji, H., Taniguchi, S., Miwa, I., et al. (1998) Ethidium bromide-induced inhibition of mitochondrial gene transcription suppresses glucose-stimulated insulin release in the mouse pancreatic beta-cell line betaHC9. *J Biol Chem* 273, 20300–20307.

8. Zylber, E., and Penman, S. (1969) Mitochondrial-associated 4 S RNA synthesis inhibition by ethidium bromide. *J Mol Biol* 46, 201–204.

9. Magda, D., Lecane, P., Prescott, J., Thiemann, P., Ma, X., Dranchak, P. K., et al. (2008) mtDNA depletion confers specific gene expression profiles in human cells grown in culture and in xenograft. *BMC Genomics* 9, 521.

10. Bacman, S. R., and Moraes, C. T. (2007) Transmitochondrial technology in animal cells. *Methods Cell Biol* 80, 503–524.

11. Moraes, C. T., Dey, R., and Barrientos, A. (2001) Transmitochondrial technology in animal cells. *Methods Cell Biol* 65, 397–412.

12. Bayona-Bafaluy, M. P., Manfredi, G., and Moraes, C. T. (2003) A chemical enucleation method for the transfer of mitochondrial DNA to rho(o) cells. *Nucleic Acids Res* 31, e98.

# Chapter 16

# Fluorescence-Activated Cell Sorting Analysis of Mitochondrial Content, Membrane Potential, and Matrix Oxidant Burden in Human Lymphoblastoid Cell Lines

**Stephen Dingley, Kimberly A. Chapman, and Marni J. Falk**

## Abstract

Fluorescence-activated cell sorting (FACS) permits specific biologic parameters of cellular populations to be quantified in a high-throughput fashion based on their unique fluorescent properties. Relative quantitation of mitochondrial-localized dyes in human cells using FACS analysis allows sensitive analysis of a variety of mitochondrial parameters including mitochondrial content, mitochondrial membrane potential, and matrix oxidant burden. Here, we describe protocols that utilize FACS analysis of human lymphoblastoid cell lines (LCL) for relative quantitation of mitochondrial-localized fluorescent dye intensity. The specific dyes described include MitoTracker Green FM to assess mitochondrial content, tetramethylrhodamine ethyl ester (TMRE) to assess mitochondrial membrane potential, and MitoSOX Red to assess mitochondrial matrix oxidant burden. Representative results of FACS-based mitochondrial analyses demonstrate the variability of these three basic mitochondrial parameters in LCLs from healthy individuals, as well as the sensitivity of applying FACS analysis of LCLs to study the effects of pharmacologic induction and scavenging of oxidant stress.

**Key words:** Mitochondria, MitoSOX Red, MitoTracker Green, TMRE, FACS

## 1. Introduction

Fluorescence-activated cell sorting (FACS) offers a high-throughput means to quantify fluorescent indicators for a variety of cell and tissue applications. It has long been used to analyze a multitude of cellular characteristics ranging from cell size to organelle abundance to specific protein levels (1). The approach first involves timed incubation of cellular suspensions with specific fluorescent dyes. Fluorescent-labeled cellular suspensions are then injected into a FACS-enabled flow cytometer. Following excitation of cells at

Lee-Jun C. Wong (ed.), *Mitochondrial Disorders: Biochemical and Molecular Analysis*, Methods in Molecular Biology, vol. 837,
DOI 10.1007/978-1-61779-504-6_16, © Springer Science+Business Media, LLC 2012

wavelengths specific to each fluorescent dye, their emitted and scattered light is recorded as they flow individually past a detector. Dye-specific data can be plotted to permit visualization of particular cellular properties. The capacity to quickly and reproducibly generate large amounts of quantifiable data, combined with an adaptability to a wide range of tissue types and fluorescent dyes, has made FACS analysis a widely used method to probe cell biology.

FACS analysis of mitochondrial biology has been utilized in a wide range of cell types (2, 3). Mitochondria-targeted fluorescent dyes are commercially available that permit targeted examination of distinct mitochondrial parameters including matrix oxidant burden (4, 5), membrane potential (6, 7), and mitochondria content (8) in living cells (9). Such fluorescent dyes have been increasingly utilized to interrogate mitochondria-specific biology in both in vitro systems and, more recently, in vivo using microscopic animal models (10, 11).

Here, we describe methods for FACS analysis of mitochondria-localized fluorescent dyes in human lymphoblastoid cell lines (LCL). Relative quantitation is performed of mean LCL fluorescence following timed incubation with MitoTracker Green FM to assess mitochondrial content, tetramethylrhodamine ethyl ester (TMRE) to assess mitochondrial membrane potential, and MitoSOX Red to assess mitochondrial matrix oxidant burden. We further describe the effect on relative matrix oxidant burden of antimycin A (AA)-induced mitochondrial oxidant stress, both alone and in combination with an antioxidant, N-acetyl-cysteine (NAC) (12). The methods described can be readily adapted to perform relative quantitation in LCLs to discern drug or toxin effects across a range of mitochondrial parameters.

## 2. Materials

### 2.1. Cell Culture and Treatment

1. RPMI 1640 Medium: 15% fetal calf serum, 2 mM L-glutamine, and 100 U/ml penicillin–streptomycin.

2. Phosphate-buffered saline (PBS) (GIBCO).

3. Dimethyl sulfoxide (DMSO).

4. 5 mM MitoSOX Red stock solution: Dilute 50 μg of MitoSOX Red with 13 μl of 100% DMSO.

5. 10 μM MitoSOX Red working solution: Dilute 4 μl of 5 mM MitoSOX Red with 2 ml of RPMI 1640.

6. 100 μM MitoTracker Green FM solution: Dilute 50 μg of MitoTracker Green FM stock with 750 μl of 100% DMSO.

7. 4 mM tetramethylrhodamine ethyl ester perchlorate (TMRE) stock solution: Dissolve 25 mg of TMRE with 12.14 ml of 100% DMSO.

8. 20 μM TMRE working solution: Dilute 4 mM of TMRE stock 1:200 in DMSO to make a 20 μM TMRE working solution.

9. Dilute 2 μl of 20 μM TMRE in 2 ml of RPMI 1640 to achieve a final concentration of 20 nM TMRE in RPMI 1640.

10. 1 mM antimycin A (AA) stock solution: Dissolve 5.4 mg of AA in 10 ml of DMSO. Store at –20°C.

11. 100 mM *N*-acetyl-cysteine (NAC) stock solution: Dissolve 163.19 mg of NAC in 10 ml of distilled water. Store at 4°C.

12. Human lymphoblastoid cell lines (LCL).

13. Multiwell cell culture plate (3 ml capacity per well).

*2.2. Fluorescence-Activated Cell Sorting Flow Cytometry*

1. 12 mm × 75 mm round bottom polystyrene tubes.

2. Dual Laser Becton Dickinson Analytical FACS Calibur Flow Cytometer equipped with a 488 nm laser and the following channels:

   (a) FL1 [530/30, 560 shortpass (SP)].

   (b) FL2 [585/42, 640 longpass (LP)].

   (c) FL3 (670 LP).

# 3. Methods

*3.1. LCL Culture Plate Preparation*

1. Collect $1 \times 10^7$ LCLs in a 15 ml conical tube for each drug and dye combination.

2. Pellet cells by centrifugation at $300 \times g$ for 1 min.

3. Discard supernatant.

4. Count LCLs.

5. Resuspend 200,000 cells/ml of RPMI 1640 (see Note 1).

6. Plate 1 ml of resuspended cells in a single well of a 24 well culture plate.

7. Plate four replicate wells for each desired drug treatment group.

8. Plate four control wells containing untreated cells to determine baseline fluorescence for FACS analysis.

*3.2. LCL Incubation with Antimycin A and N-Acetyl-Cysteine*

1. Add 5 μl of 1 mM AA stock solution to 1 ml of cells (see Note 1) in the desired wells of the culture plate to achieve a final concentration of 5 μM (see Note 2).

2. Add NAC to cells in the desired wells of the culture plate to achieve a final concentration of 5 mM (see Note 3).

*3.3. LCL Incubation with MitoTracker Green FM*

1. Add 2 μl of 100 μM MitoTracker Green FM solution to 1 ml cells in the desired wells of the culture plate to achieve a final concentration of 200 nM.

2. Incubate cells for 20 min at 37°C in a 5% $CO_2$ incubator.

3. Collect medium and cells in a 15 ml conical tube.

4. Centrifuge cells at $300 \times g$ for 1 min. Remove supernatant.

5. Wash cells by resuspending them in 1 ml of PBS maintained at 37°C.

6. Centrifuge cells at $300 \times g$ for 1 min. Remove supernatant.

7. Resuspend pelleted cells with 0.4 ml of PBS.

8. Incubate cells first for 10 min at 37°C in a 5% $CO_2$ incubator and then for 20 min at room temperature (see Note 4).

### 3.4. LCL Incubation with TMRE

1. Add 2 ml of 20 nM TMRE in RPMI 1640 solution to cells (see Note 1) in the desired wells of the culture plate to achieve a final concentration of 13.3 nM.

2. Incubate cells for 10 min at 37°C in a 5% $CO_2$ incubator.

3. Collect medium and cells in a 15 ml conical tube.

4. Centrifuge cells at $300 \times g$ for 1 min. Remove supernatant.

5. Wash cells by resuspending in 1 ml of PBS maintained at 37°C.

6. Centrifuge cells at $300 \times g$ for 1 min. Remove supernatant.

7. Repeat wash as detailed in steps 5 through 6.

8. Resuspend pelleted cells with 0.4 ml of PBS.

9. Incubate cells first for 10 min at 37°C in a 5% $CO_2$ incubator and then for 20 min at room temperature (see Note 4).

### 3.5. LCL Incubation with MitoSOX Red

1. Add 2 ml of 10 µM MitoSOX Red working solution to cells (see Note 1) in the desired wells of the culture plate to achieve a final concentration of 6.6 µM.

2. Incubate cells with MitoSOX Red at 37°C for 10 min in a 5% $CO_2$ incubator (see Note 5).

3. Collect medium and cells in a 15 ml conical tube.

4. Centrifuge cells at $300 \times g$ for 1 min. Remove supernatant.

5. Wash cells by resuspending them in 1 ml of PBS maintained at 37°C.

6. Centrifuge cells at $300 \times g$ for 1 min. Remove supernatant.

7. Resuspend pelleted cells with 0.4 ml of PBS.

8. Incubate cells first for 10 min at 37°C in a 5% $CO_2$ incubator, followed by 20 min at room temperature (see Note 4).

### 3.6. LCL Imaging by Fluorescence Microscopy

Visualization of cell fluorescence is recommended prior to proceeding with FACS analysis to verify mitochondrial localization of fluorescence signal (Fig. 1). Higher dye concentrations and longer incubation times result in nonspecific nuclear labeling, which is most easily visualized in fibroblast cell lines (Fig. 2). Thus, FACS analysis

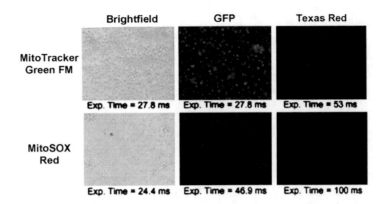

Fig. 1. Human LCL fluorescence localization with *MitoTracker Green FM* or *MitoSOX Red*. Following incubation with either *MitoTracker Green* or *MitoSOX Red*, LCLs were visualized using an Olympus phase microscope under *bright field*, a *GFP filter*, and a *Texas Red filter*. Cells showed cytoplasmic labeling and nuclear sparing, as was consistent with mitochondrial localization.

Fig. 2. Nuclear localization of fluorescent signal occurs following higher dye concentrations and/or longer exposure times. At higher dye concentrations and longer exposure times to *MitoSOX Red*, fluorescence was observed not only in the cytoplasm but also within the nucleus. This was evident in control fibroblast cell lines incubated with 10 µM MitoSOX Red for 60 min.

of cells having nuclear fluorescence should be avoided as they would not be informative for mitochondria-specific analyses (see Note 5).

1. Transfer an aliquot of washed LCLs following fluorescent dye incubation to a glass slide and cover with a cover slip.

2. Visualize LCL fluorescence following MitoSOX Red incubation with a Texas Red filter (Excitation: 560/40x, Emission: 630/75).

3. Visualize LCL fluorescence following MitoTracker Green FM incubation with a FITC/Cy2 filter (excitation: 470/40x, emission: 525/50).

**3.7. FACS Analysis of LCL Fluorescence Intensity**

1. Transfer all LCLs that remain after washing for each sample to a 12 mm×75 mm round bottom polystyrene tube.

2. Load cells into a Dual Laser Becton Dickinson Analytical FACS Calibur Flow Cytometer.

3. Obtain data using the FL1 channel for MitoTracker Green FM, the FL2 channel for TMRE, and the FL3 channel for MitoSOX Red (see Note 6).

4. Collect a total of 10,000 data events (cells) per sample (see Note 7).

5. Analyze data with desired flow cytometry analysis software. We used Cell Quest (Fig. 3) (BD Biosciences, San Jose, CA, USA) and FlowJo Software (Tree Star Inc., Ashland, OR, USA).

   1. Set the x-axis to a specific channel.

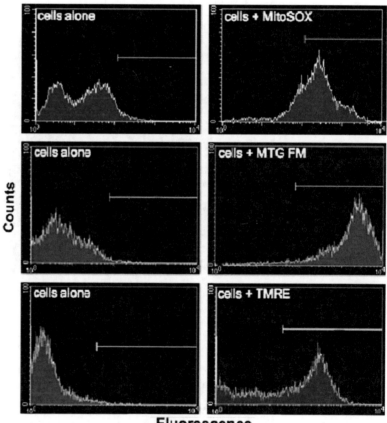

**Fluorescence**

Fig. 3. Representative histograms of LCL fluorescence data viewed in the *histogram format. Histograms* of counts versus fluorescence intensity show the clear separation in fluorescence signal between untreated LCLs and LCLs treated individually with MitoSOX Red, Mitotracker Green FM (MTG FM), or TMRE. Most control cells, which were not treated with a fluorescent dye, were excluded from the gated region. Most fluorescent dye-treated cells were captured in the gated region.

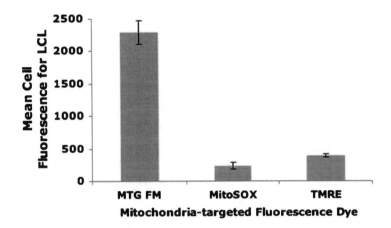

Fig. 4. Control LCL variation in mitochondria-targeted fluorescence dyes. LCLs from four control subjects were labeled with MitoTracker Green FM, MitoSOX Red, or TMRE and analyzed by FACS using the appropriate channel. *Bar height* and *error bars* indicate mean and standard deviation of fluorescence intensity for each mitochondria-targeted dye among all four LCLs.

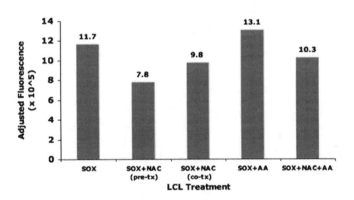

Fig. 5. Mitochondrial matrix oxidant burden of MitoSOX labeled LCLs following incubation with antimycin A and/or N-acetyl-cysteine. The baseline mitochondrial matrix oxidant burden as indicated by adjusted cell fluorescence (mean MitoSOX Red fluorescence intensity multiplied by number of cells in gated region) of LCLs that were incubated for 10 min with 6.6 μM (final concentration) MitoSOX Red was decreased with 5 mM NAC treatment for either 1 h before (preincubation) or concurrently with (coincubation) fluorescence dye exposure. Mitochondrial matrix oxidant burden increased when LCLs were concurrently treated with 5 μM AA at the time of fluorescence dye exposure. However, when LCLs were treated concurrently with both 5 μM AA and 5 mM NAC, MitoSOX fluorescence intensity was reduced to levels below baseline.

2. Set the *y*-axis to counts.

3. Gate approximately 99% of events (see Note 8).

4. Measure mean fluorescence (Fig. 4).

5. Normalize data by multiplying the mean fluorescence of gated cells by the number of cells in the gated region (Fig. 5).

## 4. Notes

1. Cell concentration was always diluted to 200,000 cells/ml so that the volume of dye and/or drug could be kept constant. Each well always contained 200,000 cells in 1 ml for our experiments.

2. Pretreatment of control LCLs with AA for 6 or 24 h prior to incubation with fluorescent dyes resulted in a progressively greater increase in MitoSOX Red fluorescence intensity than did concurrent AA treatment only during fluorescence dye incubation (data not shown).

3. No difference was seen when cells were incubated with NAC for 1 h prior to the addition of fluorescent dyes or when NAC incubation was begun concurrently with fluorescent dye incubation (Fig. 5). AA and NAC can be added together to the same culture plate well, if desired (Fig. 5).

4. LCLs should be incubated at room temperature with each fluorescent dye for 20 to 60 min. However, cell death will result if the incubation time exceeds 1 h.

5. LCLs labeled with MitoSOX Red should be imaged prior to FACS analysis to confirm diffuse cytoplasmic labeling with nuclear sparing consistent with mitochondrial localization. Cells having bright punctate nuclear labeling should not be used for FACS analysis of mitochondrial-specific parameters.

6. The FL3 channel can also be used for FACS analysis of LCLs labeled with TMRE. However, we observed better separation of labeled and unlabeled cell populations when analyzing TMRE-labeled cells in the FL2 channel.

7. Voltage compensation may need to be adjusted to obtain a scatter of data points across the y-axis. However, voltage should not be strong enough to compress treated cells close to the y-axis, which would result in data points being insufficiently separated to distinguish untreated control LCLs from fluorescent dye-treated LCLs.

8. If the observed data peak is not tight, the number of gated events can be lowered to 90%. However, it is important to keep the percentage of gated events consistent between untreated control LCLs and dye- and/or drug-treated LCLs.

## Acknowledgments

This work was funded in part by a grant from the National Institutes of Health (R03-DK082446) to M.J.F. The content is solely the responsibility of the authors and does not necessarily represent the official views of the National Institutes of Health.

## References

1. Martinez, A. O., Vigil, A., and Vila, J. C. (1986) Flow-cytometric analysis of mitochondria-associated fluorescence in young and old human fibroblasts, *Exp Cell Res 164*, 551–555.

2. Yu, D., Carroll, M., and Thomas-Tikhonenko, A. (2007) p53 status dictates responses of B lymphomas to monotherapy with proteasome inhibitors, *Blood 109*, 4936–4943.

3. Martinez-Pastor, F., Mata-Campuzano, M., Alvarez-Rodriguez, M., Alvarez, M., Anel, L., and de Paz, P. (2010) Probes and techniques for sperm evaluation by flow cytometry, *Reprod Domest Anim 45 Suppl 2*, 67–78.

4. Robinson, K. M., Janes, M. S., Pehar, M., Monette, J. S., Ross, M. F., Hagen, T. M., Murphy, M. P., and Beckman, J. S. (2006) Selective fluorescent imaging of superoxide in vivo using ethidium-based probes, *Proc Natl Acad Sci USA 103*, 15038–15043.

5. Gomes, A., Fernandes, E., and Lima, J. L. (2005) Fluorescence probes used for detection of reactive oxygen species, *J Biochem Biophys Methods 65*, 45–80.

6. Ward, M. W. (2010) Quantitative analysis of membrane potentials, *Methods Mol Biol 591*, 335–351.

7. O'Reilly, C. M., Fogarty, K. E., Drummond, R. M., Tuft, R. A., and Walsh, J. V., Jr. (2003) Quantitative analysis of spontaneous mitochondrial depolarizations, *Biophys J 85*, 3350–3357.

8. Rodriguez-Enriquez, S., Kai, Y., Maldonado, E., Currin, R. T., and Lemasters, J. J. (2009) Roles of mitophagy and the mitochondrial permeability transition in remodeling of cultured rat hepatocytes, *Autophagy 5*, 1099–1106.

9. Atkuri, K. R., Cowan, T. M., Kwan, T., Ng, A., Herzenberg, L. A., and Enns, G. M. (2009) Inherited disorders affecting mitochondrial function are associated with glutathione deficiency and hypocitrullinemia, *Proc Natl Acad Sci USA 106*, 3941–3945.

10. Dingley, S., Polyak, E., Lightfoot, R., Ostrovsky, J., Rao, M., Greco, T., Ischiropoulos, H., and Falk, M. J. (2010) Mitochondrial respiratory chain dysfunction variably increases oxidant stress in Caenorhabditis elegans, *Mitochondrion 10*, 125–136.

11. Estes, S., Coleman-Hulbert, A. L., Hicks, K. A., de Haan, G., Martha, S. R., Knapp, J. B., Smith, S. W., Stein, K. C., and Denver, D. R. (2011) Natural variation in life history and aging phenotypes is associated with mitochondrial DNA deletion frequency in Caenorhabditis briggsae, *BMC Evol Biol 11*, 11.

12. Chen, Q., Vazquez, E. J., Moghaddas, S., Hoppel, C. L., and Lesnefsky, E. J. (2003) Production of reactive oxygen species by mitochondria: central role of complex III, *J Biol Chem 278*, 36027–36031.

# Chapter 17

# Molecular Profiling of Mitochondrial Dysfunction in *Caenorhabditis elegans*

## Erzsebet Polyak, Zhe Zhang, and Marni J. Falk

### Abstract

Cellular effects of primary mitochondrial dysfunction, as well as potential mitochondrial disease therapies, can be modeled in living animals such as the microscopic nematode, *Caenorhabditis elegans*. In particular, molecular analyses can provide substantial insight into the mechanism by which genetic and/or pharmacologic manipulations alter mitochondrial function. The relative expression of individual genes across both nuclear and mitochondrial genomes, as well as relative quantitation of mitochondrial DNA content, can be readily performed by quantitative real-time PCR (qRT-PCR) analysis of *C. elegans*. Additionally, microarray expression profiling offers a powerful tool by which to survey the global genetic consequences of various causes of primary mitochondrial dysfunction and potential therapeutic interventions at both the single gene and integrated pathway level. Here, we describe detailed protocols for RNA and DNA isolation from whole animal populations in *C. elegans*, qRT-PCR analysis of both nuclear and mitochondrial genes, and global nuclear genome expression profiling using the Affymetrix GeneChip *C. elegans* Genome Array.

**Key words:** Total RNA, mitochondrial DNA, qRT-PCR, Taqman, Affymetrix GeneChip *C. elegans* Genome Array

## 1. Introduction

Primary mitochondrial disease results from mutations in either mitochondrial DNA (mtDNA)-encoded genes or in nuclear genes whose products localize to the mitochondrion (1). In addition, a host of other genetic disorders and/or pharmacologic agents can cause secondary mitochondrial dysfunction. Bidirectional nuclear-mitochondrial "cross-talk" is evident in the consistent pattern of nuclear and mitochondrial genome expression alterations that occur in the setting of mitochondrial dysfunction (2, 3). Molecular analyses offer a critical means by which to elucidate the specific

Lee-Jun C. Wong (ed.), *Mitochondrial Disorders: Biochemical and Molecular Analysis*, Methods in Molecular Biology, vol. 837,
DOI 10.1007/978-1-61779-504-6_17, © Springer Science+Business Media, LLC 2012

responses of both nuclear and mitochondrial genomes to mitochondrial dysfunction.

*Caenorhabditis elegans* offers a robust translational model animal in which to facilitate both the in vivo and in vitro study of mitochondrial disease (4). Its utility is largely rooted in the highly conserved nature of mitochondrial structure, composition, and function across evolution. Relative to wild-type (N2 Bristol) worms, a range of transgenic animals harboring mutations in nuclear-encoded respiratory chain subunits, assembly factors, enzymes, or other integral mitochondrial proteins, can be compared (5, 6). In addition, potential mitochondrial toxins and pharmacologic therapies for mitochondrial dysfunction can be systematically investigated in *C. elegans* mitochondrial mutant strains (7).

Here, we describe detailed protocols for RNA and DNA isolation from whole animal *C. elegans* populations, qRT-PCR analyses to determine the relative expression of nuclear-encoded mitochondrial genes (e.g., *SOD2* and *SOD3*) as well as relative mtDNA content (e.g., *ND4*) genes, and global nuclear genome expression profiling using the commercially available Affymetrix GeneChip *C. elegans* Genome Array. The analysis of synchronous worm populations is described to provide sufficient starting material for diverse molecular analyses, although similar methods can be applied to isolate nucleic acids from single animals. In addition, utilization of a TaqMan Gene Expression Assay for a common endogenous control gene (*drs-1*) is reported that has shown consistent performance in qRT-PCR expression analysis of primary mitochondrial dysfunction in *C. elegans* (7).

## 2. Materials

### 2.1. Total RNA Isolation from Synchronous Worm Populations

1. Nematode growth media (NGM) for worm growth plates (Bioworld) (8).
2. 10-cm Plastic culture plates.
3. OP50 *Escherichia coli* bacteria.
4. S. basal, pH 7.0: Dissolve 3.4 g of $KH_2PO_4$, 4.4 g of $K_2HPO_4$, and 5.85 g of NaCl in 800 ml of Milli-Q water. Adjust pH with NaOH and add Milli-Q water to a final volume of 1,000 ml. Sterilize by autoclaving for 30 min.
5. 1.5 ml RNase-free Eppendorf tube with tight fitting plastic pestle.
6. 15 ml Centrifuge tubes.
7. 5 M NaOH: Dissolve 200 g in approximately 800 ml of Milli-Q water and bring to a final volume of 1,000 ml.

8. Liquid bleach (Fisher Scientific) or commercially available bleach (Clorox).

9. Dimethyl sulfoxide (DMSO).

10. Kontes Pellet Pestle Cordless Motor (Fisher Scientific Inc., Pittsburgh, PA).

11. Trizol Reagent (Invitrogen Corporation, Carlsbad, CA).

12. Chloroform.

13. 70% Ethanol (EtOH).

14. RNase-free water included in RNeasy Mini Kit or Milli-Q water.

15. RNeasy Mini Kit (Qiagen, Inc., Valencia, CA).

16. NanoDrop ND-1000 Spectrophotometer.

17. Table centrifuge with capacity to accommodate 15 ml tubes at $1,300 \times g$ at room temperature.

18. Two table centrifuges to accommodate 1.7 ml Eppendorf tubes for maximum speed of $16,000 \times g$. Place one in cold room at 4°C and another one at room temperature.

19. Pharmacologic agent(s) to be studied

**2.2. DNA Isolation from Synchronous Worm Populations**

1. 10-cm NGM agar plates for worm maintenance (as per Subheading 2.1).

2. OP50 *E. coli* bacteria.

3. S. basal, pH: 7.0: Dissolve 3.4 g of $KH_2PO_4$, 4.4 g of $K_2HPO_4$, and 5.85 g of NaCl in 800 ml of Milli-Q water. Adjust pH with NaOH and add distilled water to a final volume of 1,000 ml. Sterilize by autoclaving for 30 min.

4. 15 ml Centrifuge tubes.

5. 5 M NaOH: Dissolve 200 g of 5 M NaOH in approximately 800 ml of what Milli-Q water and bring to a final volume of 1,000 ml.

6. Liquid bleach (Fisher Scientific) or commercially available bleach (Clorox).

7. Table centrifuge with capacity to accommodate 15 ml tubes at $1,300 \times g$.

8. 1.5 ml Eppendorf tubes with tight fitting plastic pestle.

9. Kontes Pellet Pestle Cordless Motor (Fisher Scientific Inc., Pittsburgh, PA).

10. QIAamp DNA Mini Kit (Qiagen, Inc., Valencia, CA).

11. NanoDrop ND-1000 Spectrophotometer.

### 2.3. qRT-PCR Analysis of Relative mtDNA Content in Synchronous Worm Populations

1. Taqman Fast Universal PCR Master mix (Applied Biosystems, Foster City, CA).

2. MicroAmp Fast Optical 96-Well Reaction Plate (Applied Biosystems, Foster City, CA).

3. Optical adhesive covers (Applied Biosystems, Foster City, CA).

4. RNase-free water or Milli-Q water.

5. TaqMan Gene Expression Assay for nuclear-encoded endogenous control gene, *drs-1* (Assay # Ce02451127_g1).

6. TaqMan Gene Expression Assay for mtDNA-encoded complex I subunit gene, *nd4* (Assay # Custom Taqman Gene Expression Assay, Applied Biosystems, Foster City, CA). Forward primer sequence: 5′GAGGCTCCTACAACAGCTAGAATAC3′. Reverse primer sequence: 5′TCATACATTGTTGTGTACAAATCTT AAACTACCT3′.

7. 7500 Fast Real-Time PCR System (Applied Biosystems, Foster City, CA).

8. 7500 Fast Real-Time PCR System Software v2.0.1 (Applied Biosystems, Foster City, CA).

### 2.4. qRT-PCR of Nuclear Gene Relative Expression in Synchronous Worm Populations

#### 2.4.1. DNase Treatment of Total RNA

1. Turbo DNA-free kit (Ambion Inc., Austin, TX).

2. Dry bath for 37°C temperature control.

3. 0.5 ml RNase-free tubes.

4. Table centrifuge (Spectrafuge 24D, Labnet).

#### 2.4.2. RT-PCR to Generate cDNA

1. High-capacity cDNA Reverse Transcription Kit (Applied Biosystems, Foster City, CA).

2. PCR tubes.

3. GeneAmp PCR system 9700 (Applied Biosystems, Foster City, CA).

#### 2.4.3. qRT-PCR to Assay Nuclear Gene Relative Expression

1. TaqMan Fast Universal PCR Master mix (Applied Biosystems, Foster City, CA).

2. MicroAmp Fast Optical 96-Well Reaction Plate (Applied Biosystems, Foster City, CA).

3. Optical adhesive covers.

4. RNase-free water or Milli-Q water.

5. TaqMan Gene Expression Assay for nuclear-encoded endogenous control gene, *drs-1* (Assay # Ce02451127_g1).

6. TaqMan Gene Expression Assay for mtDNA-encoded complex I subunit (e.g., *sod-3*, assay # Ce02404515_g1).

7. 7500 Fast Real-Time PCR System (Applied Biosystems, Foster City, CA).

8. 7500 Fast Real-Time PCR System Software v2.0.1 (Applied Biosystems, Foster City, CA).

**2.5. Affymetrix GeneChip C. elegans Genome Array Analysis**

*2.5.1. Microarray Performance*

1. GeneChip *C. elegans* Genome Array (Catalog # 900383 per five arrays, Affymetrix, Santa Clara, CA).

2. 100 ng of Total RNA from four to six biological replicates per strain/condition.

3. 2100 Bioanalyzer (Agilent Technologies, Santa Clara, CA).

*2.5.2. Microarray Platform Description*

Microarray technology permits the measurement of the expression level for thousands of transcripts simultaneously. We have previously used the GeneChip *C. elegans* Genome Array (Affymetrix, Inc.) to study the expression of most worm transcripts under different experimental conditions (2). This expression array platform measures about 22,500 known *C. elegans* transcripts. Of note, *C. elegans* Tiling 1.0R is an alternative platform that has recently been developed to interrogate the whole worm genome, which offers the benefit of being able to measure the relative expression not only of known gene transcripts but also of unknown transcripts, such as microRNAs and other small noncoding RNAs (9, 10). Both platforms use 25-base-long oligomers that are spotted on the array surface as probes to hybridize biotin-labeled cDNA generated by reverse transcription. The signal intensity of each spot after hybridization represents the abundance of cDNA, which is used to determine the expression level of the corresponding transcript.

# 3. Methods

**3.1. C. elegans Growth and Treatment**

1. *C. elegans* worm strains are described at http://www.wormbase.org and publicly available from the *Caenorhabditis* Genetics Center (University of Minnesota, Minneapolis, MN).

2. Maintain worms at 20°C on NGM agar plates spread with OP50 *E. coli* bacteria as standard nematode food source.

3. Bleach gravid adult worms to obtain sufficient eggs for purposes of generating synchronous young adult populations (8) (see Note 1).

   (a) Wash gravid worms off from NGM agar plates with 3.5 ml of sterile water into a 15 ml centrifuge tube.

   (b) Make a fresh solution of 0.5 ml 5 M NaOH and 1 ml bleach immediately prior to use.

   (c) Add bleach solution to worms in 15 ml centrifuge tube to achieve a final volume of 5 ml.

   (d) Shake tube well for several seconds and then incubate on benchtop.

(e) Repeat shaking every 2 min for a total of 10 min.

(f) Centrifuge 15 ml tube in a tabletop centrifuge for 1 min at $1,300 \times g$ to pellet the released eggs.

(g) Aspirate the supernatant to 0.1 ml.

(h) Add sterile water to 5 ml and shake well.

(i) Repeat step 3(f) and (g).

(j) Use a sterile Pasteur pipette to transfer eggs in the remaining 0.1 ml of liquid to a fresh NGM agar plate without bacteria.

4. Incubate eggs on unspread NGM agar plate (without bacteria) at 20°C overnight (see Note 2).

5. The next day, transfer L1-stage arrested animals onto fresh NGM agar plates spread with OP50 *E. coli*.

6. Maintain worms at 20°C until they have reached adult stage, as defined by the presence of eggs laid on the plate.

7. Wash worms off from NGM agar plates in 1.5 ml of S. basal, allowing adults to separate from eggs and larvae by gravity for 5 min.

8. Collect bottom layer of adults (pellet) for total RNA extraction.

### 3.2. Pharmacologic Treatment of Synchronous Young Adult or Developing Worms

*3.2.1. If Studying the Effects of a Study Drug, Add the Desired Drug or Appropriate Buffer Control to a 10-cm NGM Agar Plate After the Bacteria Lawn Has Dried and Prior to Plating Worms*

1. For control (buffer only) plates, evenly spread 10 cm bacterial-spread NGM agar plate with 500 μl of S. basal for water soluble drugs or with 20% DMSO (100 μl DMSO and 400 μl S. basal) for non-water-soluble drugs.

2. For drug treatment plate, add desired drug of interest in 500 μl total volume of S. basal or 20% DMSO, as depends on its solubility.

*3.2.2. Worms Can Be Treated with Study Drugs by Feeding Either (a) Young Adults for 24 h or (b) Developing Larvae from the L1-Arrested Stage Through the Young Adult Stage*

Pharmacologic Treatment of Young Adult Stage Worms

1. Transfer approximately 1,000 synchronous first-day young adult stage worms to NGM 10-cm plates spread with OP50 *E. coli* and desired drug concentration or buffer control (see Note 3).

2. Incubate worms on drug treatment or buffer control 10-cm NGM agar plates at 20°C for 24 h.

3. Wash worms off from NGM agar plates with 1.5 ml of S. basal into 1.5 ml Eppendorf tube.

4. Allow adults to separate from eggs and larvae by gravity for 5 min.

5. Collect bottom layer of adults (pellet) into 1.5 ml RNase-free Eppendorf tube for total RNA extraction.

**Pharmacologic Treatment of Developing Worms**

1. Transfer approximately 1,000 synchronous, L1-arrested larvae from unspread NGM agar plates (no bacteria) to drug-treated or buffer control plates (see Note 3).

2. Incubate worms on drug treatment or buffer control 10-cm NGM agar plates at 20°C for approximately 2–3 days until they reach adulthood (as defined by eggs laid on plate).

3. Wash adult worms off from NGM agar plates with 1.5 ml S. basal into 1.5 ml Eppendorf tube.

4. Allow adults to separate from eggs and larvae by gravity for 5 min.

5. Collect bottom layer of adults (pellet) in 1.5 ml RNase-free Eppendorf tube for total RNA extraction.

**3.3. Total RNA Extraction from Adult C. elegans**

1. Wash adult worm pellet with 1 ml of S. basal to remove residual bacteria.

2. Pipette off supernatant to leave worm pellet.

3. Repeat wash five to six times.

4. After the final wash, leave worm pellet with approximately 100 µl S. basal in the 1.5 ml RNase-free Eppendorf tube.

5. Add 400 µl of Trizol reagent to the worm pellet in the 1.5 ml RNase-free Eppendorf tube.

6. Grind worms for 1 min with Kontes Pellet Pestle Cordless Motor with a fitted pestle. At this point, the extraction can be stopped and the samples stored indefinitely at –80°C (see Note 4).

7. Freeze/thaw the sample by incubating first at 37°C for 10 min and then at –80°C for 10 min (see Note 5).

8. Repeat freeze/thaw cycle.

9. After the final thaw, add an additional 200 µl of Trizol reagent to the sample.

10. Incubate sample at room temperature for 5 min.

11. Add 140 µl of chloroform to the sample and shake vigorously for 15 s to mix.

12. Incubate sample at room temperature for 2 min.

13. Centrifuge sample at no more than $12,000 \times g$ for 15 min at 4°C in the cold room.

14. Remove the top aqueous phase to a fresh RNase-free 1.5 ml Eppendorf tube (see Note 6).

15. Slowly add an equal volume of 70% ethanol by pipetting.

16. The Qiagen RNeasy Mini Kit is used to finish the RNA extraction, as detailed below.

17. Transfer the mixture to a Qiagen RNeasy mini spin column.

18. Centrifuge column at maximal speed for 15 s and discard flow-through.

19. Add 700 μl of buffer RW1, centrifuge at maximal speed for 15 s, and discard flow-through.

20. Add 500 μl of buffer RPE, centrifuge at maximal speed of $16,000 \times g$ for 15 s, and discard flow-through.

21. Add an additional 500 μl of buffer RPE, centrifuge at maximal speed for 2 min, and discard flow-through.

22. Centrifuge tubes again at maximal speed for 1 min to remove any residual buffer.

23. Add 50 μl of RNase-free water to the center of the filter.

24. After 1 min, collect eluted RNA by spinning sample at maximal speed for 30 s.

25. Measure concentration and 260/280 ratio using NanoDrop spectrophotometer (see Note 7).

26. Store total RNA at –80°C until ready to use for quantitative real-time PCR or microarray expression studies.

### 3.4. DNA Isolation from Larval and Adult C. elegans

Obtain synchronous worm populations as detailed in Subheading 3.1.

### 3.4.1. Preparation of Adult Stage Worms for DNA Isolation

1. Bleach gravid adult worms (8) as described in Subheading 3.1.

2. Transfer recovered eggs to NGM plates without bacteria.

3. The following day, transfer the L1-arrested animals to OP50 *E. coli*-seeded NGM plates and leave until reach young adult stage.

4. Wash synchronous young adults off with 1.5 ml of S. basal into 1.5 ml Eppendorf tubes.

### 3.4.2. Preparation of Larval Stage Worms for DNA Isolation

1. Bleach gravid adult worms (8) as described in Subheading 3.1.

2. Transfer recovered eggs onto NGM plates without bacteria.

3. The following day, wash L1-arrested animals off from the unspread NGM agar plates with 1.5 ml of S. basal.

### 3.4.3. DNA Isolation from Adult Stage Worms

1. To isolate DNA from synchronous L1 larvae or young adult worm populations, the QIAamp DNA Mini Kit (Qiagen) is used per the recommended DNA purification from tissues protocol, with slight adaptation as detailed below for *C. elegans*.

2. Wash worms off from 10-cm NGM agar plates with 1.5 ml of S. basal into 1.5 ml Eppendorf tube.

3. Wash worms five to six times in 1 ml of S. basal in 1.5 ml Eppendorf tube to remove residual bacteria.

4. Allow worms to settle by gravity in approximately 100 μl S. basal in a 1.5 ml Eppendorf tube.

5. Grind worms for 1 min with Kontes Pellet Pestle Cordless Motor with a fitted pestle.

6. Follow the QIAamp protocol for DNA purification from tissues (Qiagen) to obtain genomic DNA as follows:

    (a) To ground worms, add 180 μl ATL lysis buffer and 20 μl of Proteinase K from the QIAamp kit.

    (b) Vortex samples.

    (c) Incubate samples for 2 h at 56°C to lyse worms, briefly vortexing twice per hour.

    (d) Follow the QIAamp protocol from step 5(b) as follows: Add 200 μl of buffer AL to samples, mix by pulse-vortexing for 15 s, then incubate at 70°C for 10 min. Briefly centrifuge the tubes to remove drops from inside the lid (taken from QIAamp DNA Mini and Blood Mini Handbook).

    (e) Add 200 μl of ethanol (96–100%) to the samples and mix by pulse-vortexing for 15 s. Briefly centrifuge the tubes to remove drops from inside the lid.

    (f) Carefully apply the mixture from above step to the QIAamp Mini spin column in a 2 ml collection tube without wetting the rim. Close the cap and centrifuge at $6,000 \times g$ (8,000 rpm) for 1 min. Place the QIAamp Mini spin column in a clean 2 ml collection tube. Discard the tube containing the filtrate.

    (g) Carefully open the QIAamp Mini spin column and add 500 μl of buffer AW1 without wetting the rim. Close the cap and centrifuge at $6,000 \times g$ (8,000 rpm) for 1 min. Place the QIAamp Mini spin column in a clean 2 ml collection tube. Discard the tube containing the filtrate.

    (h) Carefully open the QIAamp Mini spin column and add 500 μl of buffer AW2 without wetting the rim. Close the cap and centrifuge at full speed ($20,000 \times g$; 14,000 rpm or depending on the table centrifuge you have). We use Spectrafuge 24D from Labnet with maximum speed of $16,000 \times g$ for 3 min.

    (i) Place the QIAamp Mini spin column in a new 2 ml collection tube. Discard the old collection tube with the filtrate. Centrifuge at full speed for 1 min.

    (j) Place the QIAamp Mini spin column in a clean 1.5 ml microcentrifuge tube. Discard the collection tube containing the filtrate. Carefully open the QIAamp Mini spin column and add 200 μl of buffer AE. Incubate at room temperature for 1 min and then centrifuge at $6,000 \times g$ (8,000 rpm) for 1 min.

(k) Determine the genomic DNA concentration using an ND-1000 Spectrophotometer.

### 3.5. qRT-PCR Analysis of Relative mtDNA Content in C. elegans

1. Combine 40 ng of DNA per qRT-PCR reaction containing TaqMan Gene Expression Assays for the housekeeping gene, *drs-1* (Ce02451127_g1) and the mtDNA-encoded complex I subunit gene, *nd4* (custom designed TaqMan Gene Expression Assay from Applied Biosystems).

2. Add TaqMan Universal PCR Master Mix per standard Applied Biosystems protocol. Table 1 details some of the TaqMan Gene Expression Assays commonly used to study mitochondrial function in our laboratory.

3. Perform Real-time PCR on a 7500 Fast Real-Time PCR System.

4. Analyze relative gene expression using the provided software.

### 3.6. qRT-PCR Analysis of Relative Nuclear Gene Expression in C. elegans

#### 3.6.1. DNase Treatment of RNA

DNase-treat 10 μg of total RNA using TURBO DNA-free kit (Ambion) per standard protocol.

#### 3.6.2. RT-PCR to Generate cDNA from DNase-Treated RNA

Using the High-Capacity cDNA Reverse Transcription Kit and protocol from Applied Biosystems, perform RT-PCR on 1–2 μg RNA (depending on concentration) in a 20-μl reaction mixture.

#### 3.6.3. qRT-PCR

1. Use 40 ng of cDNA per qRT-PCR reaction.
   We use TaqMan Gene Expression Assays for both the housekeeping gene *drs-1* and target genes of interest (see Table 1).

2. Prepare qRT-PCR plates per standard Applied Biosystems protocol.

3. Analyze data by comparative Ct (delta-delta Ct) quantitative analysis.

### 3.7. Affymetrix GeneChip C. elegans Genome Array Analysis

#### 3.7.1. Probe Remapping

Affymetrix expression microarrays are organized by probes sets, which consist of multiple probes that map to a given transcript, where the expression level of that transcript is represented by multiple measurements from probes of the same probe set. Each Affymetrix probe set is annotated with a unique probe set identifier. However, such annotation becomes ambiguous and outdated

**Table 1**
**TaqMan gene expression assays used to study relative mtDNA content or relative expression of mitochondrial oxidative stress response**

| C. elegans gene | Gene name and function | AB assay # |
|---|---|---|
| *drs-1* | aspartyl(D) tRNA synthetase (Housekeeping gene) | Ce02451127_g1 |
| *act-3* | Actin (Housekeeping gene) | Ce02784145_s1 |
| *nd4* | ND4 (mtDNA complex I subunit) | Custom assay[a] |
| *sod-2* | SOD2 (manganese superoxide dismutase) | Ce02410777_g1 |
| *sod-3* | SOD3 (manganese superoxide dismutase) | Ce02404515_g1 |

[a]Forward primer sequence: 5′GAGGCTCCTACAACAGCTAGAATAC3′; reverse primer sequence: 5′TCATACATTGTTGTGTACAAATCTTAAACTACCT3′

over time. For example, it is not rare for the same transcript to be mapped to multiple probe sets. In addition, some probe sequences turn out to be incorrect as the worm genome becomes more accurately sequenced. Independent resources such as BRAIN-ARRAY are available that can remap and regroup original probes to provide the most up-to-date version of worm transcripts and remove probes that perform badly or have ambiguity (11). Remapping assures that each probe set is unambiguously annotated with a unique RefSeq, Entrez, or other formal gene identifier. The newest version of BRAINARRAY is what we have most recently used to map all probes on *C. elegans* Genome Array to 17,255 unique Entrez gene identifiers. Entrez gene IDs can be further associated with biological functions through gene categorization resources such as Gene Ontology (http://www.geneontology.org) and KEGG (http://www.genome.jp/kegg).

*3.7.2. Data Normalization*

Prior to analysis of relative expression levels between array groups, array data needs to be normalized to remove any systematic bias that may exist between individual arrays. Various sources of systematic bias exist, such as the amount of total RNA and the efficiency of hybridization. Robust multi-array analysis (RMA) is an algorithm commonly used to normalize and summarize array data (12). We use the RMA implemented by Bioconductor (http://bioconductor.org) to process raw, probe-level array data. The processed data is an $N \times M$ data matrix, where $N$ is the number of transcripts and $M$ is the number of arrays.

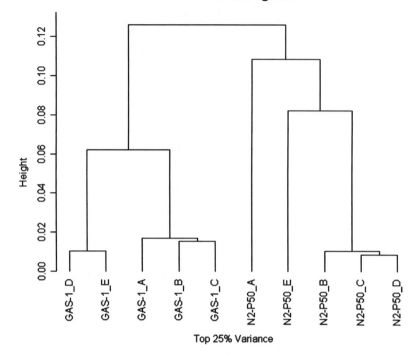

Fig. 1. Unsupervised hierarchical clustering of microarray expression data in mitochondrial complex I mutant worms. Separation of samples by an unsupervised clustering method into different groups that correlate with a known biological variable, such as mutation type, confirms that the sample groups will be distinguishable based on their gene expression profiles. Here, unsupervised clustering satisfactorily separated wild-type (N2 Bristol worms grown on OP50 *E. coli* bacteria) from complex I subunit mutant worms (*gas-1(fc21)* grown on OP50 *E. coli* bacteria) (2).

*3.7.3. Gene Expression Analysis Methods*

Both unsupervised and supervised analyses can be used to compare gene expression between sample groups. Unsupervised methods, such as hierarchical clustering and principal components analysis, evaluate the global difference of gene expression profiles without a sample grouping information. If samples in different can be separated by an unsupervised method (Fig. 1), one can conclude that sample groups will be distinguishable based on their global expression profiles. In contrast, supervised methods use the sample grouping information to identify specific genes that are differentially expressed between groups. Differential expression can be either represented by fold change or similar indices to indicate the magnitude of the group difference or by p value of a statistical test to indicate the statistical significance of the group difference. One or both types of indices can be used to sort and select differentially expressed genes between sample groups. For example, a list can be generated of all genes having at least a twofold change between groups and a p value less than 0.05. Furthermore, the percentage of false positives, called false discovery rate, can be estimated in a selected gene list.

The biological interpretation of differential expression can substantially vary for specific genes. For example, slightly altered expression of a transcription factor may dramatically alter the expression of its downstream targets. However, most group comparisons will generate hundreds to thousands of genes that have at least moderate expression changes which make it impractical for researchers to manually investigate all of their functions and relationships. Publicly available software such as Gene Set Enrichment Analysis (GSEA) allows researchers to analyze multiple genes as a set, such as genes belonging to the same KEGG metabolic pathway or Gene Ontology category (13). If most genes of a pathway analyzed by GSEA have relatively increased expression in a mutant worm strain relative to control, we can conclude that pathway is upregulated in the mutant even if the fold change of individual pathway genes may not be significant (Fig. 2). GSEA software

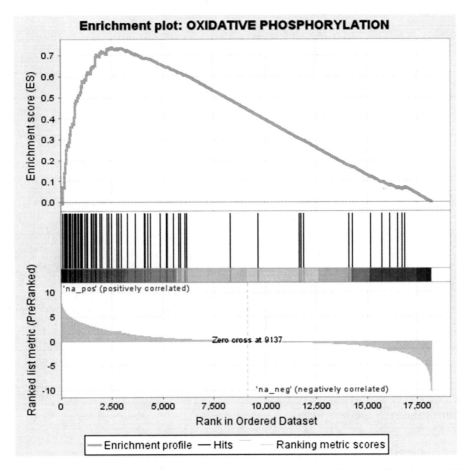

Fig. 2. Gene set enrichment analysis (GSEA) output showing relative expression of the "oxidative phosphorylation" gene cluster in mitochondrial complex I subunit (*gas-1(fc21)*) mutant worms (*left side, red highlight*) relative to wild-type N2 Bristol worms (*right side, blue highlight*) (2). *Vertical black lines* indicate the relative expression between mutant and control groups of individual gene members of the oxidative phosphorylation gene cluster, as defined in the Kyoto Encyclopedia of Genes and Genomes (KEGG). Genes that fall to the *left side* of the *green* "peak" are significantly enriched within the gene list that is upregulated in the mitochondrial mutant worms.

provides a few "standard" collections of gene sets within which genes are annotated using human gene symbols. To utilize this resource for *C. elegans*, we compile our own gene set collection of metabolic pathways from KEGG database and annotate genes with Entrez Gene IDs (2). Gene-pathway mapping information necessary to compile custom gene sets can be downloaded from ftp://ftp.genome.jp/pub/kegg/pathway/organisms/cel/cel_gene_map.tab.

## 4. Notes

1. During the bleaching process, several vigorous shaking steps are recommended. Wild-type (N2 Bristol) can tolerate this shaking, but mitochondrial mutant worms require much gentler shaking to yield viable eggs.

2. Without bacteria, worm development is arrested at the L1 larval stage. When subsequently transferred to NGM agar plates spread with OP50 *E. coli*, their development resumes simultaneously to achieve stage synchrony, as is necessary for expression analyses.

3. Estimate worm number by counting three 100-$\mu$l aliquots. Adjust worm concentration to 1,000 worms/ml of S. basal.

4. Trizol should be worked with only in a well-ventilated chemical hood.

5. Make sure that the samples are well frozen. If necessary, allow extra time for freezing. One indication that the samples are well frozen is when the sample color turns light pink.

6. When removing the top aqueous phase, carefully avoid touching the middle layer with the tip of the pipette to assure high quality of RNA.

7. For quality assurance, we also analyze RNA samples by Agilent 2100 Bioanalyzer. RNA integrity number (RIN) above 8 is acceptable (maximum RIN is 10) for downstream applications requiring nondegraded RNA, such as microarray analysis.

## Acknowledgments

This work was funded in part by grants from the National Institutes of Health (K08-DK073545, 2-P30-HD026979-21, and R01-HD065858-01A1) to M.J.F. The content is solely the responsibility of the authors and does not necessarily represent the official views of the National Institutes of Health.

## References

1. Haas, R. H., Parikh, S., Falk, M. J., Saneto, R. P., Wolf, N. I., Darin, N., and Cohen, B. H. (2007) Mitochondrial disease: a practical approach for primary care physicians, Pediatrics 120, 1326–1333.

2. Falk, M. J., Zhang, Z., Rosenjack, J. R., Nissim, I., Daikhin, E., Nissim, I., Sedensky, M. M., Yudkoff, M., and Morgan, P. G. (2008) Metabolic pathway profiling of mitochondrial respiratory chain mutants in C. elegans, Molecular genetics and metabolism 93, 388–397.

3. Peng, M., Falk, M. J., Haase, V. H., King, R., Polyak, E., Selak, M., Yudkoff, M., Hancock, W. W., Meade, R., Saiki, R., Lunceford, A. L., Clarke, C. F., and D, L. G. (2008) Primary coenzyme Q deficiency in Pdss2 mutant mice causes isolated renal disease, PLoS genetics 4, e1000061.

4. Rea, S. L., Graham, B. H., Nakamaru-Ogiso, E., Kar, A., and Falk, M. J. (2010) Bacteria, yeast, worms, and flies: exploiting simple model organisms to investigate human mitochondrial diseases, Dev Disabil Res Rev 16, 200–218.

5. Parikh, S., Saneto, R., Falk, M. J., Anselm, I., Cohen, B. H., Haas, R., and Medicine Society, T. M. (2009) A modern approach to the treatment of mitochondrial disease, Curr Treat Options Neurol 11, 414–430.

6. Rea, S. L., Ventura, N., and Johnson, T. E. (2007) Relationship between mitochondrial electron transport chain dysfunction, development, and life extension in Caenorhabditis elegans, PLoS Biol 5, e259.

7. Dingley, S., Polyak, E., Lightfoot, R., Ostrovsky, J., Rao, M., Greco, T., Ischiropoulos, H., and Falk, M. J. (2010) Mitochondrial respiratory chain dysfunction variably increases oxidant stress in Caenorhabditis elegans, Mitochondrion 10, 125–136.

8. Hope, I. A. (1999) C. elegans: A practical approach., Oxford University Press, New York.

9. Ramani, A. K., Nelson, A. C., Kapranov, P., Bell, I., Gingeras, T. R., and Fraser, A. G. (2009) High resolution transcriptome maps for wild-type and nonsense-mediated decay-defective Caenorhabditis elegans, Genome Biol 10, R101.

10. He, H., Wang, J., Liu, T., Liu, X. S., Li, T., Wang, Y., Qian, Z., Zheng, H., Zhu, X., Wu, T., Shi, B., Deng, W., Zhou, W., Skogerbo, G., and Chen, R. (2007) Mapping the C. elegans noncoding transcriptome with a whole-genome tiling microarray, Genome Res 17, 1471–1477.

11. Wang, P., Ding, F., Chiang, H., Thompson, R. C., Watson, S. J., and Meng, F. (2002) ProbeMatchDB--a web database for finding equivalent probes across microarray platforms and species, Bioinformatics 18, 488–489.

12. Irizarry, R. A., Bolstad, B. M., Collin, F., Cope, L. M., Hobbs, B., and Speed, T. P. (2003) Summaries of Affymetrix GeneChip probe level data, Nucleic Acids Res 31, e15.

13. Subramanian, A., Tamayo, P., Mootha, V. K., Mukherjee, S., Ebert, B. L., Gillette, M. A., Paulovich, A., Pomeroy, S. L., Golub, T. R., Lander, E. S., and Mesirov, J. P. (2005) Gene set enrichment analysis: a knowledge-based approach for interpreting genome-wide expression profiles, Proc Natl Acad Sci USA 102, 15545–15550.

# Part III

## Molecular Analysis of Mitochondrial Disorders

# Chapter 18

# Analysis of Common Mitochondrial DNA Mutations by Allele-Specific Oligonucleotide and Southern Blot Hybridization

**Sha Tang, Michelle C. Halberg, Kristen C. Floyd, and Jing Wang**

## Abstract

Mitochondrial disorders are clinically and genetically heterogeneous. There are a set of recurrent point mutations in the mitochondrial DNA (mtDNA) that are responsible for common mitochondrial diseases, including MELAS (mitochondrial encephalopathy, lactic acidosis, stroke-like episodes), MERRF (myoclonic epilepsy and ragged red fibers), LHON (Leber's hereditary optic neuropathy), NARP (neuropathy, ataxia, retinitis pigmentosa), and Leigh syndrome. Most of the pathogenic mtDNA point mutations are present in the heteroplasmic state, meaning that the wild-type and mutant-containing mtDNA molecules are coexisting. Clinical heterogeneity may be due to the degree of mutant load (heteroplasmy) and distribution of heteroplasmic mutations in affected tissues. Additionally, Kearns–Sayre syndrome and Pearson syndrome are caused by large mtDNA deletions. In this chapter, we describe a multiplex PCR/allele-specific oligonucleotide (ASO) hybridization method for the screening of 13 common point mutations. This method allows the detection of low percentage of mutant heteroplasmy. In addition, a nonradioactive Southern blot hybridization protocol for the analysis of mtDNA large deletions is also described.

**Key words:** Mitochondrial DNA mutation, Allele specific oligonucleotide hybridization, Southern Blot hybridization

## 1. Introduction

Mitochondrial diseases are a group of clinically and genetically heterogeneous multisystem disorders. The genes to assemble a mitochondrion are distributed between the nuclear DNA (nDNA) and mitochondrial DNA (mtDNA). It is estimated that at least 1,500 genes are required to assemble a functional mitochondrion, 37 of which are encoded by the mtDNA (1). The human mitochondrial genome, a circular double stranded DNA molecule of

Lee-Jun C. Wong (ed.), *Mitochondrial Disorders: Biochemical and Molecular Analysis*, Methods in Molecular Biology, vol. 837, DOI 10.1007/978-1-61779-504-6_18, © Springer Science+Business Media, LLC 2012

16,569 nucleotide pairs (np), encodes 13 polypeptides that are components of the electron transport chain (ETC), as well as 22 tRNAs and 2 rRNAs contributing to mitochondrial protein synthesis. The maternally inherited mtDNA is present in hundreds to thousands of copies per cell and has a very high mutation rate, in the order of 10–100 times higher than that of the nDNA. New mutations arise in cells mixed among wild-type mtDNAs (heteroplasmy) and segregate randomly during cell division (2). A variety of human diseases are directly associated with mtDNA mutations, including point mutations and large deletions. For point mutations, different percentage of mtDNA mutant can have completely distinctive clinical manifestation (3). Thus, estimation of mutant heteroplasmy levels in the relevant tissue is important in making a diagnosis (see Chapter 21 in this volume for the quantification of mutant heteroplasmy).

In this chapter, we describe the protocols for screening 13 common mtDNA point mutations by ASO (allele-specific oligonucleotide) dot blot hybridization and for the analysis of large mtDNA deletions by Southern blot (4–6). The 13 common point mutations include m.3243A > G (tRNA-Leu(UUR)) and m.3271T > C (tRNA-Leu(UUR)) for MELAS (mitochondrial encephalopathy, lactic acidosis, stroke-like episodes) (7); m.8344A > G (tRNA-Lys) and m.8356T > C (tRNA-Lys) for MERRF (myoclonic epilepsy and ragged red fibers); m.8993T > G (p.L156R, ATP6) and m.8993T > C (p.L156P, ATP6) for NARP (neuropathy, ataxia, retinitis pigmentosa) and Leigh syndromes (3); m.8363G > A (tRNA-Lys) for cardiomyopathy, MERRF, hearing loss, and Leigh syndrome (8); m.11778G > A (p.R340H, ND4), m.3460G > A (tRNA-Leu(UUR)), m.14484T > C (p.M64V, ND6), and m.14459G > A (p.A72V, ND6) for LHON (Leber's hereditary optic neuropathy) (9); and m.13513G > A (p.D393N, ND5) and m.13514A > G (p.D393G, ND5) for Leigh syndrome and MELAS (10). Large deletions in the mitochondrial genome are characteristic of Kearns–Sayre syndrome, CPEO (chronic progressive external ophthalmoplegia), Pearson marrow/pancreas syndrome, maternally inherited diabetes, and hearing loss.

## 2. Materials

### 2.1. General Guidelines

1. Prepare all solutions using ultrapure water (prepared by purifying deionized water to a sensitivity of 18 MΩ at 25°C) and analytical grade reagents.

2. Prepare and store all reagents at room temperature (unless indicated otherwise).

### 2.2. ASO Dot Blot for Point Mutation Analysis

#### 2.2.1. PCR Reagents

1. 10× buffer for Ampli*Taq*/*Taq*Gold polymerase (Applied Biosystems).
2. 5-U/μL Ampli*Taq*/*Taq*Gold polymerase (Applied Biosystems).
3. 8-mM dNTP.
4. 50-mM $MgCl_2$.
5. Sterile ultrapure water.
6. 10-μM primers listed in Table 1 for multiplex PCR.
7. 10-μM primers listed in Table 2 for the synthesis of positive controls.
8. 5-mL microtube.
9. 0.2-mL 8-strip PCR tubes with attached caps.

#### 2.2.2. Agarose Gel Electrophoresis

1. Agarose (Molecular Biology Grade).
2. 50× TAE buffer (2-M Tris–acetate, 0.05-M EDTA, pH 8.0): Combine 242-g Tris(hydroxymethyl)aminomethane, 57.1-mL acetic acid, 100-mL 0.5 M EDTA (pH 8.0), and 600-mL ultrapure water. Mix until dissolved, and use water to bring the final volume to 1 L.
3. 1× TAE buffer (40-mM Tris–acetate, 1-mM EDTA, pH 8.0): Mix 20-mL 50× TAE buffer and 980-mL water.
4. 10× loading dye: Combine 0.25-g bromophenol blue, 0.25-g xylene cyanol, and 15-g Ficoll (PM 400) in 60-mL of ultrapure water, then bring the final volume to 100 mL.
5. 2× loading dye: Diluting 1–5 of 10× loading dye with ultrapure water by mixing 1 part of 10× loading dye with four parts of ultrapure water. The volume to be prepared depends on the number of PCR products to be analyzed.
6. 0.1-μg/μL Ready-load™ 1 kb plus DNA ladder.

#### 2.2.3. ASO Probe Labeling

1. ASO probes (see Note 1): The sequences of the 20 probes are listed in Table 3. Each stock probe, received in a powder form, is prepared with ultrapure water to 100 μM. Primers are then further diluted into 10-μM and 2.5-μM working concentrations with ultrapure water. ASO cold probes for hybridization are diluted to 10 μM. The stock primers are stored at –20°C.
2. T4 polynucleotide kinase enzyme with 10× T4 kinase buffer.
3. 2.2-μM 4,500 Ci/mmol γ-$P^{32}$ ATP.

#### 2.2.4. Dot Blot and Hybridization

1. Positively charged nylon membrane.
2. 0.4-M sodium hydroxide (NaOH): Dissolve 1.6-g NaOH completely in 50-mL water, then add water to bring the final volume to 100 mL.

**Table 1**
**Primers used for the amplification of mtDNA fragments containing the corresponding point mutations**

| Primer name | Primer site | Forward (F)/ reverse (R) | Sequence (5′–3′) | Expected product size (bp) | Mutations to be analyzed |
|---|---|---|---|---|---|
| mtF3212 mtR3758 | 3,212–3,231 3,758–3,736 | F R | CACCCAAGAACAGGGTTTGT AGTAGAATGATGGCTAGGGTGAC | 546 | m.3243A>G, m.3271T>C, m.3460G>A |
| mtF8209 mtR9196 | 8,209–8,229 9,169–9,151 | F R | CATCGTCCTAGAATTAATTCC TGAAAACGTAGGCTTGGAT | 961 | m.8344A>G, m.8356T>C, m.8363G>A, m.8993T>C, m.8993T>G |
| mtF11432 mtR12170 | 11,432–11,450 12,170–12,147 | F R | ATCGCTGGGTCAATAGTAC CACAATCTGATGTTTTGGTTAAAC | 738 | m.11778G>A |
| mtF14437 mtR14686 | 14,437–14,455 14,686–14,667 | F R | AGGATACTCCTCAATAGCC CCGTGCGAGAATAATGATGT | 249 | m.14484T>C, m.14459G>A |
| mtF13317 mtR13761 | 13,317–13,337 13,761–13,741 | F R | GCACATCTGTACCACGCCTT TGCGGGGAAATGTTGTTAGT | 444 | m.13513G>A, m.13514A>G |

## Table 2
## Primers used to synthesize positive controls for the m.3271T > C, m.8356T > C, and m.13514A > G mutations

| Control | Primers | Amplicon size (bp) | Sequence (5′–3′) |
|---------|---------|--------------------|------------------|
| T3271C | T3271C (F) | 189 | ACTTAAAAC*C*TTACAGTCA |
| | mtR3758 (R) | | AGTAGAATGATGGCTAGGGTG |
| T8356C | T8356C (F) | 843 | AACACCTCT*C*TACAGTGAA |
| | mtR9169 (R) | | TGAAAACGTAGGCTTGGAT |
| A13514G | A13514G (F) | 247 | TACTCCAAAG*G*CCACATCA |
| | mtR13761 (R) | | TGCGGGGGAAATGTTGTTAGT |

The mutated nucleotide is in bold, italicized, and underlined

## Table 3
## Sequences of wild-type and mutant ASO probes

| Disease | Mutation site | S/A[a] | Sequence (5′–3′) |
|---------|---------------|--------|------------------|
| MELAS | A3243 | A | TTACCGGGCTCTGCCATCT |
| | A4343G | A | TTACCGGGCCCTGCCATCT |
| MELAS | T3271 | S | ACTTAAAACTTTACAGTCA |
| | T3271C | S | ACTTAAAACCTTACAGTCA |
| LHON | G3460 | A | GAGTTTTATGGCGTCAGCGAA |
| | G3460A | A | GAGTTTTATGGTGTCAGCGAA |
| MERRF | A8344 | A | GAGGTGTTGGTTCTCTTAAT |
| | A8344G | A | GAGGTGTTGGCTCTCTTAAT |
| MERRF | T8356 | S | AACACCTCTTTACAGTGAA |
| | T8356C | S | AACACCTCTCTACAGTGAA |
| Cardiomyopathy | G8363 | S | TTTACAGTGAAATGCCCC |
| | G8363A | S | TTTACAGTAAAATGCCCC |
| NARP | T8993 | S | CAATAGCCCTGGCCGTACG |
| | T8993C | S | CAATAGCCCCGGCCGTACG |
| | T8993G | S | CAATAGCCCGGGCCGTACG |
| LHON | G11778 | A | ATTATGATGCGACTGTGAG |
| | G11778A | A | ATTATGATGTGACTGTGAG |
| LHON | G14459 | S | ATAGCCATCGCTGTAGTATA |
| | G14459A | S | ATAGCCATCACTGTAGTATA |
| LHON | T14484 | S | AGACAACCATCATTCCC |
| | T14484C | S | AGACAACCACCATTCCC |
| Leigh | G13513 | S | TACTCCAAAGACCACATCA |
| MELAS | G13513A | S | TACTCCAAAAACCACATCA |
| | A13514G | S | TACTCCAAAGGCCACATCA |

[a]S and A indicate if the probe is from the sense (S) or antisense (A) strand

3. 0.2-M Tris–HCl, pH 7.4: Combine 121 g of Tris(hydroxymethyl) aminomethane and 600-mL water. Use concentrated HCl to adjust pH to 7.4 and then add water to a final volume of 1 L.

4. UV Stratalinker 2400 (Stratagene).

5. 20% sodium $n$-dodecyl sulfate (SDS): Dissolve 20 g of SDS into 80 mL of water by stirring. Add water until final volume is 100 mL.

6. Church hybridization buffer (1-mM EDTA, 0.5-M $Na_2HPO_4$, 7% SDS; see Note 2): First, prepare 1-M disodium phosphate ($Na_2HPO_4$, pH 7.2) by combining 71-g sodium phosphate dibasic anhydrous with 250-mL ultrapure water. Slowly add 9 mL of 85% $H_3PO_4$ until dissolved and bring final volume to 500 mL with ultrapure water. Then combine 250-mL 1-M $Na_2HPO_4$ with 1 mL of 0.5-M EDTA, pH 8.0 and 175-mL 20% SDS. Bring the final volume to 500 mL with ultrapure water.

7. 10-mg/mL salmon sperm (SSp) DNA.

8. 15-mL graduated, conical-bottom tubes with blue screw caps.

9. 20× SSC buffer (3-M NaCl, 0.3-M sodium citrate, pH 7.4): Dissolve 175-g NaCl and 88.2-g sodium citrate in 600-mL water. Adjust pH to 7.4 using HCl or NaOH. Then bring the final volume to 1 L with water.

10. 5× SSC buffer (0.75-M NaCl, 0.075-M sodium citrate, pH 7.4): To make 1-L 5× SSC, combine 250-mL 20× SSC with 750-mL ultrapure water.

11. 2× SSC buffer (0.3-M NaCl, 0.03-M sodium citrate, pH 7.4): To make 1-L 2× SSC wash solution, combine 100-mL 20× SSC with 900-mL ultrapure water.

12. CL-XPosure film, 8 × 10 in.

13. Film autoradiography cassettes with intensifying screens.

14. Kodak X-OMAT 2000 X-ray film processor (or other autoradiography film processor).

### 2.3. Southern Blot Hybridization for Large mtDNA Deletion Analysis

*2.3.1. Restriction Digestion and Agarose Gel Electrophoresis*

1. 10-U/μL *Hin*dIII with 10× buffer 2 (New England Bio labs, store at –20°C).

2. 10-U/μL *Eag*I with 10× buffer 3 (New England Bio labs, store at –20°C).

3. 10-μg/mL DIG-labeled molecular weight marker II (Roche, store at –20°C).

**2.3.2. Gel Treatment and Transferring Reagents**

1. 0.25-M HCl: Combine 980-mL water and 20 mL of 12.1-M concentrated HCl.
2. 5-M NaCl: Dissolve 292-g NaCl in 600-mL water. Then bring the final volume to 1 L with water.
3. 1.5-M NaCl, 0.5-M NaOH: Combine 50 mL of 10-M NaOH, 300 mL of 5-M NaCl, and 650-mL water.
4. 0.5-M Tris–HCl, 1.5-M NaCl, pH 8.0: Combine 500-mL 1.0-M Tris–HCl, pH 8.0, 300-mL 5-M NaCl, and 200-mL water.
5. Positively charged nylon membrane.
6. Scott multifold paper towels.
7. 3-MM chromatography paper.
8. 1-kg weight.

**2.3.3. Synthesis of DIG-Labeled Probes for Southern Blot by PCR**

1. PCR primers: The sequences of the 13 pairs of primers are listed in Table 4. Mix 100-$\mu$M primer stock solutions with sterile ultrapure water. Primers are then further diluted to 10 $\mu$M with sterile ultrapure water and stored at $-20^\circ$C.
2. PCR DIG probe synthesis kit (Roche, Catalog No. 11 636 090 910).

**2.3.4. Hybridization and Wash Reagents**

1. Hybridization tubes.
2. 50-mL graduated, conical-bottom tubes with screw caps.
3. Hybridization oven.
4. DIG Easy Hyb (Roche, Catalog number: 11603558001).
5. 2× SSC/0.1% SDS (0.3-M NaCl, 0.03-M sodium citrate, 0.1% SDS, pH 7.4): Combine 100-mL 20× SSC, 5-mL 20% SDS, and 895-mL ultrapure water.
6. 0.5× SSC/0.1% SDS (75-mM NaCl, 7.5-mM sodium citrate, 0.1% SDS, pH 7.4): Combine 25-mL 20× SSC, 5-mL 20% SDS, and 970-mL ultrapure water.
7. DIG wash and block buffer set (Roche, Catalog number: 11585762001): #1 Washing buffer 10×, #2 Maleic acid buffer 10×, #3 Blocking solution 10×, #4 Detection buffer 10×.

**2.3.5. DIG Luminescent Detection Materials**

1. DIG luminescent detection kit (Roche, Catalog Number: 11 363 514 910) with 150-mU/mL anti-DIG-AP antibody, blocking reagents, and CSPD.
2. CL-XPosure Film, 8 × 10 in.
3. Development Folders.
4. Development Cassettes.

**Table 4**
**Sequences for primers used for the synthesis of probes for Southern blot hybridization**

| Primer pair | Primer name | Primer site | Forward (F)/ reverse (R) | Sequence (5'–3') | Expected product size (bp) |
|---|---|---|---|---|---|
| 1 | mtF1351 mtR3135 | 1,351–13,568 3,135–3,118 | F R | GCA AGA AAT GGG CTA CAT TGT CCT TTC GTA CAG GGA | 1,785 |
| 2 | mtF3085 mtR4917 | 3,085–3,102 4,917–4,897 | F R | ATC CAG GTC GGT TTC TAT GCT TAC GTT TAG TGA GGG A | 1,842 |
| 3 | mtF4881 mtR6016 | 4,881–4,899 6,016–5,996 | F R | CCC ATC TCA ATC ATA TAC C CGA ATA AGG AGG CTT AGA G | 1,135 |
| 4 | mtF5960 mtR7282 | 5,960–5,977 7,282–7,262 | F R | CCT ATT ATT CGG CGC ATG GAA TGA GCC TAC AGA TGA T | 1,322 |
| 5 | mtF7234 mtR8600 | 7,234–7,251 8,600–8,580 | F R | CCG ATG CAT ACA CCA CAT AGA ATG ATC AGT ACT GCG G | 1,366 |
| 6 | mtF8295 mtR9169 | 8,295–8,314 9,169–9,149 | F R | CAC TGT AAA GCT AAC TTA GC TGA AAA CGT AGG CTT GGA T | 874 |
| 7 | mtF9104 mtR10629 | 9,104–9,125 10,629–10,609 | F R | TCA CAA TTC TAA TTC TAC TGA C GCA CAA TAT TGG CTA AGA G | 1,206 |
| 8 | mtF10551 mtR11757 | 10,551–10,568 11,757–11,735 | F R | TCC TCC CTA CTA TGC CTA TTT GAG TTT GCT AGG CAG AAT | 1,206 |
| 9 | mtF11688 mtR13086 | 11,688–11,705 13,086–13,065 | F R | CCG GCG CAG TCA TTC TCA TTC CTG CTA CAA CTA TAG TG | 1,398 |
| 10 | mtF12949 mtR13738 | 12,949–12,966 13,738–13,718 | F R | AAC GCT AAT CCA AGC CTC TGA GAA ATC CTG CGA ATA G | 789 |
| 11 | mtF13695 mtR15185 | 13,695–13,712 15,185–15,166 | F R | CAT TAA ACG CCT GGC AGC GGC GGA TAG TAA GTT TGT | 1,490 |
| 12 | mtF15119 mtR16543 | 15,119–15,137 16,543–16,524 | F R | GCA ACA GCC TTC ATA GGC T CGT GTG GGC TAT TTA GGC | 1,424 |
| 13 | mtF16411 mtR1424 | 16,411–16,429 1,424–1,405 | F R | CGT GAA ATC AAT ATC CCG C ATC CAC CTT CGA CCC TTA | 1,583 |

# 3. Methods

Carry out all procedures at room temperature, unless otherwise indicated.

### 3.1. ASO Dot Blot Hybridization for Point Mutation Analysis

*3.1.1. Multiplex PCR to Amplify the Five mtDNA Fragments Containing the 13 Common Point Mutations*

1. Dilute extracted total genomic DNA sample to a final concentration of 50 ng/µL using DNA hydration buffer.

2. Using 10-µM working concentrations of ASO PCR primers listed in Table 1, assemble a MasterMix containing all reagents except DNA samples (Table 5) in a 5-mL microtube on ice. Make sufficient amount of MasterMix for the number of samples being tested. Vortex gently and aliquot 59 µL of MasterMix into each PCR reaction tube; add 1 µL of 50-ng/µL total genomic DNA sample to be tested to each well (see Note 3).

3. Seal the tubes and run PCR under the following conditions:

   Stage 1, 1 cycle: 95°C, 5 min

   Stage 2, 36 cycles: 95°C, 45 s; 55°C, 45 s; 72°C, 2 min

   Stage 3, 1 cycle: 72°C, 5 min

   Stage 4, 1 cycle: 4°C, hold

   PCR product can be stored at 4°C overnight or at −20°C until ready to be used.

4. Genomic DNA and/or synthetic controls for each point mutation: Known patient-positive controls from previous runs are used as quality controls. For rare mutations without authentic patient controls, such as m.3271T>C, m.8356T>C, and

## Table 5
## Components of the multiplex PCR

| Reagent | 1× Reaction volume (µL) | Final concentration |
|---|---|---|
| Sterile ultrapure water | 35.4 | |
| *Taq* polymerase 10× buffer | 6 | 1× |
| dNTP (8 mM) | 2.4 | 0.32 mM |
| MgCl$_2$ (50 mM) | 3 | 2.5 mM |
| The 10 forward and reverse primers (listed in Table 1, 10 µM) | 1.2 µL each (12 µL in total) | 0.2 µM each |
| *Taq* polymerase (5 U/µL) | 0.24 | 0.02 U/µL |
| Genomic DNA (50 ng/µL) | 1 | 0.83 ng/µL |
| Total volume | 60 | |

**Table 6**
**Components for PCR synthesis of positive controls for m.3271T > C, m.8356T > C, and m.13514A > G**

| Reagent | 1 Reaction volume (µL) | MasterMix for 3 reactions | Final concentration |
|---|---|---|---|
| Normal control genomic DNA (50 ng/µL) | 2 | 6 | 2 ng/µL |
| 10× FastStart buffer | 5 | 15 | 1× |
| dNTP (8 mM) | 1 | 3 | 0.16 mM |
| 5× GC-Rich PCR reaction buffer | 10 | 30 | 1× |
| Forward primer (listed in Table 2, 2.5 µM) | 4 | | 0.2 µM |
| Reverse primer (listed in Table 2, 2.5 µM) | 4 | | 0.2 µM |
| FastStart Taq polymerase (5 U/µL) | 0.4 | 1.2 | 0.04 U/µL |
| Sterile ultrapure water | 23.6 | 70.8 | |
| Total volume | 50 | | |

m.13514A > G, synthetic positive controls are made by PCR (11). The PCR setup is listed in Table 6. Aliquot 42-µL MasterMix without primer sets into 0.2-mL 8-strip PCR tubes. Each primer set is then added separately to the appropriate tube. Attached caps to the tubes are used to seal the reaction before placing in the PCR machine under the following conditions:

Stage 1, 1 cycle: 95°C, 5 min

Stage 2, 35 cycles: 95°C, 45 s; 55°C, 45 s; 72°C, 1 min

Stage 3, 1 cycle: 72°C, 5 min

Stage 4, 1 cycle: 4°C, hold

The resulting PCR products are ready for blotting and can be kept frozen at −20°C until ready for use.

5. Verification of successful multiplex PCR amplification by agarose gel electrophoresis: Follow the standard protocol for agarose gel electrophoresis. Add 2 µL of each PCR product to 2 µL of 2× loading dye before loading and run the PCR products on 1.5% agarose gel in 1× TAE. Use 1 kb plus DNA markers for sizing. To ensure proper spacing of all five PCR fragments, run the gel for 30 min at 150 V (Fig. 1).

*3.1.2. Dot Blot*

1. Prepare 20 blots named as listed in Table 7. Use positively charged nylon membrane and draw circles with a fine-point pen as a guide for dotting PCR product (see Note 4). On each membrane, draw circles for each PCR product, a normal

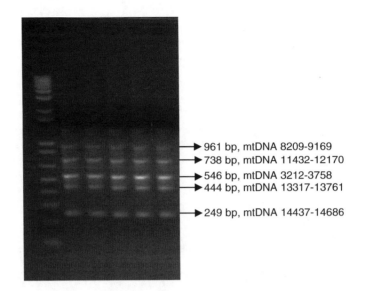

Fig. 1. Verification of successful multiplex PCR amplification by 1.5% agarose gel electro-phoresis.

control, two patient-positive controls specific to that mutation, and an NTC control (Fig. 2).

2. Dot 2 μL of PCR product onto each circle using a multichannel pipette. Use a single channel pipette to dot two specific patient-positive controls (PC1 and PC2) to their respective position (Fig. 2) (see Note 5).

3. After dotting all samples, let the membranes air-dry for at least 15 min.

4. In a small tray, soak membranes in 0.4-M NaOH for 10 min to denature.

5. Discard excess 0.4-M NaOH in proper biohazard container and rinse membranes with distilled water.

6. Neutralize membranes in 0.2-M Tris–HCl, pH 7.4 for 10 min.

7. Using UV Stratalinker 2400 for cross-linking the denatured PCR product to the membrane at 1,200-μJ for 30 s.

*3.1.3. Synthesis of Probes (Note 6)*

1. Make a MasterMix for the 24 labeling reactions according to Table 8.

2. Aliquot 2.75-μL MasterMix into separate 8-strip PCR tubes, then add 1 μL of 2.5-μM specific ASO probe (Table 3) into the corresponding tube.

3. Run labeling reaction with the following conditions:

   Stage 1, 1 cycle: 37°C, 40 min

   Stage 2, 1 cycle: 65°C, 10 min

   Stage 3, 1 cycle: 4°C, hold

**Table 7**
**Amounts of probes used for hybridization of the 20 blots used to analyze the 13 common mtDNA point mutations**

| | Blot name | Cold probe (10 μM) | Hybridization amount | Hot probe (2 μM) | Hybridization amount |
|---|---|---|---|---|---|
| 1 | m.3243nl | A3243G | 5 μL | 3243nl | 2.5 μL |
| 2 | m.3243A>G | 3243nl | 10 μL | A3243G | 2.5 μL |
| 3 | m.3271T>C | 3271nl | 2.5 μL | T3271C | 5 μL |
| 4 | m.3460G>A | 3460nl | 10 μL | G3460A | 2.5 μL |
| 5 | m.8344nl | A8344G | 5 μL | 8344nl | 2.5 μL |
| 6 | m.8344A>G | 8344nl | 5 μL | A8344G | 2.5 μL |
| 7 | m.8356T>C | 8356nl | 2.5 μL | T8356C | 5 μL |
| 8 | m.8363G>A | 8363nl | 2.5 μL | G8363A | 6 μL |
| 9 | m.8993nl | T8993C, T8993G | 2 μL | 8993nl | 8 μL |
| 10 | m.8993T>C | 8993nl, T8993G | 2 μL | T8993C | 8 μL |
| 11 | m.8993T>G | 8993nl, T8993C | 2 μL | T8993G | 8 μL |
| 12 | m.11778nl | G11778A | 5 μL | 11778nl | 2.5 μL |
| 13 | m.11778G>A | 11778nl | 5 μL | G11778A | 2.5 μL |
| 14 | m.14459nl | G14459A | 2.5 μL | 14459nl | 5.5 μL |
| 15 | m.14459G>A | 14459nl | 2.5 μL | G14459A | 5.5 μL |
| 16 | m.14484nl | T14484C | 5 μL | 14484nl | 2.5 μL |
| 17 | m.14484T>C | 14484nl | 5 μL | T14484C | 2.5 μL |
| 18 | m.13513nl | G13513A, A13514G | 5 μL | 13513nl | 2.5 μL |
| 19 | m.13513G>A | 13513nl, A13514G | 5 μL | G13513A | 2.5 μL |
| 20 | m.13514A>G | 13513nl, G13513A | 5 μL | A13514G | 2.5 μL |

4. Add 33.75-μL ultrapure water to each reaction to bring the total final volume to 37.5 μL. Hot ASO probes can be stored at −20°C for up to 4 weeks until ready to use.

*3.1.4. Pre-hybridization and Hybridization*

1. Place each membrane into a labeled 15-mL conical tube, rolled up lengthwise with the dot blots facing inward (see Note 7).

2. Pre-warm 50 mL of pre-aliquoted Church buffer and sheared 10-mg/mL salmon sperm (SSp) DNA to 65°C±1°C until the Church buffer is homogenous and the SSp is completely thawed.

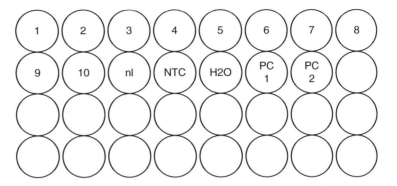

Fig. 2. Layout of the dot blots. "nl" stands for normal control. NTC is no-template control. PC1 and PC2 are two different patients with the mutation.

## Table 8
## Components of the probe labeling reactions

| Reagent | 1 Reaction volume (μL) | MasterMix for 25 reactions(μl) | Final concentration |
|---|---|---|---|
| Specific ASO primer, 2.5 μM | 1.0 | | 0.67 μM |
| Ultrapure water | 1.25 | 31.25 | |
| 10× T4 kinase buffer | 0.375 | 9.375 | 1× |
| 10× T4 kinase enzyme | 0.375 | 9.375 | 1× |
| γ-P$^{32}$ ATP | 0.75 | 18.75 | |
| Total volume | 3.75 | | |

3. Add 500-μL 10-mg/mL SSp to 50 mL of Church buffer to make a final concentration of 100 μg/mL.

4. Invert several times to mix and aliquot approximately 2 mL to each conical tube containing the ASO membranes (see Note 8).

5. Pre-hybridize the membranes in a hybridization oven at 65°C±1°C for 20 min.

6. Thaw, vortex, and spin down both the 10-μM cold ASO probes and the 2.5-μM hot ASO probes (see Note 9).

7. Following Table 7, add cold ASO probes and then hot probes to the appropriate tube (see Note 10).

8. Hybridize at 65°C±1°C for 20 min and then turn the temperature down to 34°C±1°C. Leave the membranes rotating for at least 3 h to overnight.

1. Empty the radioactive hybridization solution out of the conical tubes carefully. In order to remove the residual probes, fill each tube with 5× SSC, rinse the membrane, and pour off the 5× SSC. Wash the membrane with 5× SSC for another time and pour off the 5× SSC.

2. Fill the tube containing the membrane with 2× SSC with the volume depending on the age of the probes being used. If the probe's half-life has not been reached yet, fill the tube ½–¾ full (6–9 mL) with 2× SSC. After 2 weeks, however, only fill the tubes ¼–½ full (3–6 mL) to prevent overwashing.

3. Seal the tubes with 2× SSC and place them in the hybridization oven at 37°C±1°C for 20 min.

4. Lay out the membranes facedown on smoothed-out plastic wrap. Align the blots hybridized to the wild-type probe and corresponding mutant probes side by side. Place an old piece of film (or other form of support) on top, tighten the excess plastic wrap around the support to keep the membranes in place, and flip the constructed item so the membranes appear faceup and with only the plastic wrap covering their surface.

5. In a dedicated dark room and using film cassettes with image enhancers, place the supported membranes inside the cassettes, making sure that the film is sandwiched between the plastic-covered membranes and the image enhancer.

6. To keep any light from penetrating the film during exposure, place the cassettes in an opaque film guard bag and store them at –80°C±10°C.

7. A short exposure is done after approximately 6 h, as well as a longer exposure that lasts at least 16 h to overnight.

8. DNA samples containing homoplasmic wild-type sequence should produce a positive signal when hybridized to the normal probe and show no signal with mutant probe (lanes 3–5, 8, and 11 in Fig. 3). DNA samples containing heteroplasmic mutations will produce positive signals with both the wild-type and the mutant probes (lanes 1–2, 6–7). The synthetic control homoplasmic for the mutant only hybridizes to the mutant probe (lane 10).

### 3.2. Southern Blot Hybridization for the Detection of Large mtDNA Deletions

*3.2.1. Restriction Enzyme Digestion*

1. Dilute genomic DNA from blood sample or muscle specimen to 500 ng/μL and 50 ng/μL, individually (see Note 11).

2. Set up *Hind*III and *Eag*I digestions according to Table 9. Make a MasterMix for all the components except for genomic DNA for *Hind*III and *Eag*I, respectively. Aliquot 35 μL into each tube and add 5-μL diluted genomic DNA. Use 2.5-μg total genomic DNA for blood samples and 0.25-μg total genomic DNA for muscle samples. The total volume for restriction enzyme digestion is 40 μL (Table 9). Incubate at 37°C for 5 h or overnight.

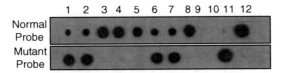

Fig. 3. Example of allele-specific oligonucleotide dot blot hybridization results. *Lanes 1* and *2*: heteroplasmic mutant controls; *lanes 3–5* and *8*: samples with wild-type mtDNA; *lanes 6–7*: samples with heteroplasmic mutant; *lane 9*: non-template control; *lane 10*: homoplasmic mutant control; *lane 11*: homoplasmic wild-type control; *lane 12*: water control.

## Table 9
## *Hin*dIII and *Eag*I digestions of total genomic DNA

| Reagent | Volume (μL) | Reagent | Volume (μL) |
|---|---|---|---|
| [a]Total genomic DNA | 5 | [a]Total genomic DNA | 5 |
| 10× NEB buffer 2 | 4 | 10× NEB buffer 2 | 4 |
| *Hin*dIII (20 U/μL) | 1 | *Eag*I (10 U/μL) | 1 |
| WFI quality sterile water | 30 | WFI quality sterile water | 30 |
| Total volume | 40 | Total volume | 40 |

[a] Total amount of DNA used in the digestion depends on the tissue type. For blood samples, a total of 2.5-μg DNA is used (5 μL of 500 ng/μL); for muscle samples, 0.25-μg DNA is used (5 μL of 50 ng/μL)

*3.2.2. Agarose Gel Electrophoresis*

1. Spin down the DNA digests and add 4.5 μL of 10× loading dye.
2. Load the DNA digests on to 0.8% agarose gel in 1× TAE containing 0.4 μg/mL of ethidium bromide.
3. Use 5-μL 10-ng/μL DIG-labeled molecular weight marker II mixture as molecular weight markers.
4. Run the gel overnight at ~47 V in 1× TAE buffer.
5. Stop electrophoresis when the front dye is about 3–4 cm from the end of the gel.
6. Photograph the gel under UV light. Label the picture clearly.

*3.2.3. Agarose Gel Transfer and DNA Fixation*

1. Depurinate the DNA in the gel by rocking it gently in 0.25-M HCl for 10 min.
2. Pour off the 0.25-M HCl and rinse the gel with distilled water.
3. Soak the gel in denaturation solution (1.5-M NaCl, 0.5-M NaOH) for 30 min to denature the double-stranded DNA.
4. Pour off the denaturation solution and rinse the gel with distilled water, followed by renaturation with neutralization solution (0.5-M Tris–HCl, 1.5-M NaCl, pH 8.0) for 15 min.
5. Repeat the neutralization step, (Subheading 3.2.3, step 4).

6. Pour off renaturation solution. Soak the gel in 20× SSC for 15 min twice.

7. Cut out one piece of positively charged nylon membrane and two pieces of 3-MM chromatography paper according to the gel size and label with a fine-point pen with quick-drying ink.

8. Rinse the membrane and the filter paper with distilled water briefly.

9. Soak the membrane and filter paper with 20× SSC for 5 min.

10. Using a tray as secondary containment, lay a piece of plexi-glass horizontally across. Fill the bottom of the tray with 20×SSC. Lay a large sheet of filter paper over the plexi-glass so that the paper touches the bottom of the tray.

11. Wet the filter paper with 20× SSC. Smooth out any bubbles by gently rolling the 10-mL pipette horizontally on top of the gel.

12. Lay the gel upside down on top of the wetted filter paper. Smooth out the gel to remove all bubbles.

13. Place the membrane on top of the gel and lay the two wetted pieces of 3-MM filter paper on top of the membrane.

14. Place a stack of paper towels (6–10 cm) on top and then a 1-kg weight on top and make sure it is level (Fig. 4).

15. Let the DNA transfer from the gel to the positively charged nylon membrane for at least 6 h to overnight.

16. Wash the membrane with 2× SSC for 10 min.

17. Place wet membranes in UV cross-linker. Set the UV cross-linker at 1,200 μJ for 30 s. The membrane is ready for pre-hybridization, Subheading 3.2.5.

*3.2.4. Synthesis of DIG-Labeled Probes*

1. Dilute the total genomic DNA from a normal control to 40 ng/μL with ultrapure water.

2. Make a MasterMix on ice of all the reagents except the primers for the 13 labeling reactions according to Table 10.

3. Aliquot 48 μL into each PCR tube and then add 1 μL each for corresponding forward and reverse primers (Table 4).

4. PCR conditions:

Stage 1, 1 cycle: 95°C, 2 min

Stage 2, 35 cycles: 95°C, 30 s; 56°C, 30 s; 72°C, 2 min

Stage 3, 1 cycle: 72°C, 7 min

Stage 4, 1 cycle: 4°C, hold

*3.2.5. Hybridization and Wash*

1. Place the cross-linked membrane from Subheading 3.2.3, step 17 in the hybridization tube. Add 25-mL DIG Easy Hyb and pre-hybridize in the hybridization oven at 42°C±1°C for 30 min.

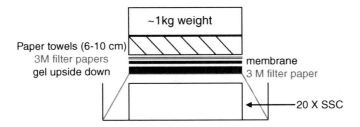

Fig. 4. Capillary transfer setup.

## Table 10
## Components of Southern blot probe labeling reactions

| Reagent (PCR DIG probe synthesis kit) | 1 Reaction volume (µL) | MasterMix for 14 reactions (µL) | Final concentration |
|---|---|---|---|
| 10× PCR buffer (with Mg²⁺) (vial 3) | 5 | 70 | 1× |
| PCR DIG label mix (vial 2) | 2.5 | 35 | |
| 2-mM dNTP (vial 4) | 2.5 | 35 | 0.1 mM |
| Forward primer (10 µM, listed in Table 4) | 1 | | 0.2 mM |
| Reverse primer (10 µM, listed in Table 4) | 1 | | 0.2 mM |
| Enzyme mix (3.5 U/µL, vial 1) | 0.75 | 10.5 | 0.0525 U/µL |
| Genomic DNA (40 ng/µL) | 1 | 14 | 0.8 ng/µL |
| Sterile ultrapure water | 36.25 | 507.5 | |
| Total volume | 50 | | |

2. In a 1.5-mL microfuge tube, add 100 µL of DIG Easy Hyb containing 2-µL DIG-labeled probe mixture.

3. Denature the mixture by boiling for 5 min and immediately place on ice for 5 min.

4. In a new 50-mL conical tube, add 25-mL DIG Easy Hyb and then the denatured DIG-labeled probe mixture to make hybridization solution. Mix well but avoid foaming.

5. Pour off the pre-hybridization solution from the hybridization tube and add in the hybridization solution. Hybridize at 42°C±1° C for 16 h or overnight (see Note 12).

6. Pour off the hybridization solution and add 100-mL 2× SSC/0.1% SDS in the hybridization tube. Wash for 5 min at room temperature.

7. Repeat the 2× SSC/0.1% SDS wash for 5 min at room temperature.

8. Wash for 15 min in 100-mL high-stringency buffer (0.5× SSC/0.1% SDS) in a tray at 57.5°C±1°C twice.

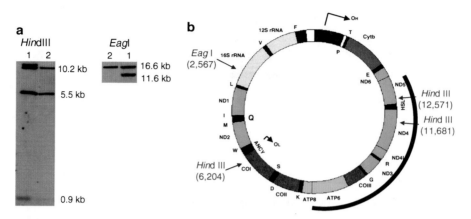

Fig. 5. Example of Southern blot hybridization results. (**a**) *Lane 1*: sample with a heteroplasmic deletion of ~5 kb in the mitochondrial genome; *lane 2*: wild-type control. (**b**) *Hind*III has three recognition sites in the mtDNA and *Eag*I linearizes the mitochondrial genome. The common ~5 kb deletion abolishes two *Hind*III sites.

*3.2.6. DIG Illuminescence Detection*

1. Rinse the membrane in 50–200-mL wash buffer (0.1-M maleic acid, 0.15-M NaCl, 0.3% Tween 20, pH 7.5) for 5 min.

2. Incubate the membrane in 50–200-mL blocking solution in a tray for 2–3 h at room temperature.

3. Add 4 μL of 150-mU/mL anti-DIG-AP into 20-mL blocking solution to make antibody solution. Mix well and pour to a small tray. Remove the membrane from the blocking solution and put it into the tray containing the antibody solution.

4. Incubate the membrane in antibody solution for 30 min at room temperature.

5. Remove the membrane from the antibody solution and put it into a tray containing 200-mL washing buffer. Wash for 15 min at room temperature.

6. Pour off the washing buffer and pour in 200-mL fresh washing buffer. Wash the membrane for another 15 min at room temperature.

7. Equilibrate the membrane for 2–5 min in 50–100-mL detection buffer (0.1-M Tris–HCl, pH 9.5, 0.1-M NaCl). Incubate at room temperature for 5 min.

8. Squeeze out excess liquid and seal the edges of the development folder.

9. Incubate the damp membrane for 15 min at 37°C±1°C.

10. Expose to X-ray film for 10–30 min at room temperature.

11. Results are depicted in Fig. 5 (see Note 13). *Hind*III has three cleavage sites in the mitochondrial genome, thus generating three *Hind*III fragments for wild-type mtDNA: 10.2, 5.5, and 0.9 kb. *Eag*I has only one cutting site in the mitochondrial

genome. It linearizes the mitochondrial genome to form a 16.6-kb band corresponding to the wild-type. MtDNA heteroplasmic for the common ~5 kb large deletion (spanning two *Hin*dIII recognition sites) is expected to show an extra band of 11.6 kb in both *Hin*dIII and *Eag*I blots.

## 4. Notes

1. Optimal ASO probes should be 15–21 nucleotides in length with the point mutation of interest localized approximately in the middle of the probe. If the nucleotide is a G in the mutant, an antisense probe should be made to reduce the nonspecific base pairing.

2. Church buffer may precipitate at room temperature. Warm the Church buffer to 60°C or until precipitations completely dissolved prior to use.

3. If total genomic DNA, which contains both nDNA and mtDNA, is used as the template in the PCR amplification, 50 ng should be utilized. If the starting material is plasmid, much less template should be used. In addition to each clinical sample being tested, be sure to include all positive controls for the 13 mutations being screened, as well as a known normal control and a no-template control (NTC), which only contains the MasterMix. Positive controls should undergo PCR with every ASO batch to ensure no degradation of PCR product, and NTC ensures that no contamination is present in the PCR reagents.

4. The ink used should be quick-drying or the lines will rinse away during washing.

5. Two positive patient controls, one with high and the other with low heteroplasmy levels for the mutant, should be included in each run. A strong signal for the one with high mutant load indicates an overall successful run. A visible dark dot for the one with low heteroplasmy level of the mutation ensures the sensitivity for the run.

6. When working with γ-P$^{32}$ ATP, it is important to follow basic radiation safety protocol and to work behind a Plexiglas shield in a radiation-certified containment area.

7. This prevents the face of the membrane from sticking to the walls of the tube, which can cause uneven hybridization and washing.

8. With the cap on tap the tube against a hard surface to open the rolled membrane up as far as it can: this will help maintain even hybridization and washing.

9. Normal protocol recommends the addition of a standard volume of 5 µL cold probe and 2.5 µL of hot probe to each 2-mL hybridization solution. Experimental results show that some mutations have stronger signals than others. To obtain even signal intensity of different probes, different amount of each probe may be used (Table 7). If the resulting signal is too strong, which may result in high background, addition of more cold probes can help minimize the background. If the signal is weak, addition of more hot probes can enhance the signal. $\gamma$-P$^{32}$ ATP has a half-life of approximately 14 days. By the third and fourth week of using the same probes, the overall signal may become weak, and thus, more hot probes may be used.

10. When pipetting, make sure to guide the tip down to the residual hybridization solution at the bottom of the tube: if the tip touches the membrane it can create uneven distribution of probes leading to uneven signal intensity throughout the membrane.

11. Different amounts of total genomic DNA are used in this analysis depending on the tissue type since mtDNA content varies among different tissues. We use 2.5-µg total genomic DNA for blood samples and 0.25-µg total genomic DNA for muscle samples.

12. Southern blot hybridization employing human DNA probe is generally carried out at 65°C in 10× SSC buffer with no formamide or 42°C with 50% formamide. The DIG Easy Hyb contains high concentration of urea to lower the hybridization temperature as though it contains 50% formamide. Therefore, the hybridization step is carried out at 42°C.

13. Southern blot hybridization is a low-cost and sensitive way to detect mtDNA deletions. However, this methodology cannot precisely map the breakpoints for the deletion. Alternatively, mtDNA deletions can be analyzed on oligonucleotide array comparative genomic hybridization (aCGH) with the MitoMet® platform to determine the location, size, and heteroplasmy level of the mtDNA deletion (12).

## References

1. Wallace DC. Mitochondria as chi. Genetics. 2008 Jun;179(2):727–35.

2. Wallace DC. Why do we still have a maternally inherited mitochondrial DNA? Insights from evolutionary medicine. Annu Rev Biochem. 2007;76:781–821.

3. Tatuch Y, Christodoulou J, Feigenbaum A, Clarke JT, Wherret J, Smith C, et al. Heteroplasmic mtDNA mutation (T—G) at 8993 can cause Leigh disease when the percentage of abnormal mtDNA is high. Am J Hum Genet. 1992 Apr;50(4):852–8.

4. Wong LJ. Comprehensive molecular diagnosis of mitochondrial disorders: qualitative and quantitative approach. Ann N Y Acad Sci. 2004 Apr;1011:246–58.

5. Wong LJ, Senadheera D. Direct detection of multiple point mutations in mitochondrial DNA. Clin Chem. 1997 Oct;43(10):1857–61.

6. Wong LJ, Cobb BR, Chen TJ. Molecular analysis of mitochondrial DNA point mutations by polymerase chain reaction. Methods Mol Biol. 2006;336:135–43.

7. Ciafaloni E, Ricci E, Shanske S, Moraes CT, Silvestri G, Hirano M, et al. MELAS: clinical features, biochemistry, and molecular genetics. Ann Neurol. 1992 Apr;31(4):391–8.

8. Shtilbans A, Shanske S, Goodman S, Sue CM, Bruno C, Johnson TL, et al. G8363A mutation in the mitochondrial DNA transfer ribonucleic acidLys gene: another cause of Leigh syndrome. J Child Neurol. 2000 Nov;15(11): 759–61.

9. Brown MD, Trounce IA, Jun AS, Allen JC, Wallace DC. Functional analysis of lymphoblast and cybrid mitochondria containing the 3460, 11778, or 14484 Leber's hereditary optic neuropathy mitochondrial DNA mutation. J Biol Chem. 2000 Dec 22;275(51):39831–6.

10. Brautbar A, Wang J, Abdenur JE, Chang RC, Thomas JA, Grebe TA, et al. The mitochondrial 13513 G>A mutation is associated with Leigh disease phenotypes independent of complex I deficiency in muscle. Mol Genet Metab. 2008 Aug;94(4):485–90.

11. Liang MH, Johnson DR, Wong LJ. Preparation and validation of PCR-generated positive controls for diagnostic dot blotting. Clin Chem. 1998 Jul;44(7):1578–9.

12. Chinault AC, Shaw CA, Brundage EK, Tang LY, Wong LJ. Application of dual-genome oligonucleotide array-based comparative genomic hybridization to the molecular diagnosis of mitochondrial DNA deletion and depletion syndromes. Genet Med. 2009 Jul;11(7):518–26.

# Chapter 19

# Sequence Analysis of the Whole Mitochondrial Genome and Nuclear Genes Causing Mitochondrial Disorders

**Megan L. Landsverk, Megan E. Cornwell, and Meagan E. Palculict**

## Abstract

The diagnosis of mitochondrial disorders has increased considerably over the past few years. However, the genetics are complex, as the causative mutations can be in either the mitochondrial or the nuclear genome. Identification of the molecular defects in the causative genes is the key to a definitive diagnosis of the disease. Here, we describe PCR-based sequence analysis of the entire mitochondrial genome and a group of nuclear genes known to cause mitochondrial disorders.

**Key words:** Mitochondrial Whole Genome, Mitochondrial DNA Depletion Syndrome, PCR, Sequence Analysis

## 1. Introduction

Mitochondrial disorders are a highly heterogeneous group of diseases. Since mitochondria are the energy-producing organelles of the cell, virtually all organ systems may be affected if there is a mitochondrial defect. However, tissues with a high energy demand such as the brain and muscle are more susceptible. Depending on the mutation and the disease-causing gene, mitochondrial disorders may be autosomal recessive, autosomal dominant, X-linked, or maternally inherited. Identification of disease-causing mutations and how they are inherited is important for proper patient management and to facilitate genetic testing of at-risk relatives (1, 2).

The human mitochondrial DNA is a small, circular, double-stranded molecule 16.6 kb in size. It contains a total of 37 genes encoding two ribosomal RNAs (rRNAs), 22 transfer RNAs (tRNAs), and 13 polypeptides which encode components of the respiratory chain/oxidative phosphorylation system (3). Cells typically contain hundreds to thousands of copies of mitochondrial

Lee-Jun C. Wong (ed.), *Mitochondrial Disorders: Biochemical and Molecular Analysis*, Methods in Molecular Biology, vol. 837, DOI 10.1007/978-1-61779-504-6_19, © Springer Science+Business Media, LLC 2012

DNA (mtDNA). Therefore, both normal and mutated mtDNA may coexist in a particular tissue, a condition known as heteroplasmy. Some mitochondrial mutations are more common than others and are associated with particular clinical phenotypes (2, 4). These mutations can be quickly and easily screened for using techniques other than sequence analysis (see Chapter 18 "Analysis of Common Mitochondrial DNA Mutations by Allele specific Oligonucleotide and Southern Blot Hybridization"). However, if the screening test is negative and there is a clear indication of maternal inheritance, sequence analysis of the entire mitochondria genome is warranted (2). Here, we describe sequence analysis of the entire mitochondrial genome in a clinical laboratory setting.

While some mutations in mtDNA are clearly linked to mitochondrial disease, the vast majority of defects causing mitochondrial deficiencies are found in nuclear genes. An estimated 1,500 nuclear-encoded proteins are targeted to the mitochondria (5). However, mutations have been reported in less than 200 of these genes (2). Identification of these changes is usually performed by PCR amplification of the coding regions of candidate genes followed by Sanger sequencing analysis. The selection of candidate nuclear genes for diagnostic molecular testing is generally based on the patient's clinical presentation and history (1). For some disorders, a group of candidate nuclear genes may be tested. For example, the amount of mtDNA in an affected tissue can be quickly determined using qPCR methods (see Chapter 22 "Measurement of Mitochondrial DNA Copy Number"). If there is a suggestion of a mtDNA depletion, a group of relevant nuclear genes can be tested first. Mitochondrial respiratory chain enzyme analysis (ETC) can also give an indication of pertinent nuclear genes to be analyzed (see Chapter 4 "Biochemical Analyses of Electron Transport Chain Complexes by Spectrophotometry"). Here, we discuss a group of genes that are most commonly responsible for the mtDNA depletion syndrome and are frequently sequenced together in a clinically available panel (*POLG, DGUOK, TK2,* and *SUCLA2*). Because panels such as this one are often analyzed simultaneously for a single individual, it is optimal to have PCR primers that amplify products of a similar size under the same conditions and use the same master mix. However, conditions may have to be adjusted due to the presence of repetitive DNA regions, GC-rich regions, or copy number variations. Therefore, we also describe steps taken to minimize nonspecific primer amplification due to repetitive DNA sequences, to avoid the presence of single nucleotide polymorphisms (SNPs) in primer regions, and to increase PCR amplification in GC-rich regions.

## 2. Materials

### 2.1. PCR

1. Extracted genomic DNA from patient tissue samples (6).

2. FastStart Taq DNA polymerase 5 U/µl kit: FastStart Taq polymerase 5 U/µl, 10× PCR reaction buffer + 20 mM MgCl$_2$, 10×PCR reaction buffer with no MgCl$_2$, 25 mM MgCl$_2$, 5× GC-rich solution (Roche Applied Science, Indianapolis, IN, USA) (see Note 1).

3. 8 mM Deoxynucleotides (dNTPs).

4. PCR primers. Primers for mitochondrial whole genome sequence analysis are listed in Table 1. Primers for the mitochondrial DNA depletion panel are listed in Table 2.

5. TE buffer: 10 mM Tris-HCl pH 7.5, 1 mM EDTA pH 8.0.

6. DNA hydration buffer (Qiagen Inc., Valencia, CA, USA).

7. Sterile water.

8. 96-Well PCR plates and strip caps.

9. PCR thermocycler.

### 2.2. Agarose Gel

1. Agarose.

2. Agarose gel electrophoresis buffer (1× TAE): 40 mM Tris–acetate, 1 mM EDTA pH 8.0.

3. Ethidium bromide (EtBr) 1% solution.

4. Horizontal electrophoresis apparatus with chambers and combs.

5. Microwave oven.

6. DNA size standard: 1 kb plus DNA ladder at 1 µg/µl (Invitrogen, Carlsbad, CA, USA).

7. Power supply device.

8. 2× Agarose gel loading dye: 0.125% bromophenol blue, 0.125% xylene cyanol, and 7.5% Ficoll 400.

### 2.3. Big Dye

1. ABI prism Big Dye terminator cycle sequencing version 3.1 kit: Big Dye® terminator v3.1 ready reaction mix, M13 (-21) primer, pGEM control DNA, 5× sequencing buffer (Applied Biosystems, Foster City, CA, USA).

2. Standard M13 forward and reverse sequencing primers at 2.5 µM stock concentration. M13F: 5′-TGT AAA ACG ACG GCC AGT-3′ M13R: 5′-CAG GAA ACA GCT ATG ACC-3′.

### 2.4. PCR Product and Big Dye Reaction Purification

1. ExcelaPure 96-well UF PCR purification plates and Performa DTR V3 96-well Big Dye purification plates (Edge Biosystems, Gaithersburg, MD, USA) (see Note 2).

**Table 1**
**Primer sequences for whole mitochondrial genome sequencing**

| Primer set | Forward primer | Sequence (5′–3′) | Reverse primer | Sequence (5′–3′) | PCR product size (bp) | Master mix | PCR program |
|---|---|---|---|---|---|---|---|
| 1 | F467 | CCCATACTACTAATCTCATC | R1173 | GGTCCTTGAGTTTTAAGCT | 706 | FS | 55 |
| 2 | F1095 | TAGCCCTAAACCTCAACAGT | R1895 | GCTTTGGCTCTCCTTGCAA | 800 | FS | 55 |
| 3 | F1819 | TAATATAGCAAGGACTAACCC | R2623 | TCCCTATTTAAGGAACAAGTG | 813 | FS | 55 |
| 4 | F2551 | GTGACACATGTTTAACGGCC | R3353 | GCGATTAGAATGGGTACAATG | 802 | FS | 55 |
| 5 | F3268 | AACTTTACAGTCAGAGGTTC | R4085 | ACAAAATATGTTGTGTAGAGTTC | 817 | FS | 55 |
| 6 | F4013 | CCCTCACCACTACAATCTT | R4821 | ACCTCTGGGACTCAGAAGT | 808 | FS | 55 |
| 7 | F4707 | ACTCTCCGGACAATGAAC | R5629 | GATTTGCGTTCAGTTGATG | 922 | FS | 55 |
| 8 | F5460 | ACACTCATCGCCCTTACCA | R6194 | TATGCGGGGAAACGCCATAT | 734 | FS | 55 |
| 9 | F6061 | TCTACAACGTTATCGTCACAG | R6783 | TAAATATATGGTGTGCTCACTC | 722 | FS | 55 |
| 10 | F6698 | GAACCATTTGGATACATAGGTA | R7510 | AAGTCATGGAGGCCATGGG | 812 | FS | 55 |
| 11 | F7411 | ACACATTCGAAGAACCCGTA | R8218 | TAGGACGATGGGCATGAAAC | 808 | FS | 55 |
| 12 | F8128 | AACCACTTTCACCGCTACACG | R8966 | ATGGTTTCGATAATAACTAGTATG | 839 | FS | 55 |

| 13 | F8865 | GATTATAGGCTTTCGCTCTAA | R9698 | AAGCAGTGCTTGAATTATTTGG | 834 | FS | 55 |
| 14 | F9603 | ACACATCCGTATTACTCGCA | R10411 | TATACCAATTCGGTTCAGTCT | 808 | FS | 55 |
| 15 | F10341 | ATCATCATCCTAGCCCTAAG | R11150 | CCAAGGTGGGGATAAGTG | 809 | FS | 55 |
| 16 | F11040 | TCTACCTCTCTATACTAATCTC | R11757 | TTTGAGTTTGCTAGGCAGAAT | 718 | FS | 55 |
| 17 | F11608 | ATCGGTCATTGCATACTCTTC | R12360 | GGTTATAGTAGTGTGCATGG | 752 | FS | 55 |
| 18 | F12239 | ACTCATGCCTCCATGTCTA | R13028 | GGGTGGAGACCTAATTGGG | 789 | FS | 55 |
| 19 | F12949 | AACGCTAATCCAAGCCTC | R13761 | TGCCGGGGAAATGTTGTTAG | 812 | FS | 55 |
| 20 | F13650 | CACCCTTACTAACATTAAACGAA | R14504 | TAATTTATTTAGGGGAATGATG | 854 | FS | 55 |
| 21 | F14437 | AGGATACTCCTCAATAGCC | R15236 | TTCATTGAACTAGGTCTGTCC | 799 | FS | 55 |
| 22 | F15132 | TAGGCTATGTCCTCCCGTG | R15964 | TTTCTCTGATTTGTCCTTGG | 813 | FS | 55 |
| 23 | F15843 | TACCAACTATCTCCCTAATTGA | R32 | TGAGTGGTTAATAGGGTGATAGA | 758 | FS | 55 |
| 24 | F16472 | GGGTAGCTAAAGTGAACTG | R610 | AGTGTATTGCTTTGAGGAGG | 683 | FS | 55 |

**Table 2**
**Primer sequences for genes in mitochondrial DNA depletion panel**

| Amplicon | Forward (F) reverse (R) | Sequence (5′–3′) | PCR product size (bp) | Master mix | PCR program |
|---|---|---|---|---|---|
| POLG_Ex1a | F | ACAGGACGTGTCTCTCC | 517 | FS_GC | R55 |
| | R | GTGCTGGTCCAGGTTGTCC | | | |
| POLG_Ex1b | F | GCAAGTGCTATCCTCGGAG | 605 | FS_GC | R55 |
| | R | CACATCAGCGCTCCCTACG | | | |
| POLG_Ex1c | F | GGAATGATGATTATGGATACACCT | 557 | FS_GC | R55 |
| | R | CTTCTGCAGGTGCTCGAC | | | |
| POLG_Ex1d | F | GCTGCACGAGCAAATCTTC | 447 | FS_GC | R55 |
| | R | CCTGCTTATGTCCCCAACC | | | |
| POLG_Ex2 | F | GTAGCTGTTTGAGTTAGGAGC | 494 | FS_GC | R55 |
| | R | CACAGCTGGTCAACAGATGC | | | |
| POLG_Ex3 | F | CAAGCATGAACAAGCATGAGG | 470 | FS_GC | R55 |
| | R | AGTCCCAGGATGAGATCTGG | | | |
| POLG_Ex4–5 | F | CAGAAGTCCCAGAGGAAAGC | 640 | FS_GC | R55 |
| | R | AGCCTAGAAAAGCTAAGGTCC | | | |
| POLG_Ex6–7a | F | AGCTGTGCCATGTCAGTGG | 707 | FS_GC | R55 |
| | R | GGAAGACAATCAGGAGCAGGA | | | |
| POLG_Ex6–7b | F | CAGCTCAGGGATTGGGC | 572 | FS_GC | R55 |
| | R | CTCAATCACAGGACCTTCCC | | | |
| POLG_Ex8a | F | AAGGATTGCTCCAGCCTTCT | 358 | FS_GC | R55 |
| | R | AGTGTGACTGAATGGCAGCA | | | |
| POLG_Ex8b | F | CCTGCTCCTGATTGTCTTCC | 220 | FS_GC | R55 |
| | R | GTCCTGAGAATGGGAGCAAGG | | | |

| | F/R | Sequence | | | |
|---|---|---|---|---|---|
| POLG_Ex9 | F | GGGTGGGACATTGTGA | 413 | FS_GC | R55 |
| | R | ACCCAAACTCTTTCCACTAGC | | | |
| POLG_Ex10-11 | F | GAATTGTGGAAGGCACTAGC | 583 | FS_GC | R55 |
| | R | CTGGCTGGGAAGAACTAGG | | | |
| POLG_Ex12 | F | CTGATGACGACAGTTTCAGG | 302 | FS_GC | R55 |
| | R | GAAAGTCTCAGGTGTGTCAC | | | |
| POLG_Ex13 | F | GGGCTCAGTGTTGGGAGG | 353 | FS_GC | R55 |
| | R | GACCCAGGCTGGCCCTCTGTGGGA | | | |
| POLG_Ex14-15 | F | TGGAGTATAGCAGTCCTGG | 518 | FS_GC | R55 |
| | R | ACTAGAGTCCTGCCTGACC | | | |
| POLG_Ex16-17a | F | GCCATCCCCTCAGGAAAGG | 653 | FS_GC | R55 |
| | R | GGACAGTAAAGCAGGCCTCG | | | |
| POLG_Ex16-17b | F | GTGGCCATCTCTGGAACTGT | 740 | FS_GC | R55 |
| | R | TGGGCAGGAGATAGAACAGA | | | |
| POLG_Ex18-19 | F | TGGAAGTGATATGTGAACATTCC | 594 | FS_GC | R55 |
| | R | CTACAAACATTGGTAAGGTCC | | | |
| POLG_Ex20 | F | CAGAGTGAAGCTTCTCTTGG | 389 | FS_GC | R55 |
| | R | GTCAAAACTGACCAGTCTGG | | | |
| POLG_Ex21 | F | ACAGTGCTGGACCTTCACC | 487 | FS_GC | R55 |
| | R | ACATTCACTCTGGACACAGG | | | |
| POLG_Ex22a | F | CATGTAACATTACCGTTCGTGG | 464 | FS_GC | R55 |
| | R | TTTACCCAACAAGCAACAATGC | | | |
| POLG_Ex22b | F | TTCGTGGCAATTGTTCTCAA | 324 | FS_GC | R55 |
| | R | TGAAAAATGGCTGGCCTTAG | | | |
| TK2_E1 | F | TCAGTGCTGGTATTATTTGC | 610 | F_GC | 55 |
| | R | CTCCATCCCAGAACCAAAGC | | | |

(continued)

**Table 2**
**(continued)**

| Amplicon | Forward (F) reverse (R) | Sequence (5'–3') | PCR product size (bp) | Master mix | PCR program |
|---|---|---|---|---|---|
| TK2_E2 | F | CCAGGGAGTGAGCATAAAC | 316 | F_GC | 55 |
| | R | GAAATTGGTAACAGAGGTGG | | | |
| TK2_E3 | F | TGAAAGTCTTGGACTGAACC | 322 | F_GC | 55 |
| | R | AAATTACACCTGTGGCTTGC | | | |
| TK2_E4 | F | GATAGAGTTGTGACCTCACG | 351 | F_GC | 55 |
| | R | TTCTCCAACTCAGTTAAGAGC | | | |
| TK2_E5 | F | CCAACCCTGCCTGTGTAGG | 525 | F_GC | 55 |
| | R | GATGCTTCCGGCTGCTGGTC | | | |
| TK2_E6 | F | ATTCTCATATTCAGCAAAGTCC | 349 | F_GC | 55 |
| | R | CTGCCATGAGGATTCGTGG | | | |
| TK2_E7 | F | TGGTACCCTTGGTGACTCC | 305 | F_GC | 55 |
| | R | AAGTGCCTCACCCACTGC | | | |
| TK2_E8 | F | GAGCTGTGCAGACCTTCC | 321 | F_GC | 55 |
| | R | TCACCCTCGATACACTTGG | | | |
| TK2_E9 | F | CGCTGTCCACTTGAGCCAG | 376 | F_GC | 55 |
| | R | ACCAGCCTTCCCACCTTCC | | | |
| TK2_E10 | F | GCTTCCATCTGTCTGTCTCC | 314 | F_GC | 55 |
| | R | TGAGACCATTAGGAAAATCAAG | | | |
| SUCLA2_E1 | F | CGCCAGGGAGCTGGTCTA | 407 | F_GC | 65–55 |
| | R | CAGCACTCCCAGGCAAGTC | | | |
| SUCLA2_E2 | F | GCTGTAATTCAGGTACTCTG | 408 | FS | 65–55 |
| | R | GCTAAAAGTTGTTATCAAACC | | | |

| | | | | | |
|---|---|---|---|---|---|
| SUCLA2_E3-4 | F | CTATGTCTGGCATGCTGCTC | 627 | FS | 65–55 |
| | R | GCTCATGACATACAAACAGA | | | |
| SUCLA2_E5 | F | ATGTGGGTAGAGAGTAATAG | 321 | FS | 65–55 |
| | R | CTTCAGTTGTACGGTTATCT | | | |
| SUCLA2_E6a | F | GACTCATGTTTATCATTTAAG | 354 | FS | 65–55 |
| | R | CTACATAGGTAGAAGTTTAAC | | | |
| SUCLA2_E6b | F | GGTAGCTGCTATTTCAGTAATGTTTG | 466 | FS | 65–55 |
| | R | GGGTCTAATAGTGTTTTGTTTTAATGG | | | |
| SUCLA2_E7-8 | F | GGGCAGTTTTAATTCCCAAA | 975 | FS | 65–55 |
| | R | AAAATCTACACTGGCCACACAA | | | |
| SUCLA2_E9 | F | TCATGAAGCACATATACTAG | 281 | FS | 65–55 |
| | R | TAACAAACTGAACTCCATG | | | |
| SUCLA2_E10 | F | TGGAAGTCTCCTCTGTTGAG | 291 | FS | 65–55 |
| | R | GCTGAAACACAAATTATTAGCTG | | | |
| SUCLA2_E11a | F | GAGTGTACATTAATTTTGCAC | 229 | FS | 65–55 |
| | R | GAACAATAACACAGAAACACAG | | | |
| SUCLA2_E11b | F | CAGAATAGTCCAGATGTTTAAATGG | 300 | FS | 65–55 |
| | R | GATGGCAATTACAATCTCCACA | | | |
| DGK_E1 | F | ACGATCTCCTTACGTCAACG | 403 | FS | 55 |
| | R | AGATACACAGTGCACTTTGC | | | |
| DGK_E2 | F | TAGCAGCCTTCTCCTTCAGC | 273 | FS | 55 |
| | R | CAAACTGGGCTACTTTACTCACTG | | | |
| DGK_E3a | F | AGGTAGGGGTGTGTGTGGAG | 347 | FS | 55 |
| | R | GCAGGATCTGTGCAAGATC | | | |
| DGK_E3b | F | CCTTTTGTGGAAGGAAAC | 424 | FS | 55 |
| | R | GTGACCATAGCAGTGCAGGA | | | |

(continued)

**Table 2**
**(continued)**

| Amplicon | Forward (F) reverse (R) | Sequence (5'–3') | PCR product size (bp) | Master mix | PCR program |
|---|---|---|---|---|---|
| DGK_E4 | F | CTCTCTCGTGCCTTTCATTCC | 454 | FS | 55 |
|        | R | GTTCCTACATCAACAGCAATC |     |    |    |
| DGK_E5 | F | AAAGGCATGGCTTGTAATGC | 348 | FS | 55 |
|        | R | TCCAGTATATTCTGAACCTTCC |     |    |    |
| DGK_E6 | F | AGTCATGTTGAATTTAGATCTGT | 363 | FS | 55 |
|        | R | ATCTGCACAAGATTCCATAGG |     |    |    |
| DGK_E7 | F | GAGACTACTGGATTCTATTCC | 376 | FS | 55 |
|        | R | GTCTAACAATAATTAACAAACC |     |    |    |

2. 96-Well optical reaction plates.

3. Sterile water.

4. Centrifuge capable of holding 96-well plates.

5. 96-Well plate shaker.

6. Clear polymer seals for 96-well plates.

**2.5. Sequencing and Analysis**

1. ABI 3730XL DNA analyzer.

2. Sequence analysis software such as Mutation Surveyor, DNA variant analysis software (Soft Genetics, State College, PA, USA).

# 3. Methods

**3.1. Primer Design for Nuclear Genes**

1. Identify your gene of interest by entering the name into the UCSC genome browser (http://genome.ucsc.edu/). If there is more than one transcript to choose from, select the NM number that corresponds to your protein of interest. Click on the transcript of interest and under "Links to sequence" click on genomic sequences. Hit submit using default conditions and save the results as a word file. Check the "Mask Repeats" box and highlight "to N" and save the results in a second word file.

2. Using ENSEMBL genome browser (http://www.ensembl.org/Homo_sapiens/Info/Index), identify your gene/transcript of interest and ensure that it matches the one chosen in UCSC genome browser. Click on the transcript of interest and under transcript-based displays click on "Exons" to get the sequence of each exon of the gene of interest. Once the individual exons of the transcript of interest are clear, click on the Gene tab at the top of the page, and under gene-based displays, click on "sequence." Make sure that sequence variations are marked. If they are not, click on "Configure this page," and under "show variations," highlight "yes and show links." Save the ENSEMBL results as a word file. Save each exon ±200–300 bp on either side individually for further analysis.

3. Before primer design, check for SNPs in the region surrounding each exon from the ENSEMBL file. In the world file, mark SNPs near the exon with < >. Compare the ENSEMBL file with the results from "mask repeats" in step 1 above and mark the regions resulting in "N" with < >. Set the PCR target boundaries at ±60–70 bp on either side of the target exon and mark using [ ] ensuring that the target PCR product is between 200 and 600 bp in size (see Note 3). Copy and paste these results into Primer3 (http://frodo.wi.mit.edu/primer3/). Using default conditions, click on "pick primers." A list of complementary primers that can be used to PCR the exon of

interest should appear (see Note 4). For each exon of an entire gene, chose primer pairs that are close to the same melting temperature (Tm) to ensure that all of the exons will amplify under similar PCR conditions. Repetitive regions, such as homopolymer tracks, will affect sequence quality. Therefore, if they lie in the region of interest, chose a second primer pair to ensure clean 2× coverage of the exon of interest (see Note 5).

4. In the UCSC genome browser, click on "PCR," enter the top choice forward, and reverse primers from Primer3 in step 3 for in silica PCR. The generated output should be a single PCR product between 200 and 600 bp in size. If not, pick an alternate primer pair or change the target sequence in step 3 and repeat.

5. If only one PCR product is generated, copy that PCR product and paste into "Blat" under UCSC genome browser and hit submit. Ensure that there is only one result with the expected PCR product size, and less than ten additional regions with matching genomic sequence (see Note 6).

6. Once all of the above conditions are met, mark the primer sequences in the word file containing all of the exons and save.

7. For universal M13 sequencing, add the M13 forward and reverse primer sequences listed in Subheading 2.3 item 2 to the 5′ end of the forward and reverse primers respectively.

*3.2. PCR Setup*

1. Suspend primer oligonucleotides in TE buffer to a 100-μM stock solution (see Note 7, and Tables 1 and 2).

2. Make 2.5 μM working solution by adding 5 μl of the 100 μM forward primer stock and 5 μl of the 100 μM reverse primer into 190 μl of TE buffer (see Note 8).

3. Dilute the DNA samples to be used for mitochondrial whole genome analysis to 2 ng/μl and those to be used for nuclear gene analysis to 50 ng/μl with DNA hydration buffer (see Note 9).

4. Prepare 1.1× volume of master mix based on conditions and number of amplicons specific to the gene(s) of interest (Tables 1–3), keeping reagents on ice. Include a no template control (NTC) for each amplicon (see Note 10).

5. To each tube, or well of a 96-well plate, add 1 μl of DNA, 4 μl of working primer solutions, and 20 μl of master mix. Add only working primer stocks and master mix to NTC tubes/wells, no DNA.

6. Quick spin (30 s) the tubes or plate at room temperature in a centrifuge at 14,000×*g* and ensure that there are minimal bubbles in samples. If there are many bubbles, place back into the centrifuge and spin for an additional 1–2 min.

## Table 3
**Master mix for mitochondrial DNA whole genome, _DGUOK_, and _SUCLA2_ amplicons 2–11 and _POLG_, _TK2_, and _SUCLA2_ amplicon 1**

| Components | Volume (μl) | Final conc. |
|---|---|---|
| _Master mix for mitochondrial DNA whole genome, DGUOK, and SUCLA2 amplicons 2–11_ | | |
| Sterile water | 16.8 | |
| 10× FastStart buffer (Roche) | 2.5 | 1× |
| dNTP (8 mM) | 0.5 | 0.16 mM |
| FastStart enzyme (5 U/ul) (Roche) | 0.2 | 0.04 U/μl |
| Mast mix total volume | 20 | |
| Forward and reverse primer mix (2 μl of 2.5 mM ea) | 4 | 0.2 mM ea |
| DNA template (50 ng/μl) | 1 | 2 ng/μl |
| Total reaction volume | 25 | |
| _Master mix for POLG, TK2, and SUCLA2 amplicon 1_ | | |
| Sterile water | 11.8 | |
| 10× FastStart buffer (Roche) | 2.5 | 1× |
| 5× GC-rich solution (Roche) | 5 | 1× |
| dNTP (8 mM) | 0.5 | 0.16 mM |
| FastStart enzyme (5 U/μl) (Roche) | 0.2 | 0.04 U/μl |
| Mast mix total volume | 20 | |
| Forward and reverse primer mix (2 μl of 2.5 mM ea) | 4 | 0.2 mM ea |
| DNA template (50 ng/μl) | 1 | 2 ng/μl |
| Total reaction volume | 25 | |
| Sterile water | 11.8 | |

_3.3. PCR Amplification_

1. Place the tubes or plate into a thermocycler machine and program PCR conditions based on size of PCR products and primer melting temperatures. Use the appropriate thermocycling conditions for the targeted genes (see Note 11 and Tables 1, 2, and 4).

_3.4. Analysis of PCR Product on Agarose Gel_

1. Clean the appropriate casting tray, place on a flat surface, and ensure that it is level (see Note 12). Place appropriate sized comb(s) in tray, teeth down (see Note 13). Ensure that the bottom of the comb(s) does not touch the bottom of the casting tray.

2. Prepare 1.5% agarose gel for analysis. The volume of gel required will depend on the size of the gel box being used. For 100 ml total volume, weigh out 1.5 g of agarose. Pour agarose into 500-ml flask containing a large stir bar. Add 100 ml of 1× TAE to the flask and microwave until the agarose is dissolved and liquid is clear. A common approach is to microwave on

**Table 4**
**PCR conditions for mitochondrial DNA whole genome, *POLG*, *TK2*, and *DGUOK*, and *SUCLA2***

| PCR conditions for mitochondrial DNA whole genome, *POLG*, *TK2*, and *DGUOK* |
| --- |
| 1. 95°C, 5min |
| 2. 95°C, 45 s |
| 3. 55°C, 45 s |
| 4. 72°C, 60 s |
| Repeat 2–4 for 35 cycles |
| 5. 72°C, 5 min |
| 6. 4°C, hold |
| PCR conditions for *SUCLA2* |
| 1. 95°C, 5 min |
| 2. 94°C, 45 s |
| 3. 65°C, 45 s |
| 4. 72°C, 2 min |
| Repeat 2–4 for ten cycles |
| 5. 94°C, 45 s |
| 6. 55°C, 45 s |
| 7. 72°C, 2 min |
| Repeat 5–7 for 25 cycles |
| 5. 72°C, 5 min |
| 6. 4°C, hold |

high for 2 min, remove the flask and swirl, and place back into the microwave for an additional 0.5–1 min until clear. The flask will be hot, and take proper precautions for handling hot liquids.

3. Place hot flask on magnetic stir plate and stir at medium speed, and avoid introducing bubbles into the gel. Stir until the flask can be comfortably touched, approximately 20 min. Alternatively, run the flask under cold water while swirling to cool faster.

4. Add 5 μl of 1% ethidium bromide solution to the flask (5 μl/100 ml of liquid gel), and stir/swirl for an additional 1–2 min.

5. Pour liquid slowly into the tray, taking care not to introduce bubbles or drop stir bar into tray. If bubbles persist near the comb(s), use a pipette tip to remove. Allow gel to cool until it is completely solidified, approximately 30 min.

6. Prepare gel tank by filling with 1× TAE up to ~1 cm below the electrode attachment point.

7. Place solidified gel into the tank, and adjust buffer height so that it completely covers the gel and is about 2 mm from

the top arch between the comb teeth. Carefully remove the comb(s).

8. Mix 2 μl of 2× loading dye and 2 μl of each PCR product (see Note 14).

9. Load 5 μl of 1-kb ladder in the first lane and the 4 μl sample + dye mix from the above step to each subsequent lane. If loading multiple genes, or multiple individuals for one gene, load a ladder in between for ease of analysis.

10. Place lid on gel tank, and connect red and black electrodes to appropriate outlets on power supply. Make sure that samples run towards the red cathode. Set the voltage on the power supply to 200 V and run for 15 min.

11. When gel is ready, remove it from the tank and analyze using a gel-imaging system (see Note 15).

12. Ensure that all PCR products have clear, bright, single bands of the expected size and that NTC lanes have no PCR products present. If any of the PCR products have nonspecific band(s), do not amplify or have clear primer-dimers, repeat that PCR reaction.

*3.5. Purification of PCR Products for Sequencing Analysis*

1. The following purification steps are performed using Edge Biosystems, ExcelaPure 96-well UF PCR purification plates. If an alternate system is used, please follow manufacturer's instructions.

2. Dilute the remaining PCR product to 100 μl total volume with sterile water.

3. Add the PCR products to the purification plate (see Note 16). To avoid confusion when purifying multiple reactions, maintain the location of each well from the PCR reaction plate to the purification plate as much as possible.

4. Centrifuge the plate at $1,400 \times g$ for 5 min. Discard the eluate.

5. Add another 100 μl to each well of the plate, and centrifuge again at $1,400 \times g$ for 5 min. Discard the eluate.

6. Add 100 μl of sterile water to each well and seal the plate. Secure the plate to a shaker and shake at ~300 rpm on a Gyrotory shaker model G2 or at ~600 rpm on a VWR microplate shaker for 5 min (see Note 17).

7. After 5 min, transfer the reconstituted product to a designated 96-well plate. These samples can be frozen and stored until ready to sequence.

*3.6. Big Dye Setup*

1. Prepare Big Dye master mix (see Table 5) for both M13 forward and M13 reverse sequencing for all nuclear gene samples. For mtDNA samples, use the same PCR primers that were

**Table 5**
**Master mix for Big Dye reaction and PCR conditions for Big Dye reaction for all samples**

| Components | Volume (μl) | Final conc. |
|---|---|---|
| *Master mix for Big Dye reaction, all samples* | | |
| Sterile water | 5.75 | |
| Big Dye terminator | 0.25 | |
| Big Dye 5× buffer | 2 | 1× |
| Primer, M13 or mitochondrial specific (2.5 μM) | 1.5 | 0.36 μM |
| DNA from PCR purification | 1 | |
| Total volume | 10.5 | |
| *PCR conditions for Big Dye reaction, all samples* | | |
| 1. 96°C, 2 min | | |
| 2. 96°C, 10 s | | |
| 3. 5°C, 45 s | | |
| 4. 60°C, 4 min | | |
| Repeat 2–4 for 35 cycles | | |
| 5. 4°C, hold | | |

used in Subheading 3.2. Do not include NTCs in Big Dye reactions.

2. To each tube, or well of a 96-well plate, add 9.5 μl of master mix and 1 μl of purified PCR product.

3. Place in thermocycler and run using the conditions listed in Table 5.

**3.7. Big Dye Purification**

1. After Big Dye program is complete, briefly spin down the samples.

2. Prepare a Performa DTR V3 96-well P/N 80808 purification plate (Edge Biosystems), by spinning the plate at $850 \times g$ for 3 min. If alternate purification plates are used, please follow manufacturer's instructions.

3. Place a 96-well optical reaction collection plate underneath the Edge plate, and add the entire Big Dye reaction sample to the 96-well Edge plate.

4. Spin the plates together at $850 \times g$ for 5 min. Eluate should collect in the bottom plate. Discard the Edge plate. Seal the collection plate with heat-sealing film.

5. Place the collection plate in a thermocycler and heat at 96°C for 2–5 min to denature.

6. Place the samples on ice for 2–5 min. At this point, the samples are ready for sequencing (see Note 18).

**3.8. Sample Sequence Analysis**

1. Sequence samples using ABI 3730XL DNA analyzer following manufacturer's instructions.

2. Using available software for sequence analysis, ensure that every amplicon of each sample has clean 2× coverage (see Note 19). If the sequence does not have clean 2× coverage, the sample may not be clean enough after Big Dye purification, therefore repeat steps 3.6–3.8 and reanalyze.

3. Compare the sequence traces for the human mitochondrial genome using the revised Cambridge reference sequence (RCRS) and the reported polymorphisms and mutations listed in http://mitomap.org database as reference. For target nuclear genes, use the reference sequence chosen in Subheading 3.1. Make note of heterozygous, homozygous, and hemizygous changes in nuclear genes and heteroplasmic or homoplasmic changes in the mitochondrial genome (see Note 20).

4. Classify changes based on current American College of Medical Genetics guidelines and Chapter 23 "Determination of the clinical significance of an unclassified variant" (7).

# 4. Notes

1. While most commercially available Taq polymerases can be used in this protocol, we use the FastStart Taq polymerase kit because it is a hot start polymerase that has high processivity and fidelity, maximizes specificity, and has given us the best reproducible results. In addition, the inclusion of the GC-rich solution in the kit provides ease of master mix preparation when both +GC-rich and –GC-rich master mixes are required at the same time.

2. Because of the volume of samples and lot reproducibility, we use a commercially available product. Alternate purification methods such as homemade plates and single column purification kits are also available.

3. For some exons, it may not be possible to set the optimal PCR product boundaries at ±60–70 bp around the exon due to genomic repeat regions or the size of the exon itself. It is also possible to amplify more than one exon in a single amplicon if the exons are close enough together. Since the optimal size of the PCR products is between 200 and 600 bp and the coding region of the exon is the primary region of interest, the target PCR product boundaries can be adjusted, and additional PCR primer sets for any given region can be designed. For example, the first coding exon of the *POLG* gene is approximately 660 bp, is flanked by repetitive regions, and has a $(CAG)n$ poly Q track in the middle. Multiple primer sets are required to get clean 2× coverage of the entire exon (see Fig. 1).

Primers

```
ttggggacgcagtaaatgctcaaggaatgatgattatggatacacctattacatatatggtaaaat      Ex1-3F
aacgctttatatcatctgtctcctttasgatttggggtggaaggcaggcatggtcaaaccc
atttcactgacaggagagccagagacaggacgtgtctctccacgt<y>ttccagccagtaaaa      Ex1-1F
ga[agccaagctggagcccaaagccaggtgttctgactcccagcgtgggggtccctgcacc
aacATGAGCC<G>CCTGCTCTGGAGGAAGGTGGCCGGCGCCACCGTCGGGCCAGGG<Y>CG
GTTCCAGCTCCGGGGCGCTGGGTCTCCAGCTCCGTCCCCGCGTCCGACCCCAGCGACGGGC
AGCGGCGGC<R>GC<R>GCAGCAGCAGCAGCAGCAGCAGCAGCAGCAACAGCAGCCTCAGC
AGCCGCAAGTGCTATCCTCGGAGGGCGGGCAGCTGCGGCACAACCCATTGGACATCCAG      Ex1-2F
ATGCTCTCGAGAGGGCTGCACGAGCAAATCTTCGGGCAAGGAGGGGAGATGCCTGGCGA      Ex1-4F
GGCCGCGGTGCGCCGCAGCGTCGAGCACCTGCAGAAGCACGG<K>CTC<K>GGGGGCAG      Ex1-3R
CCAGCCGTGCCCTTGCCCGACGTGGAGCTGCGCCTGCCGCCCCTCTACGGGGACAACCTG      Ex1-1R
GACCAGCACTTCCGCCTCCTGGCCCAGAAGCAGAGCCTGCCCTACCTGGAGGCGGCCAAC
TTGCTGTTGCAGGCCCAGCTGCCCCCGAAGCCCCCGGCTTGGGCCTGGGCGGAGGGCTGGA
CCCGGTACGGCCCCGAGGGG<G>AGGCCGTACCCGTGGCCATCCCCGAGGAGC<R>GGCCC
TGGTGTTCGACGTGGAGGTCTGCTTGGCAGAGGGAACTTGCCCCACATTGGCGGTGGCCAT
ATCCCCCTCGGCCTGgtaagtaggg<k>cagggttggggacataagcaggcatgggggcccagc      Ex1-4R
ttaatagt]ttgtttcagtgaacattttctgaggtcctgttacg<k>gctgggtgctcacg
tagggagcgctgatgtgttgaattag<v>actagacccctgtttatgtgggactcactttcctg      Ex1-2R
```

Fig. 1. Primer design for POLG exon 1. For clean 2× coverage of the entire exon, multiple PCR amplicons are required. *Bold capital letters* denote the coding exon; *lower case letters* denote surrounding intronic regions. SNPs are marked with < >, the target region is marked with [ ], and repetitive regions are *underlined*. Primers are marked in *bold*, *italic*, and *alternate font*. Note, some primers have been placed in repetitive regions due to sequence constraints.

4. If no primer sets are listed for the region of interest, adjust the PCR target boundaries and try again. Additionally, conditions such as primer melting temperatures can be adjusted.

5. We have often found that for greater than five of the same nucleotide in a row, a second amplicon is required for clean 2× coverage. A minimum of clean 2× coverage is required for diagnostic sequencing. For research purposes, the quality threshold may be lowered, such that 1× coverage may suffice.

6. A large number of results generally indicate homology to other regions of the genome. This can result in nonspecific primer annealing and a decreased PCR efficiency. For this reason, PCR primers in repeat regions should be avoided. However, alternative primer sites may not be possible, and primers may have to be placed in repetitive regions (see Fig. 1).

7. We let the oligo sit for at least 10 min with occasional vortexing to ensure complete resuspension.

8. Working primer dilutions can be stored at −20°C for up to 2 years or at 4°C for 1 month. Working primers should not be frozen and thawed more than five times.

9. The required DNA concentration for mitochondrial sequencing is lower than that for nuclear gene sequencing due to the increased copy number of mitochondrial DNA in each sample. The total volume of the final DNA dilution for nuclear genes depends on the number of amplicons per gene and the number of genes being tested per individual. For most genes, between 50 and 100 μl final volume is enough.

10. For example, the mitochondrial DNA depletion syndrome panel (*POLG*, *DGUOK*, *TK2*, and *SUCLA2*) has a total of 51 amplicons (23, 8, 10, and 11 respectively). However, we have found that while *POLG* and *TK2* amplification work best using FastStart Taq polymerase supplemented with a buffer for GC-rich regions, *DGUOK* amplification works well using FastStart Taq polymerase, and *SUCLA2* works the best using FastStart Taq polymerase for every amplicon except exon 1 which requires GC-rich buffer supplementation.

11. Again, different genes are best amplified under different conditions. For *POLG*, *DGUOK*, and *TK2*, we use a 55° annealing temperature. For *SUCLA2*, we use a step-down PCR program from 65 to 55°. Any standard thermocycler can be used.

12. A gel box of the appropriate size should be chosen based on the number of amplicons to be analyzed. Cast the gel according to manufacturer's instructions.

13. For a number of systems, multiple combs can be used simultaneously. We use four, 25-well combs, for a total of 100 amplicons run simultaneously.

14. To save on reagent costs, rather than using tubes to mix the samples and loading dye, samples can be mixed on paraffin wax strips.

15. While a UV light box and camera can suffice for this step, for ease of record keeping and image accessibility, a system in which digital images can be saved is optimal.

16. Alternately, if cross contamination is a concern, add 77 μl of sterile water to each well of the purification plate first, and then, add the remaining 23 μl of PCR product to the sterile water for a total volume of 100 μl.

17. When using an alternate plate shaker, optimize shaker speed and time to ensure sufficient agitation. When resuspending the purified PCR product manually, ensure that the sample is adequately mixed by vigorously pipetting up and down at least 20 times.

18. In order to maintain sample quality, no plate should be left at room temperature for longer than 12–15 h. This is to avoid dye break down due to extended exposure to humidity in the atmosphere. In case of a prolonged wait time (over 8 h), plates should be heat-sealed and stored away from light at 4°C.

19. ABI sequence analysis software assigns a quality score (QV) to each base call. QV is a per-base estimate of the basecaller accuracy and quality. Typical high quality bases will have QV 20–50. We use mutation surveyor software which compares the patient sample sequence for each gene to a reference sequence, for sequence analysis. Lane quality, which is a measurement of the

average signal to noise ratio for the region of interest in an amplicon, is carefully examined for background noise. Ideally, a minimum of one clean forward and one clean reverse sequence covering the same region is required for a sample to pass quality assessment. However, due to regions such as poly As at the 3′ or 5′ end of particular exons, two reverse or two forward traces from two different amplicons may suffice.

20. For the mitochondrial genome, if two nucleotides peaks are observed at a single position, make a note of the occurrence of heteroplasmy. Heteroplasmy levels cannot easily be determined based on peak heights from sequence traces. If quantification is desired, use ARMS PCR (see Chapter 21 "Quantification of mtDNA Mutation Heteroplasmy"). For nuclear genes, two nucleotide peaks at a single position indicate a heterozygous change.

## References

1. Wong, L. J., Scaglia, F., Graham, B. H., and Craigen, W. J. (2010) Current molecular diagnostic algorithm for mitochondrial disorders, *Mol Genet Metab 100*, 111–117.

2. Wong, L. J. (2010) Molecular genetics of mitochondrial disorders, *Dev Disabil Res Rev 16*, 154–162.

3. Wallace, D. C., Lott, M. T., Brown, M. D., and Kerstann, K. (2001) Mitochondria and neuro-opthalmologic diseases, In *The Metabolic and Molecular Bases of Inherited Disease* (Scriver, C. R., Beaudet, A. L., Sly, W. S., and Valle, D., Eds.), pp 2425–2509, McGraw-Hill, New York.

4. Wong, L. J. (2007) Diagnostic challenges of mitochondrial DNA disorders, *Mitochondrion 7*, 45–52.

5. Calvo, S., Jain, M., Xie, X., Sheth, S. A., Chang, B., Goldberger, O. A., Spinazzola, A., Zeviani, M., Carr, S. A., and Mootha, V. K. (2006) Systematic identification of human mitochondrial disease genes through integrative genomics, *Nat Genet 38*, 576–582.

6. Venegas, V., Wang, J., Dimmock, D., and Wong, L. J. (2011) Real-time quantitative PCR analysis of mitochondrial DNA content, *Curr Protoc Hum Genet Chapter 19*, Unit 19, 17.

7. Richards, C. S., Bale, S., Bellissimo, D. B., Das, S., Grody, W. W., Hegde, M. R., Lyon, E., and Ward, B. E. (2008) ACMG recommendations for standards for interpretation and reporting of sequence variations: Revisions 2007, *Genet Med 10*, 294–300.

# Chapter 20

# Utility of Array CGH in Molecular Diagnosis of Mitochondrial Disorders

## Jing Wang and Mrudula Rakhade

## Abstract

Array comparative genomic hybridization (aCGH) is a powerful clinical diagnostic tool that can be used to evaluate copy number changes in the genome. Targeted aCGH provides a much higher resolution in targeted gene regions to detect copy number changes within single gene or single exon. A custom-designed oligonucleotide aCGH platform (MitoMet®) has been developed to provide tiled coverage of the entire 16.6-kb mitochondrial genome and high-density coverage of a set of nuclear genes associated with metabolic and mitochondrial related disorders, for quick evaluation of copy number changes in both genomes (1). The high-density probes in mitochondrial genome on the MitoMet® array allow estimation of mtDNA deletion breakpoints and deletion heteroplasmy (2). This technology is particularly useful as a complementary diagnostic test to detect large deletions in genes related to mitochondrial disorders.

**Key words:** aCGH, MitoMet® array, Mitochondrial Disorder

## 1. Introduction

Large exonic or whole gene deletions and duplications have been reported to be a frequent cause of many diseases such as DMD, GLDC, RETT, and OTC (3, 4). Traditionally, these large deletions were detected by various methods including multiplex ligation-dependent probe amplification (MLPA) (5), restriction fragment analysis on Southern blots (6), fluorescent in situ hybridization (FISH) (7), and multiplex PCR or quantitative real-time PCR (1, 8). However, each of these methods has limitations with size of detection (7), high false positive and false negative rates, especially in the detection of duplications (5, 9). These procedures are often tedious, and the interpretation of results is usually not straightforward.

Oligonucleotide probes on microarrays corresponding to sequences throughout the entire genome have now been shown

Lee-Jun C. Wong (ed.), *Mitochondrial Disorders: Biochemical and Molecular Analysis*, Methods in Molecular Biology, vol. 837, DOI 10.1007/978-1-61779-504-6_20, © Springer Science+Business Media, LLC 2012

to give quantitative hybridization responses under standardized conditions, allowing rapid and relatively inexpensive analysis of chromosomal copy number variation as a clinical test (10–12). Most of the applications of this technology have been designed to detect major chromosomal copy number changes affecting >100 kb (12, 13) in the whole genome. This approach has been applied to higher resolution analysis of specific genes of interest with high-density probe coverage in array design (14).

The custom-designed array is an oligonucleotide array CGH targeted to genes involved in mitochondrial and metabolic disorders. This array contains tiled coverage of the entire 16.6-kb mitochondrial genome and high-density coverage on nuclear genes responsible for mitochondrial DNA (mtDNA) biogenesis, maintenance of mitochondrial deoxynucleotide pools, mitochondrial transcription and translation factors, and respiratory chain complex subunits and complex assembly. MitoMet® array can be used to detect deletions in both nuclear and mitochondrial genomes (1, 2, 15–17).

## 2. Materials

### 2.1. Equipment

1. High-resolution microarray scanner bundle (includes scanner, PC and LCD monitor, barcode reader and feature extraction software. Agilent Technologies, Inc., Santa Clara, CA).
2. NanoDrop ND-1000 UV–VIS spectrophotometer (Thermo Fisher Scientific, Inc., Wilmington, DE).
3. Microcentrifuge.
4. Incubator (set to 37°C).
5. Hybridization oven (set to 65°C).
6. PCR machine with heated lid.
7. Magnetic stirring hot plate and magnetic stir bars.
8. SpeedVac concentrator.
9. Nitrogen-purged low-humidity storage desiccators (see Note 1).

### 2.2. Reagents and Materials (see Note 2)

1. Agilent oligonucleotide based aCGH microarray ($8 \times 60$ K) (custom design) (see Note 3).
2. Agilent oligo aCGH hybridization kit (including $2 \times$ oligo aCGH hybridization solution and $10 \times$ blocking agent. Agilent Technologies, Inc., Santa Clara, CA).
3. Agilent oligo aCGH wash buffer 1 and wash buffer 2 (Agilent Technologies, Inc., Santa Clara, CA).
4. Stabilization and drying solution (Agilent Technologies, Inc., Santa Clara, CA).

5. Alu I (10 U/μl). Store at –20°C.

6. Rsa I (10 U/μl). Store at –20°C.

7. Human Cot-1 DNA (1 mg/ml in 10 mM Tris–HCl, pH 7.4, 1 mM EDTA). Store at –20°C.

8. 99.8% Acetonitrile.

9. BioPrime array CGH genomic labeling module (Invitrogen Corporation, Carlsbad, CA), contains:

   (a) exo-Klenow fragment of DNA polymerase I. Store at –20°C.

   (b) 2.5× Random primers. Store at –20°C.

   (c) 10× dCTP nucleotide mix (1.2 mM dATP, dGTP, and dTTP, 0.6 mM of dCTP in 10 mM Tris–HCl pH 8.0, 1 mM EDTA). Store at –20°C.

10. 1× TE buffer (pH 8.0): 10 mM Tris–HCl pH 8.0, 1 mM EDTA.

11. Cyanine 5-dCTP and cyanine 3-dCTP. Both are supplied at a concentration of 1.0 mM in 10 mM Tris–HCl, pH 7.6, 1.0 mM EDTA. (PerkinElmer Life and Analytical Sciences, Boston, MA).

12. Amicon Centricon filter devices (YM-30 filter) (Millipore Corporation, Billerica, MA).

## 3. Methods

Figure 1 shows the workflow of the procedures of oligo aCGH described here.

### 3.1. Restriction Digestion of Genomic DNA

1. Label two 0.2 ml 8-tube strips, one for patient's DNA and one for gender-matched reference DNA (see Note 4). Add 0.5 μg of total DNA from each corresponding individual and appropriate amount of nuclease-free water to make a final volume of 15 μl in each tube (see Note 5).

2. Prepare digestion master mix according to Table 1. The volume of digestion mater mix for each sample is 6.0 μl. Calculate total amount of digestion master mix needed based on number of samples. Keep all reagents on ice during preparation.

3. Add 6.0 μl of digestion master mix to each tube containing 15 μl of gDNA sample. Total volume now is 21 μl.

4. Place the tubes on ice as soon as digestion master mix was added. Mix well by flicking the bottom of the tubes. Quick spin in a microcentrifuge to drive the contents off the walls and lid.

5. Put the strips in a programmed PCR machine set for 37°C for 2 h and then 65°C for 20 min to inactivate the endonucleases.

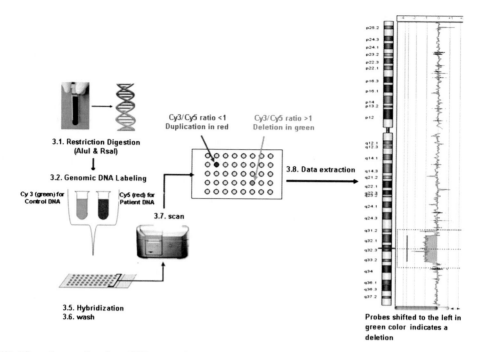

Fig. 1. Workflow diagram for oligo aCGH processing.

## Table 1
## Digestion master mix component

| Reagents | 1× Reaction (μl) | Each slide (8× reaction) | Final concentration |
|---|---|---|---|
| H$_2$O | 3.3 | 26.4 | – |
| 10× Buffer C | 2.1 | 16.8 | 1× |
| BSA (10 U/μl) | 0.1 | 0.8 | 0.34 U/μl |
| Alu I (10 U/μl) | 0.25 | 2.0 | 0.86 U/μl |
| Rsa I (10 U/μl) | 0.25 | 2.0 | 0.86 U/μl |
| *Total volume* | *6.0* | *48.0* | |

Alternatively, this step can be carried out in temperature-adjusted water bath or heating blocks. Store fragmented DNA on ice until ready for labeling.

***3.2. Genomic DNA Labeling***

Cy3-dCTP and Cy5-dCTP are light sensitive. Minimize the light exposure through the labeling procedures. This step should be performed in a dark room.

1. Quick spin the tubes with digested DNA (21 μl) in a microcentrifuge to drive the contents off the walls and lid.

2. Add 20 µl of 2.5× random primer solution (supplied by BioPrime array CGH genomic labeling module) into each post digestion sample tube.

3. Transfer the tubes to a programmed PCR machine set at 95°C for 5 min, and then hold at 4°C for at least 5 min until ready for labeling. Alternatively, this step can be carried out in a 95°C heating block for 5 min, and then chill on ice for 5 min.

4. Prepare labeling master mix on ice. For each reaction, add 5 µl of 10× dCTP nucleotide mix and 1 µl of exo-Klenow fragment of DNA polymerase I. The total volume of labeling master mix for each reaction is 6.0 µl. Calculate total amount of labeling master mix needed based on total sample number.

5. Add 6.0 µl labeling master mix to each tube containing digested DNA.

6. Add 3 µl of Cy3-dCTP to control tubes or 3 µl of Cy5-dCTP to patient tubes. The final volume is 50 µl.

7. Put the strips in a programmed PCR machine set for 37°C for 2 h, and then 65°C for 20 min to inactivate the enzyme. This step can be carried out in temperature-adjusted water bath or heating blocks (see Note 6).

**3.3. Removal of Excess Unreacted Cy3-dCTP and Cy5-dCTP (see Note 7)**

This step should be performed in a dark room.

1. Transfer samples to new 1.5-ml tubes.

2. Add 430 µl 1× TE (pH 8.0) to each labeling tube (total volume 480 µl).

3. Label the Amicon filter tubes and the 1.5-ml collection vials provided by the Amicon kit accordingly. Add the diluted labeled gDNA samples from Subheading 3.3, step 2 above to Amicon filter tubes.

4. Centrifuge at 14,000×g for 10 min at room temperature. Discard flow through.

5. Add 480 µl 1× TE (pH 8.0) to each filter tube.

6. Centrifuge at 14,000×g for 10 min at room temperature. Discard flow through.

7. Place the filter tube in an inverted direction into a new 1.5-ml sample retentive vial supplied by Amicon. Refer to Amicon manufactory instruction for details.

8. Centrifuge 1,000×g for 1 min at room temperature.

9. Collect the purified Cy3- or Cy5-labeled gDNA samples. The volume is around 15 µl. Take one ~1.5 µl of purified labeled gDNA to determine the yield and degree of labeling (see Note 8).

**3.4. Preparation of Hybridization Samples**

This step should be performed in a dark room.

1. Combine 8 μl of Cy3-labeled reference DNA and 8 μl of Cy5-labeled patient DNA.

2. Prepare Agilent 10× blocking agent: add 1,350 μl distilled water to a vial with lyophilized blocking agent (supplied with the kit); place at room temperature for 60 min before use. The remaining Agilent 10× blocking agent can be stored at –20°C for next use.

3. Prepare hybridization master mix solution. For each reaction, mix 2 μl of human Cot-1 DNA (1 mg/ml), 4.5 μl of 10× block agent, and 22.5 μl of 2× hybridization buffer. The total hybridization master mix volume for each reaction is 29 μl. Make sufficient amount of hybridization master mix based on number of samples.

4. Pipette up and down a few times to mix reagents followed by a quick spin in a microcentrifuge to drive the contents off the walls and lid.

5. Add 29 μl hybridization master mix to the combined Cy3/Cy5-labeled gDNA yield in Subheading 3.4, step 1 (16 μl). The total volume is 45 μl. Pipette up and down to mix and quick spin in a microcentrifuge.

6. Put the samples in a programmed PCR machine set for 95°C for 3 min, and then 37°C for 30 min. This step can be carried out in temperature-adjusted heating blocks.

7. Quick spin in a microcentrifuge to drive the contents off the walls and lid. Place tubes in 37°C incubator while setting up chamber assembly. Keep the tube in dark with foil.

**3.5. Assembly of Hybridization Chamber**

This step should be performed in a dark room (see Note 9).

1. Load a clean gasket slide into the Agilent SureHyb chamber base, with the gasket label facing up and aligned with the rectangular section of the chamber base. Ensure that the gasket slide is level and seated properly within the chamber base.

2. Slowly dispense 42 μl hybridization sample mixture onto the gasket well according to hybridization worksheet (Table 2); be careful to dispense each sample into the correct gasket well.

## Table 2
## Sample layout worksheet

| Agilent | DNA sample: #1 | DNA sample: #2 | DNA sample: #3 | DNA sample: #4 |
|---------|----------------|----------------|----------------|----------------|
| Agilent | DNA sample: #5 | DNA sample: #6 | DNA sample: #7 | DNA sample: #8 |

Active slide number: _____

3. Place the slide "Agilent"-labeled barcode side down onto the SureHyb gasket slide, so the numeric barcode side is facing up and the "Agilent"-labeled barcode is facing down. The label on the gasket slide should be lined up with the label on the array slide. Assess that the sandwich pair is properly aligned.

4. Place the SureHyb chamber cover onto the sandwiched slides and slide the clamp assembly onto both pieces.

5. Hand tighten the clamp onto the chamber.

6. Vertically rotate the assembled chamber to wet the slides and assess the mobility of the bubbles. Tap the assembly on the palm of your hand if necessary to move stationary bubbles.

7. Place assembled slide chamber in the rotator rack in a hybridization oven set to $65°C \pm 2°C$. Be sure that the rotator is balanced both side-to-side and front-to-back. Set the hybridization rotator to 18 rpm.

8. Hybridize at $65°C \pm 2°C$ for 20–24 h.

**3.6. Washing the Slides**

1. Prepare five washing dishes according to the following procedure (see Note 10):

   Dish 1: Completely fill dish 1 with room temperature wash buffer 1.

   Dish 2: Fill 250–260 ml of room temperature wash buffer 1 with stir bar on a magnetic stir plate.

   Dish 3: Fill 300 ml of pre-warmed (37°C) wash buffer 2 into pre-warmed dish with stir bar. Dish 3 needs to be placed on a magnetic plate with heating element set to 37°C (see Note 11).

   Dish 4: Fill with 99.8% acetonitrile. (Acetonitrile is toxic and flammable and must be used in a vented fume hood.)

   Dish 5: Fill with room temperature stabilization and drying solution (see Note 12).

2. With the sandwich completely submerged in wash buffer 1 (dish 1), disassemble the sandwich open from the barcode end only. Do this by slipping one of the blunt ends of the forceps between the slides and then gently turn the forceps upwards or downwards to separate the slides. Let the gasket slide drop to the bottom of the staining dish.

3. Remove the microarray slide and put into the slide rack in dish 2 containing wash buffer 1. Minimize exposure of the slide to air. Stir using setting 4 for 5 min.

4. Transfer slide rack to dish 3 containing pre-warmed wash buffer 2 (37°C). Stir using setting 4 for 1 min (see Note 11).

5. Remove the slide rack from dish 3 and tilt the rack slightly to minimize the wash buffer carryover. Immediately transfer the

slide rack to dish 4 containing acetonitrile. Stir using setting 4 for 30 s.

6. Transfer the slide rack to dish 5 filled with stabilization and drying solution. Stir using setting 4 for 30 s.

7. Slowly remove the slide rack, trying to minimize droplets on the slides. It should take 5–10 s to remove the slide rack.

8. Immediately put the slide with barcode side facing up in slide holder with an ozone-barrier slide cover on top of the array. Scan slide immediately to minimize impact of environmental oxidants on signal intensities. If necessary, store slides in original slide boxes in a $N_2$ purge box in the dark.

### 3.7. Scan

1. Place assembled slide holders into scanner carousel.

2. Verify scan settings as below:
   Scan region is set to scan area ($61 \times 21.6$ mm).
   Scan resolution ($\mu$m) is set to 5.
   Dye channel is set to red and green.
   Green PMT is set to 100%.
   Red PMT is set to 100%.

3. Select settings for array barcode.

4. Click Scan Slot *m-n* on the Scan Control main window. The letter *m* represents the start slot where the first slide is located, and the letter *n* represents the end slot where the last slide is located.

### 3.8. Data Extraction

1. Save multiple ".tif" images.

2. Open the Agilent Feature Extraction (FE) program (Agilent Feature Extraction software v9.5).

3. Import ".tif" files into Feature Extraction and convert image file to ".txt" files.

4. Interpret data with Agilent software (Agilent Genomic Workbench 5.0) or compatible analytic software (Fig. 2).

## 4. Notes

1. Do not store microarray slides in open air after opening the foil wrap. Store the microarray slides at room temperature under a $N_2$ purge box in the dark.

2. Reagents can be stored at room temperature unless otherwise indicated. Acetonitrile and stabilization and drying solution should be stored in flammable cabinet and disposed as flammable solvent in separate collection bottles for disposal by environmental safety.

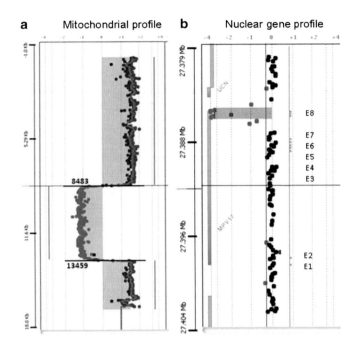

**Fig. 2.** Examples of targeted oligo aCGH results. (**a**) Mitochondrial profile. Highly tiled probes covered the entire mitochondrial. The *red dots* represent the undeleted region, whereas the *green dots* represent the deleted region. A 5-kb large mtDNA deletion was detected by targeted aCGH. The approximate deletion breakpoints were at mtDNA positions m.8483 and m.13495. (**b**) Nuclear gene profile of *MPV17* (2p23.3). The entire *MPV17* gene and its 5′UTR and 3′UTR regions were covered by densely spaced probes. The *black dots* represent the undeleted probes, and the *green dots* represent the deleted probes. The log ratio of −4 indicates a homozygous deletion. A homozygous deletion involving exon 8 of the *MPV17* gene was detected.

3. The aCGH array described in this protocol is a custom-designed array (Agilent Technologies, Inc., Santa Clara, CA). Each glass slide is formatted with eight high-definition 60,000 arrays (8×60 K format). In each array, there are about 6,400 probes to cover the entire mitochondrial genome. Nuclear genes involving mtDNA depletion syndromes, mitochondrial OXPHOS, and respiratory chain complex assembly were densely covered by 47,600 probes. The average probe space in targeted gene region is ~250 bp per probe. In addition, about 6,000 backbone probe were also included in this array with an average probe space of 500 kb. Other custom-designed targeted aCGH from different venders can also be used for this purpose.

4. Every array CGH experiment requires a normal DNA (the "reference DNA") that is labeled and co-hybridized to the array along with the patient sample. We recommend using gender-matched reference DNA for each sample. A male and a female

reference DNA samples should be prepared by your laboratory. Same male and female reference DNA should be used for all patient samples to ensure the consistent calls for CNV.

5. Genomic DNA (gDNA) quality is critical for good experimental results. gDNA extracted from blood is highly recommended to ensure the quality of the result. Use NanoDrop ND-1000 UV–VIS spectrophotometer (or equivalent) to assess both gDNA concentration and purity. High-quality genomic DNA samples should have A260/A280 ratio within 1.8 to 2.0 and A260/A230 ratio >2.0. Genomic DNA integrity can be checked by using agarose gel electrophoresis or BioAnalyzer. High-quality DNA will appear as a high-molecular-weight band in agarose gel, without a smear of lower-molecular-weight bands suggesting degradation or shearing of the DNA samples. The presence of RNAs can be detected as a cloud fluorescence in the low-molecular-weight region of the gel. If gDNA quality is insufficient, re-purification is recommended.

6. At this point, samples may be covered with aluminum foil and stored at $-20°C \pm 2°C$ up to 1 week if experiment cannot be processed. However, it is not recommended.

7. Non-reacted Cy3 and Cy5 molecules can interfere with the subsequent microarray experiment and increase background noise if they are not efficiently removed prior to hybridization.

8. Use NanoDrop ND-1000 UV–VIS spectrophotometer to measure the yield and degree of the labeling. From the main menu, select "MicroArray Measurement," and from the Sample Type menu, select "DNA-50"; use 1.5 μl of 1× TE to set blank of the instrument. Use 1.5 μl-labeled gDNA sample for quantification. Measure the absorbance at 260, 550, and 650 nm for DNA, Cy3, and Cy5, respectively. Degree of labeling = $(340 \times pmol/\mu l \, dye)/(ng/\mu l \, gDNA \times 1,000)$. It is important to match the degree of labeling of Cy3/Cy5-labeled DNA for every sample. Ideal signal intensity from a Cy3- or Cy5-labeled DNA should be in the range of 9–12 pmol/μl. The discrepancy of signal intensity between Cy3- and Cy5-labeled gDNA should be less than 3 pmol/μl. If the labeling signal intensity is too low, SpeedVac can be used to concentrate labeled DNA. However, if there is insufficient labeled DNA after concentrated by SpeedVac, it is recommended to repeat the labeling procedure.

9. Keep the temperature of hybridization sample mixtures as close to 37°C as possible. Microarray slides should be handled carefully. Each microarray is printed on the side of the glass slide containing the "Agilent"-labeled barcode. This side is called the "active side." The numeric barcode is on the "inactive side" of the glass slide. The hybridization sample mixture is applied

directly to the gasket slide and not to the microarray slide. Then the active side of the microarray slide is put on top of the gasket slide to form a "sandwich slide pair." To avoid damaging the microarray, always handle glass slides carefully by their edges. Never touch the surfaces of the slides. Never allow the microarray surface to dry out during the hybridization process and washing steps. It requires practice and experience to get the chambers very tight, but not too tight to break the slide.

10. Each of dishes one to three can be used to wash a maximum of five slides. Replace with fresh solution if there are more than five slides. Each of dishes 4 and 5 can be used for 20–25 slides wash. Place active face of slides toward the wall of the dish. Transfers slides fast to avoid liquid residue.

11. Pre-warm wash buffer 2 for dish 3 in a 37°C±2°C incubator overnight, or at least 3 h before washing.

12. Inspect the stabilization and drying solution for precipitation prior to use. If precipitate is present, warm the solution in a 37°C±2°C water bath until the precipitate returns to solution. Warm stabilization and drying solution only in a closed container and ensure that the container is no more than 70% full, to allow for vapor expansion. Allow the solution to equilibrate to room temperature before use.

## References

1. Wong, L. J., Dimmock, D., Geraghty, M. T., Quan, R., Lichter-Konecki, U., Wang, J., Brundage, E. K., Scaglia, F., and Chinault, A. C. (2008) Utility of oligonucleotide array-based comparative genomic hybridization for detection of target gene deletions. *Clin Chem* 54, 1141–8.

2. Chinault, A. C., Shaw, C. A., Brundage, E. K., Tang, L. Y., and Wong, L. J. (2009) Application of dual-genome oligonucleotide array-based comparative genomic hybridization to the molecular diagnosis of mitochondrial DNA deletion and depletion syndromes. *Genet Med* 11, 518–26.

3. Mendell, J. R., Buzin, C. H., Feng, J., Yan, J., Serrano, C., Sangani, D. S., Wall, C., Prior, T. W., and Sommer, S. S. (2001) Diagnosis of Duchenne dystrophy by enhanced detection of small mutations. *Neurology* 57, 645–50.

4. Shchelochkov, O. A., Li, F. Y., Geraghty, M. T., Gallagher, R. C., Van Hove, J. L., Lichter-Konecki, U., Fernhoff, P. M., Copeland, S., Reimschisel, T., Cederbaum, S., Lee, B., Chinault, A. C., and Wong, L. J. (2009) High-frequency detection of deletions and variable rearrangements at the ornithine transcarbamylase

(OTC) locus by oligonucleotide array CGH. *Mol Genet Metab* 96, 97–105.

5. Schouten, J. P., McElgunn, C. J., Waaijer, R., Zwijnenburg, D., Diepvens, F., and Pals, G. (2002) Relative quantification of 40 nucleic acid sequences by multiplex ligation-dependent probe amplification. *Nucleic Acids Res* 30, e57.

6. Monaco, A. P., Bertelson, C. J., Middlesworth, W., Colletti, C. A., Aldridge, J., Fischbeck, K. H., Bartlett, R., Pericak-Vance, M. A., Roses, A. D., and Kunkel, L. M. (1985) Detection of deletions spanning the Duchenne muscular dystrophy locus using a tightly linked DNA segment. *Nature* 316, 842–5.

7. Bendavid, C., Kleta, R., Long, R., Ouspenskaia, M., Muenke, M., Haddad, B. R., and Gahl, W. A. (2004) FISH diagnosis of the common 57-kb deletion in CTNS causing cystinosis. *Hum Genet* 115, 510–4.

8. Sieber, O. M., Lamlum, H., Crabtree, M. D., Rowan, A. J., Barclay, E., Lipton, L., Hodgson, S., Thomas, H. J., Neale, K., Phillips, R. K., Farrington, S. M., Dunlop, M. G., Mueller, H. J., Bisgaard, M. L., Bulow, S., Fidalgo, P., Albuquerque, C., Scarano, M. I., Bodmer, W., Tomlinson, I. P., and Heinimann, K. (2002)

Whole-gene APC deletions cause classical familial adenomatous polyposis, but not attenuated polyposis or "multiple" colorectal adenomas. *Proc Natl Acad Sci USA* 99, 2954–8.

9. Schrijver, I., Rappahahn, K., Pique, L., Kharrazi, M., and Wong, L. J. (2008) Multiplex ligation-dependent probe amplification identification of whole exon and single nucleotide deletions in the CFTR gene of Hispanic individuals with cystic fibrosis. *J Mol Diagn* 10, 368–75.

10. Boone, P. M., Bacino, C. A., Shaw, C. A., Eng, P. A., Hixson, P. M., Pursley, A. N., Kang, S. H., Yang, Y., Wiszniewska, J., Nowakowska, B. A., del Gaudio, D., Xia, Z., Simpson-Patel, G., Immken, L. L., Gibson, J. B., Tsai, A. C., Bowers, J. A., Reimschisel, T. E., Schaaf, C. P., Potocki, L., Scaglia, F., Gambin, T., Sykulski, M., Bartnik, M., Derwinska, K., Wisniowiecka-Kowalnik, B., Lalani, S. R., Probst, F. J., Bi, W., Beaudet, A. L., Patel, A., Lupski, J. R., Cheung, S. W., and Stankiewicz, P. (2010) Detection of clinically relevant exonic copy-number changes by array CGH. *Hum Mutat* 31, 1326–42.

11. Ou, Z., Kang, S. H., Shaw, C. A., Carmack, C. E., White, L. D., Patel, A., Beaudet, A. L., Cheung, S. W., and Chinault, A. C. (2008) Bacterial artificial chromosome-emulation oligonucleotide arrays for targeted clinical array-comparative genomic hybridization analyses. *Genet Med* 10, 278–89.

12. Dhami, P., Coffey, A. J., Abbs, S., Vermeesch, J. R., Dumanski, J. P., Woodward, K. J., Andrews, R. M., Langford, C., and Vetrie, D. (2005) Exon array CGH: detection of copy-number changes at the resolution of individual exons in the human genome. *Am J Hum Genet* 76, 750–62.

13. Lu, X., Shaw, C. A., Patel, A., Li, J., Cooper, M. L., Wells, W. R., Sullivan, C. M., Sahoo, T., Yatsenko, S. A., Bacino, C. A., Stankiewicz, P., Ou, Z., Chinault, A. C., Beaudet, A. L., Lupski, J. R., Cheung, S. W., and Ward, P. A. (2007) Clinical implementation of chromosomal microarray analysis: summary of 2513 postnatal cases. *PLoS One* 2, e327.

14. Brunetti-Pierri, N., Paciorkowski, A. R., Ciccone, R., Mina, E. D., Bonaglia, M. C., Borgatti, R., Schaaf, C. P., Sutton, V. R., Xia, Z., Jelluma, N., Ruivenkamp, C., Bertrand, M., de Ravel, T. J., Jayakar, P., Belli, S., Rocchetti, K., Pantaleoni, C., D'Arrigo, S., Hughes, J., Cheung, S. W., Zuffardi, O., and Stankiewicz, P. (2010) Duplications of FOXG1 in 14q12 are associated with developmental epilepsy, mental retardation, and severe speech impairment. *Eur J Hum Genet*.

15. Compton, A. G., Troedson, C., Wilson, M., Procopis, P. G., Li, F. Y., Brundage, E. K., Yamazaki, T., Thorburn, D. R., and Wong, L. J. (2010) Application of oligonucleotide array CGH in the detection of a large intragenic deletion in POLG associated with Alpers Syndrome. *Mitochondrion*.

16. Lee, N. C., Dimmock, D., Hwu, W. L., Tang, L. Y., Huang, W. C., Chinault, A. C., and Wong, L. J. (2009) Simultaneous detection of mitochondrial DNA depletion and single-exon deletion in the deoxyguanosine gene using array-based comparative genomic hybridisation. *Arch Dis Child* 94, 55–8.

17. Zhang, S., Li, F. Y., Bass, H. N., Pursley, A., Schmitt, E. S., Brown, B. L., Brundage, E. K., Mardach, R., and Wong, L. J. (2010) Application of oligonucleotide array CGH to the simultaneous detection of a deletion in the nuclear TK2 gene and mtDNA depletion. *Mol Genet Metab* 99, 53–7.

# Chapter 21

# Quantification of mtDNA Mutation Heteroplasmy (ARMS qPCR)

## Victor Venegas and Michelle C. Halberg

## Abstract

Pathogenic mitochondrial DNA (mtDNA) mutations are usually present in heteroplasmic forms that vary in concentration among different tissues. Manifestation of clinical phenotypes depends on the degree of mtDNA mutation heteroplasmy (mutation load) in affected tissues. It is therefore important to quantify the degree of mutation heteroplasmy in various tissues. In this chapter, we outline the design of allele refractory mutation system (ARMS)-based quantitative PCR (qPCR) analysis of common mtDNA point mutations, a cost-effective and sensitive single-step method to simultaneously detect and quantify heteroplasmic mtDNA point mutations.

**Key words:** Mitochondrial DNA point mutations, MtDNA, ARMS qPCR, MtDNA heteroplasmy, MtDNA mutations, Quantitative PCR

## 1. Introduction

The mitochondrial respiratory chain (RC) is the essential final common pathway for aerobic metabolism to produce ATP. Thus, tissues and organs with a high energy demand are more susceptible to mitochondrial dysfunction (1, 2). There are hundreds to thousands of mitochondria per cell, and each mitochondrion contains 2–10 copies of the mitochondrial DNA (mtDNA) molecule, which is a 16.6 kb circular double-stranded DNA, containing 37 genes: 2 rRNA, 22 tRNA, and 13 protein subunits essential for respiratory chain complexes. Oxidative DNA damage and a high mutation rate are observed in the mtDNA as a consequence of limited DNA repair mechanisms, proximity to the site of reactive oxygen species production, and a lack of protective histones.

Pathogenic mtDNA mutations are usually heteroplasmic, meaning that both mutant and wild-type mtDNA forms are coexisting. Single-cell and hybrid-cell studies have shown that the

Lee-Jun C. Wong (ed.), *Mitochondrial Disorders: Biochemical and Molecular Analysis*, Methods in Molecular Biology, vol. 837, DOI 10.1007/978-1-61779-504-6_21, © Springer Science+Business Media, LLC 2012

mutant mtDNA must exceed a critical threshold level before a cell expresses a biochemical abnormality of the mitochondrial respiratory chain. The clinical expressivity of mtDNA disorders depends on the level of heteroplasmy in affected tissues that can vary among matrilineal relatives in the same family and among different tissues within the same individual. Furthermore, the mutant mtDNA threshold sufficient to cause certain symptoms or disease varies among different tissues and among different mutations. The variability of mutant heteroplasmy and tissue distribution can explain the degree of heterogeneity of the clinical phenotype. Therefore, it is important to identify the deleterious mtDNA mutation and to determine the degree of mutant heteroplasmy in various tissues in patients and matrilineal relatives of the affected individual for proper patient management and genetic counseling of family members.

While there are studies that show the mutant mtDNA is uniformly high when multiple tissues are analyzed, such as the m. T8993G mutation, the degree of mutant load in blood does not always correlate with the clinical phenotype (3). The skewed segregation of mutant mtDNA between different tissues is thought to arise during early embryonic development. The shared embryonic origin of specific tissues, as in hair bulbs and the brain, calls for analysis of mutant heteroplasmy in a variety of noninvasive tissues such as urine sediment cells, buccal mucosa cells, and hair follicles (4). It is important to analyze any available muscle specimen, skin fibroblasts, or autopsy or biopsy tissues to establishing a clinical correlation.

Methods widely used to detect heteroplasmic mutations include restriction fragment length polymorphism (RFLP), direct sequencing, TaqMan allelic discrimination analysis, and PCR/allele-specific oligonucleotide (ASO) dot blot hybridization. However, the ability of these methods to quantify heteroplasmy is limited (5). The detection limit for RFLP–ethidium bromide (EtBr) gel analysis is about 10% (5). Although the last-cycle hot PCR-RFLP quantification may be more sensitive (6), the use of radioactive materials is a disadvantage.

Here, we describe ARMs RT qPCR procedures to quantitatively screen for common mtDNA point mutations associated with mitochondrial syndromes (Table 1). ARMS real-time (RT) qPCR is a cost-effective method that measures the relative amount of wild-type and mutant mtDNA in a single step.

It requires as little as 0.1 ng of total cellular DNA per reaction and can be used to quantify mutation heteroplasmy in various noninvasive tissues such as hair follicles, buccal swab, and urine sediment. The ARMS primers designed by introducing a mismatched nucleotide immediately 5' to the mutation site greatly increase the amplification specificity of allele-specific modified primers (7). The specific nucleotide at the 3' end of the modified ARMS primer discriminates the wild-type and the mutant alleles.

# Table 1
## ARMS primer sequences and qPCR conditions

| Forward primer | Sequence (5'-3') | Reverse primer | Sequence (5'-3') | Size (bp) | Annealing temperature (°C) |
|---|---|---|---|---|---|
| ARMS-A3243-1m2 | CAGGGTTTGTTAAGATGGCAtA | mtR3319 | TGGCCATGGGTATGTTGTTA | 97 | 63 |
| ARMS-A3243G-1m2 | CAGGGTTTGTTAAGATGGCAtG | mtR3319 | TGGCCATGGGTATGTTGTTA | 97 | |
| ARMS T3271-1m-qz1 | CGGTAATCGCATAAAACTTAAAAgT | mtR3319 | TGGCCATGGGTATGTTGTTA | 59 | 63 |
| ARMS T3271C-1m-qz1 | CGGTAATCGCATAAAACTTAAAAgC | mtR3319 | TGGCCATGGGTATGTTGTTA | 59 | |
| ARMS-G3460-R-1m | CTACTACAACCCTTCGCTGAaG | mtR3553-33 | GAG CGA TGG TGA GAG CTA AGG | 94 | 58 |
| ARMS-G3460A-R-1m | CTACTACAACCCTTCGCTGAaA | mtR3553-33 | GAG CGA TGG TGA GAG CTA AGG | 94 | |
| ARMS-R-A8344-1m2 | GGGCATTTCACTGTAAAGAGGTGTTGaT | mtF8127 | AACCACTTTCACCGCTACACG | 130 | 60 |
| ARMS-R-A8344G-1m2 | GGGCATTTCACTGTAAAGAGGTGTTGaC | | AACCACTTTCACCGCTACACG | 130 | |
| ARMS-T8356-1m | GATTAAGAGAACCAACACCTCgT | mtR8448-23 | ATTTTTAGTTGGGTGATGAGGAATA | 95 | 58 |
| ARMS-T8356C-1m | GATTAAGAGAACCAACACCTCgC | mtR8448-23 | ATTTTTAGTTGGGTGATGAGGAATA | 95 | |
| ARMS-G8363-1m | GAGAACCAACACCTCTTTACAGgG | mtR8448-23 | ATTTTTAGTTGGGTGATGAGGAATA | 88 | 58 |
| ARMS-G8363A-1m | GAGAACCAACACCTCTTTACAGgA | mtR8448-23 | ATTTTTAGTTGGGTGATGAGGAATA | 88 | |
| ARMS-T8993-1m | TACTCATTCAACCAATAGCCaT | mtR9046-9026 | TTAGGTGCATGAGTAGGTGGC | 75 | 60 |
| ARMS-T8993C-1m2 | TACTCATTCAACCAATAGCCaC | mtR9046-9026 | TTAGGTGCATGAGTAGGTGGC | 75 | |
| ARMS-T8993G-1m2 | TACTCATTCAACCAATAGCCaG | mtR9046-9026 | TTAGGTGCATGAGTAGGTGGC | 75 | |
| ARMS-G11778-1m | CTACGAACGCACTCACAGTaG | mtR11859 | AGGTTAGCGAGGCTTGCTAG | 104 | 60 |
| ARMS-G11778A-1m | CTACGAACGCACTCACAGTaA | mtR11859 | AGGTTAGCGAGGCTTGCTAG | 104 | |

(continued)

**Table 1**
**(continued)**

| Forward primer | Sequence (5'-3') | Reverse primer | Sequence (5'-3') | Size (bp) | Annealing temperature (°C) |
|---|---|---|---|---|---|
| ARMS-G14459-1m | GGATACTCCTCAATAGCCATaG | mtR14558-34 | GGTGTGTTATTATTCTGAATTTTGG | 120 | 60 |
| ARMS-G14459A-1m | GGATACTCCTCAATAGCCATaA | mtR14558-34 | GGTGTGTTATTATTCTGAATTTTGG | 120 | |
| ARMS-T14484-1m | GTAGTATATCCAAAGACAACCtT | mtR14558-34 | GGTGTGTTATTATTCTGAATTTTGG | 95 | 60 |
| ARMS-T14484C-1m | GTAGTATATCCAAAGACAACCtC | mtR14558-34 | GGTGTGTTATTATTCTGAATTTTGG | 95 | |
| ARMS-G13513-1m | CTCACAGGTTTCTACTCCAAtG | R13636 | GACCTGTTAGGGTGAGAAGAA | 143 | 60 |
| ARMS-G13513A-1m | CTCACAGGTTTCTACTCCAAtA | R13636 | GACCTGTTAGGGTGAGAAGAA | 143 | |
| ARMS-A13514-1m2 | CTCACAGGTTTCTACTCCAAAtA | R13636 | GACCTGTTAGGGTGAGAAGAA | 143 | 60 |
| ARMS-A13514G-1m2 | CTCACAGGTTTCTACTCCAAAtG | R13636 | GACCTGTTAGGGTGAGAAGAA | 143 | |

## 2. Materials

### 2.1. Pre-PCR

1. DNA samples: DNA extracted from various clinical specimens or cloned plasmid DNA controls (8).

2. DNA Hydration Solution for diluting DNA samples: 10 mM tris (hydroxymethyl) aminomethane hydrochloride, 0.1 mM ethylenediaminetetraacetic acid (EDTA) in water, pH 8.0.

3. NanoDrop ND-1000 UV–Vis spectrophotometer for DNA quantification.

4. Specific ARMS primers for wild-type and mutant alleles (Table 1).

5. 100 μM ARMS primer stock solution: Dissolve the powder primer (usually around 30–50 nmol) with an appropriate amount of sterile water [water for injection (WFI) quality], usually ten times the number of nmoles in microliter volume. For example, if there are 35 nmol, add 350 μl of sterile water to the tube to make a 100 μM stock solution. 6.5 μM ARMS primer working solution: Add 5 μl of 100 μM stock solution to 95 μl of WFI quality sterile water.

6. Dilute positive control (PC) and normal control (NC) plasmid DNA to about $2 \times 10^{-2}$ ng/μl for an initial ARMS qPCR experiment; the plasmid DNA may require further dilution to reach a $C_T$ value at around 27–29 (see Note 1).

7. iTaq SYBR Green Supermix with ROX (Bio-Rad).

8. 96-Well, low-profile PCR Fast plates (Axygen PCR-96-LP-AB-C) (see Note 2).

9. Adhesive film for real-time qPCR (Axygen) (see Note 2).

### 2.2. Post-PCR

1. ABI Prism 7900HT sequence detector system (Applied Biosystems).

2. Sequence Detection System (SDS) data analysis software (version 2.2).

## 3. Methods

### 3.1. ARMs qPCR Primer Design

1. Identify your target mtDNA point mutation of interest. http://www.mitomap.org/MITOMAP/HumanMitoSeq.

2. The terminal 3′-nucleotide of the ARMS qPCR primer must be allele specific (see Note 3). The modified primer can be either forward or reverse.

3. The ARMS forward and reverse primers should have similar melting temperature (Tm). The Tm for ARMS primers should

be designed around 62–72°C based on the formula $4(G+C)+2(A+T)$. The annealing temperature is determined by the lowest Tm of a primer set.

4. Introduce a mismatch to the penultimate nucleotide to the 3′ terminus of the primer containing either the wild-type or mutant point mutation at the 3′ end (see Note 3).

5. Adjust the length of the primers to maintain a Tm of 62–72°C (see Note 4).

6. ARMS primers for quantification of mtDNA common mutations are listed in Table 1.

### 3.2. Generation of Positive Control and Normal Control Plasmid DNA

The wild-type and mutant plasmid controls, designated as normal control (NC) and positive control (PC) respectively, are generated by cloning the specific PCR products into plasmid as follows:

1. Choose the target DNA sequence similar to ARMS qPCR primers as listed in Table 1 except that the NC and PC DNA PCR primers do not contain the mismatched 3′ penultimate nucleotide residue, and the mutation nucleotide is sitting at the middle of the target primer DNA sequence rather than the 3′ end of the primer (9) (see Note 5).

2. Clone the PCR products into pCR2.1-TOPO plasmid according to manufacturer's protocols (Invitrogen).

3. Isolate plasmid DNA using a Qiagen Plasmid Miniprep Kit per manufacturer's specifications and protocols.

4. Quantify plasmid DNA controls using the NanoDrop ND-1000 UV–Vis spectrophotometer using methods previously described (8).

5. Verify that the insert contains target wild-type or mutant mtDNA by restriction enzyme digestion and sequencing of the insert DNA using M13 universal primers provided with the kit.

### 3.3. Pre-qPCR

1. Quantify concentration of DNA samples using NanoDrop ND-1000 UV–Vis spectrophotometer. Serially dilute the DNA samples using DNA hydration solution, beginning with a dilution of 2 ng/μl from the stock DNA and further diluting down to 0.1 ng/μl or less, depending on the mutation and tissue type (see Note 6).

2. Using 5 μM ARMS primer working solutions to create two master mixes: the modified primer containing the point mutation mixed with its paired primer and the modified wild-type primer mixed with its paired primer.

3. Preparation of master mix for ten reactions: To a 1.5 ml microcentrifuge tube, mix 50 μl of iTaq SYBER Green Supermix with ROX, 10 μl of 5 μM forward, and 10 μl of 5 μM reverse

primer (final concentration of 0.5 μM for each primer). Depending on the number of PCRs to be set up, this can be scaled up or down. Always remember to make the master mix 10% more than the actual volume needed for the reactions.

4. Pipette 7 μl of master mix to each of the wells in a 96-well Fast plate. Add 3 μl of diluted DNA, PC, NC, or WFI quality sterile water as a no template control (NTC) (final concentration will vary based on the dilution of DNA and the mtDNA copy number in the tissue source the DNA sample is extracted from) (see Note 7).

5. The qPCR is set up for plating on a 96-well Fast plate and ran on a Fast plate block in the ABI Prism 7900HT sequence detector system. Each DNA sample is run in triplicates for both the wild-type and mutant ARMS master mixes. Each well contains a final volume of 10 μl (see Note 8).

6. Once the master mixes and DNA template are plated, the 96-well Fast plate is covered with an adhesive plate sealer to prevent evaporation. Centrifuge the sealed plate for at least 5 min at $2,000 \times g$ or higher to ensure all bubbles are removed prior to inserting the plate in the ABI Prism 7900HT.

**3.4. PCR and Post-qPCR**

1. The program used in association with the 7900HT is SDS, version 2.2. For ARMS qPCR, the standard format is used with four stages (manually adding the dissociation stage to the preset format). The conditions for the qPCR are shown in Table 2.

2. The annealing temperature will vary depending on the lowest melting temperature (Tm) of the ARMS primer sets of the mutation being run. For m.3243A > G, the annealing temperature is 63°C.

3. Using the standard format with all 96 wells selected, the program takes approximately 1 h and 15 min to run. When

**Table 2**
**ARMs qPCR conditions**

| Stage | Number of cycles | Time | Temperature (°C) |
|---|---|---|---|
| 1 | 1 | 2 min | 50 |
| 2 | 1 | 20 s | 95 |
| 3 | 40 | 15 s | 95 |
|  |  | 30 s | 58–63 |
| 4 | 1 | 15 s | 95 |
|  |  | 20 s | 50 |
|  |  | 15 s | 95 |

complete, the resulting data will show the $C_T$ values of each individual well.

4. The resulting data can be exported in .txt format to be analyzed, and the used 96-well plate containing PCR product can be disposed of properly.

**3.5. Analysis of Data**

1. Each triplicate $C_T$ value for a DNA sample must be within a deviation of no more than $0.5\,C_T$ unit from each other; if not, any outliers should be deleted and an average taken from the remaining duplicate $C_T$ values. If all the $C_T$ values of the triplicate differ from each other in greater than $0.5\,C_T$ unit, the triplicate measurements should be repeated. The large deviation is most likely due to pipetting errors or bubble formation.

2. The SDS data analysis software will generate a spreadsheet with the $C_T$ value for each qPCR. To calculate the percentage of mutation heteroplasmy, first find $\Delta C_T$, that is, $C_{T\ \text{wild-type}} - C_{T\ \text{mutant}}$. Then use the following formula to calculate the mutation load:

$$\text{Mutant heteroplasmy level (\%)} = 1 / [1 + (1 / 2)^{\Delta C_T}] \times 100\%$$

3. To validate results from an ARMS qPCR run, PC should show >99%, whereas NC should show <0.1%. The NTC should show "undetermined" $C_T$ values, indicating that the $C_T$ is greater than 40 threshold cycles in the qPCR. Occasionally, $C_T$ values of 35 or higher will show up in the results, which indicate negligible contamination. Consistent $C_T$ values <35 in the NTC wells suggest that either the DNA hydration buffer or the ARMS primer solutions are contaminated, and the experiments should be repeated with fresh dilutions.

**3.6. Assessment of PCR efficiency**

1. Serially dilute PC and NC control plasmid DNA over six log units, around a $10^6$ fold range.

2. Carryout ARMS qPCR for the serially diluted PC and NC control plasmid DNA as described in Subheading 3.2.

3. To generate amplification plots, take the average of the triplicate $C_T$ results for each PC and NC control plasmid dilution and plot them on a semilog graph (Fig. 1). Optimal concentrations of DNA template should generate a linear exponential increase of PCR product.

4. To verify that there is only one PCR product present in each reaction well, a single sharp peak should be observed in the dissociation curve (Fig. 2). Additional peaks would indicate the presence of primer dimer product or nonspecific PCR amplification.

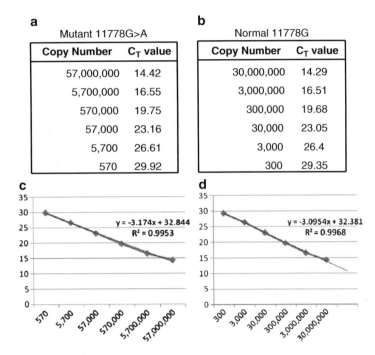

Fig. 1. Standard curve. Plasmid DNA containing normal 11778G (negative control) or mutant 11778G > A (positive control) was diluted to indicated copy number, negative control (**a**) and positive control (**b**). The standard curve is generated by log (copy number) vs. $C_T$ value, negative control (**c**) and positive control (**d**).

*3.7. Quality Assessment of the ARMS qPCR Quantification of mtDNA Point Mutations*

1. Make a series of heteroplasmic mixtures as listed in Table 3. Make the PC and NC plasmid DNA solutions in the same concentration. Mix 0, 5, 10, 15, … parts of PC and 100, 95, 90, 85,… parts of NC DNA solutions to make 0, 5, 10,….% of mutant heteroplasmy mixtures (Table 3) (see Note 9). An example of mixture of various proportions is shown in Table 3.

2. Carry out ARMS qPCR analysis for each mixture as described in Subheading 3.3 (Table 2).

3. Calculate the percentage of mutation heteroplasmy by $\Delta C_T$ $(C_{T\ wild\text{-}type} - C_{T\ mutant})$.

   The mutant heteroplasmy level is $(\%) = [1/(1 + (1/2)\Delta C_T)] \times 100\%$ (see Note 10) Plot the observed and calculated percentages of mutant control DNA plasmid (Fig. 3). If the accurate concentration of the mutant and wild-type control DNA is mixed, the observed percentage of mutant heteroplasmy should coincide with the expected percentage as shown in Fig. 3.

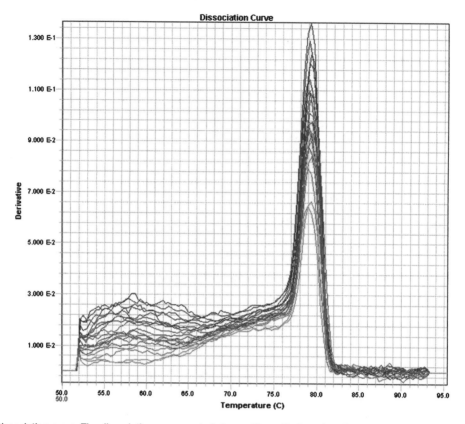

Fig. 2. Dissociation curve. The dissociation curve analysis is used to verify there is only one PCR product present, represented by a single sharp peak. Additional peaks would indicate the presence of primer dimer product or nonspecific PCR amplification.

## 4. Notes

1. DNA template should be adjusted to a concentration that would give a $C_T$ value at around 27–29. This range demonstrates the highest inter- and intra-run reproducibility.

2. The 96-well plate must be compatible with the ABI Prism 7900HT sequence detector system which requires Fast Optical PCR plates. The single 96-well piece polypropylene plates have consistent well-to-well thickness for PCR process uniformity. The plates should be sealed with optically clear adhesive film to prevent cross contamination and sample evaporation.

3. ARMS primers are specifically designed based on the absence of 3′ exonucleolytic proofreading activity associated with *Taq* DNA polymerase. In addition, the terminal 3′-nucleotide must be allele specific. Along with a mismatched nucleotide immediately 5′ to the mutation site, the designed primer will further

**Table 3**
**Quality assessment of the ARMs qPCR quantification of mtDNA point mutations**

| NC9%/PC9% | N1 | N2 | N3 | Average | P1 | P2 | P3 | Average | $\Delta C_T$ | $2^{(ddCT)}$ | Mutation rate (%) |
|---|---|---|---|---|---|---|---|---|---|---|---|
| N100/P0 | 27.86 | 27.66 | 27.63 | 27.72 | UD | UD | UD | #DIV/0! | #DIV/0! | #DIV/0! | #DIV/0! |
| N95/P5 | 27.72 | 27.96 | 27.96 | 27.88 | 31.93 | 32.03 | 31.95 | 31.97 | −4.092 | 0.058 | 5.53 |
| N90/P10 | 27.79 | 27.79 | 28.06 | 27.88 | 30.38 | 31.00 | 30.77 | 30.72 | −2.837 | 0.139 | 12.27 |
| N85/P15 | 28.25 | 28.22 | 28.00 | 28.16 | 30.52 | 30.06 | 30.42 | 30.33 | −2.177 | 0.220 | 18.09 |
| N80/P20 | 28.13 | 28.02 | 28.05 | 28.06 | 29.78 | 29.94 | 30.31 | 30.01 | −1.945 | 0.259 | 20.60 |
| N75/P25 | 28.36 | 28.55 | 28.13 | 28.34 | 29.44 | 29.99 | 29.85 | 29.76 | −1.416 | 0.374 | 27.25 |
| N60/P40 | 28.79 | 28.91 | 28.54 | 28.75 | 29.24 | 29.10 | 28.99 | 29.11 | −0.361 | 0.778 | 43.77 |
| N50/P50 | 28.92 | 28.93 | 28.49 | 28.78 | 29.12 | 29.42 | 29.06 | 29.20 | −0.421 | 0.746 | 42.74 |
| N40/P60 | 28.79 | 29.64 | 29.40 | 29.28 | 28.79 | 28.77 | 28.52 | 28.69 | 0.583 | 1.498 | 59.97 |
| N25/P75 | 29.73 | 29.45 | 29.88 | 29.69 | 27.94 | 28.33 | 28.35 | 28.20 | 1.483 | 2.796 | 73.65 |
| N20/80 | 30.34 | 30.30 | 30.60 | 30.41 | 27.85 | 28.12 | 28.38 | 28.12 | 2.293 | 4.902 | 83.05 |
| N15/P85 | 31.20 | 30.57 | 30.64 | 30.80 | 27.74 | 27.79 | 27.83 | 27.78 | 3.018 | 8.100 | 89.01 |
| N10/P90 | 31.96 | 31.29 | 31.61 | 31.62 | 27.88 | 27.82 | 27.72 | 27.81 | 3.813 | 14.06 | 93.36 |
| N5/P95 | 32.25 | 31.62 | 31.64 | 31.84 | 27.37 | 27.76 | 27.90 | 27.68 | 4.161 | 17.89 | 94.70 |
| N0/P100 | 37.61 | 38.30 | 36.54 | 37.49 | 27.49 | 27.52 | 27.77 | 27.59 | 9.892 | 950.3 | 99.894 |
| No template | UD | UD | 36.89 | 36.89 | UD | UD | UD | #DIV/0! | #DIV/0! | #DIV/0! | #DIV/0! |

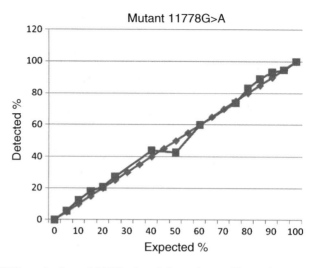

Fig. 3. ARMS results for m.11778G>A mutation using positive and negative plasmid controls in Table 3.

increase the specificity of target DNA amplification. Either the forward or the reverse primer can be the modified as long as they work optimally to discriminate between the normal and mutated target sequences.

4. Large primers with a corresponding high Tm allow the use of high annealing temperatures to improve specificity and reduce the chance of mispriming elsewhere on DNA template.

5. During the initial experimental design, the generation of wild-type and mutant control DNAs is necessary for the validation of the protocols and the assessment of the efficiency and specificity of the ARMS qPCR primers. The plasmid controls are then used in each experiment to ensure that every ARMS reaction is successful.

6. Optimal dilution of genomic DNA varies by mutation and tissue type. For m.3243A>G, DNA from blood yields an ideal threshold cycle $(C_T)$ value of 27 when it is diluted down to 0.3–0.6 ng. For other tissue types – including muscle, urine, and buccal – a concentration of 0.03–0.3 ng is best. Hair follicle samples contain higher mtDNA copy number, thus, should be further diluted.

7. For every qPCR run, there is a PC and an NC, which are 100% mutant and 100% wild-type plasmid controls, respectively. When the control plasmid DNAs are analyzed with their corresponding ARMS primers, PC yields a >99% mutation load while NC yields a <0.1% mutant load. When ARMS qPCR is performed with total cellular DNA samples, it is important to establish a correct linear correlation between the expected and

observed percentage heteroplasmy using PC and NC plasmid DNAs. Only when the standard curve of the control mixtures can be validated that the accurate measurements of heteroplasmy in clinical samples can be ensured. In addition, an NTC control is always performed to ensure that there is no contamination. Cloned PC and NC controls are diluted to 50 ng/$\mu$l as a stock solution before being further diluted into working concentrations. Dilutions vary among different mtDNA samples from different tissue type to obtain both an optimal $C_T$ and 100% wild-type and mutant results respectively. Plasmid control samples should be diluted to $2 \times 10^{-2}$ ng/$\mu$l and diluted further to reach a $C_T$ value around 27–29.

8. The SYBR Green dye is used to detect the target DNA concentration by preferentially binding to double-stranded DNA PCR products, creating a complex that absorbs blue light ($\lambda_{max} = 488$ nm) and emits green light ($\lambda_{max} = 522$ nm); this allows the targeted and amplified DNA sequence to be detected and quantified. SYBR Green is light sensitive, so once the master mixes are made, all proceeding steps should be performed in a low-light setting. When plating, it is also important to expunge the pipette slowly to help prevent bubbles in the reaction; bubbles during the PCR cycles can adversely affect the spectrophotometer reading of the SYBR Green and yield variable results from the PCR.

9. When conducting ARMS qPCR quantification, the control heteroplasmic mixture experiment is an important procedure for quality assessment. These procedures establish a correlation of $C_T$ with the degree of heteroplasmy in control DNA mixtures.

10. Remove outlier $C_T$ values within the triplicate experiments from the final calculation. If the discrepancy of $C_T$ values among the triplicates is >0.5, repeat the measurements.

# References

1. Wallace, D. C. (1999) Mitochondrial diseases in man and mouse, *Science 283*, 1482–1488.

2. Smeitink, J., van den Heuvel, L., and DiMauro, S. (2001) The genetics and pathology of oxidative phosphorylation, *Nat Rev Genet 2*, 342–352.

3. Enns, G. M., Bai, R. K., Beck, A. E., and Wong, L. J. (2006) Molecular-clinical correlations in a family with variable tissue mitochondrial DNA T8993G mutant load, *Mol Genet Metab 88*, 364–371.

4. Ashley, M. V., Laipis, P. J., and Hauswirth, W. W. (1989) Rapid segregation of heteroplasmic bovine mitochondria, *Nucleic Acids Res 17*, 7325–7331.

5. Bai, R. K., and Wong, L. J. (2004) Detection and quantification of heteroplasmic mutant mitochondrial DNA by real-time amplification refractory mutation system quantitative PCR analysis: a single-step approach, *Clin Chem 50*, 996–1001.

6. Moraes, C. T., Ricci, E., Bonilla, E., DiMauro, S., and Schon, E. A. (1992) The mitochondrial tRNA(Leu(UUR)) mutation in mitochondrial encephalomyopathy, lactic acidosis, and stroke-like episodes (MELAS): genetic, biochemical, and morphological correlations in skeletal muscle, *Am J Hum Genet 50*, 934–949.

7. Newton, C. R., Graham, A., Heptinstall, L. E., Powell, S. J., Summers, C., Kalsheker, N., Smith, J. C., and Markham, A. F. (1989)

Analysis of any point mutation in DNA. The amplification refractory mutation system (ARMS), *Nucleic Acids Res 17*, 2503–2516.

8. Wang, J., Venegas, V., Li, F., and Wong, L. J. Analysis of mitochondrial DNA point mutation heteroplasmy by ARMS quantitative PCR, *Curr Protoc Hum Genet Chapter 19*, Unit 19 16.

9. Liang, M. H., Johnson, D. R., and Wong, L. J. (1998) Preparation and validation of PCR-generated positive controls for diagnostic dot blotting, *Clin Chem 44*, 1578–1579.

# Chapter 22

# Measurement of Mitochondrial DNA Copy Number

Victor Venegas and Michelle C. Halberg

## Abstract

Mitochondrial disorders are complex and heterogeneous diseases that may be caused by molecular defects in either the nuclear or mitochondrial genome. The biosynthesis and maintenance of the integrity of the mitochondrial genome is solely dependent on a number of nuclear proteins. Defects in these nuclear genes can lead to mitochondrial DNA (mtDNA) depletion (Spinazzola et al. Biosci Rep 27:39–51, 2007). The mitochondrial DNA (mtDNA) depletion syndromes (MDDSs) are autosomal recessive disorders characterized by a significant reduction in mtDNA content. These genes include *POLG, DGUOK, TK2, TYMP, MPV17, SUCLA2, SUCLG1, RRM2B*, and *C10orf2*, all nine genes have mutations reported to cause various forms of MDDSs. In this chapter, we outline the real-time quantitative polymerase chain reaction (qPCR) analysis of mtDNA content in muscle or liver tissues.

**Key words:** mtDNA copy number, mtDNA content, mtDNA qPCR, Quantification of mtDNA content, mtDNA depletion

## 1. Introduction

Human mitochondrial DNA (mtDNA) is a small, 16.6-kb circular double-stranded DNA encoding 22 tRNA, 2 rRNA, and 13 polypeptides that encode protein subunits of the electron transport chain complexes I, III, IV, and V. Proteins involved in mtDNA replication, transcription, translation, and repair are encoded by the nuclear DNA. Molecular defects in the genes responsible for mtDNA biogenesis, and the maintenance of mtDNA integrity can lead to a series of disorders in which there is a significant reduction in the cellular mtDNA content, known as mtDNA depletion syndrome (Table 1) (2, 3). These genes include *TYMP, POLG, DGUOK, TK2, SUCLA2, MPV17, SUCLG1, RRM2B*, and *C10orf2* (1, 4–9). In particular, mutations in *TK2* and *SUCLA2*

Lee-Jun C. Wong (ed.), *Mitochondrial Disorders: Biochemical and Molecular Analysis*, Methods in Molecular Biology, vol. 837, DOI 10.1007/978-1-61779-504-6_22, © Springer Science+Business Media, LLC 2012

**Table 1**
**Mitochondrial DNA depletion disorders**

| Gene | Symbol | Clinical presentation | References |
|------|--------|----------------------|------------|
| Thymidine phosphorylase | *TYMP* | Mitochondrial neurogastrointestinal encephalomyopathy (MNGIE) | (16) |
| Thymidine kinase 2 | *TK2* | Myopathic depletion syndrome | (12) |
| Succinyl-CoA ligase, ADP binding, beta subunit | *SUCLA2* | Encephalomyopathy, Leigh-like encephalopathy, dystonia, deafness | (10, 11) |
| DNA polymerase gamma | *POLG* | Encephalopathy, hepatic impairment | (9) |
| Deoxyguanosine kinase | *DGUOK* | Mitochondrial hepatoencephalopathy, hepatic and neurologic symptoms during infancy | (5, 7) |
| MPV17, a mitochondrial inner membrane protein | *MPV17* | Infantile hepatic failure | (8) |
| TWINKLE, a DNA helicase | *PEO1* | Hepatoencephalopathy | (6, 7) |
| Succinate-CoA ligase, GDP binding, alpha subunit | *SUCLG1* | Lactic acidosis, encephalomyopathy | (13, 14) |
| P53-inducible ribonucleotide reductase subunit M2 B | *RRM2B* | Infantile encephalopathy, Leigh-like and MNGIE-like diseases | (17, 18) |

are responsible for the severe myopathic form of mtDNA depletion syndrome (10–13). Mutations in *POLG*, *DGUOK*, *MPV17*, and *TWINKLE* are responsible for the hepatocerebral form of mtDNA depletion syndrome (5–8, 14). Deficiency in *SUCLG1* causes encephalomyopathy (15), while deficiency of thymidine phosphorylase (TP) causes MNGIE (mitochondrial neurogastrointestinal encephalomyopathy) (16). Mutations in p53-inducible ribonucleotide reductase, *RRM2B* gene cause mtDNA depletion in multiple organs including muscle (17, 18). Quantitative PCR analysis of the mtDNA content in affected tissues revealed severe reduction of mtDNA copy number (5, 8, 18, 19).

MtDNA depletion is the most common cause of multisystemic mitochondrial disease and may be identified by a characteristic pattern of deficiency of complexes I, III, and IV, with relative preservation of complex II activity. In addition, muscle or liver specimens may also show increased CS activity and mitochondrial proliferation and/or abnormal mitochondria on electron microscopy (8). Analysis of mtDNA content is required for confirmation of mtDNA depletion. While gel electrophoresis and Southern blot analysis of the nuclear DNA (nDNA) and mtDNA simultaneously followed by quantitative analysis of the relative quantities of these components can also be used, the approach is labor intensive and only semiquantitative (20). Furthermore, the semiquantitative technique requires micrograms of purified DNA, significantly higher than the amount (nanograms) of DNA needed for PCR analysis. Here we describe the development of a quantitative real-time PCR approach that allows quick screening for mtDNA contents in affected tissues using a SYBR green assay.

## 2. Materials

### 2.1. Pre-PCR

1. DNA samples: DNA extracted from various clinical specimens (21).

2. DNA hydration solution for diluting DNA samples: 10-mM tris (hydroxymethyl) aminomethane hydrochloride, 0.1-mM ethylenediaminetetraacetic acid (EDTA) in sterile water (Water for Injection (WFI) Quality), pH 8.0.

3. NanoDrop ND-1000 UV–Vis spectrophotometer for DNA quantification.

4. Specific qPCR primers for mitochondrial tRNA$^{Leu(UUR)}$ gene, alternate primers for mtDNA 16S rRNA gene, and nuclear ß-2-microglobulin (ß2M) gene listed in Table 2.

5. 100-μM ARMS primer stock solution: Dissolve the powder primer (usually around 30–50 nmol) in appropriate amount of WFI quality sterile water, usually ten times the number of nmoles in μl volume. For example, if there are 35 nmol, add 350 μl of sterile water to the tube to make a 100-μM stock solution.

6. 5-μM ARMS primer working solution: Add 5 μl of 100 μM stock solution to 95 μl of WFI quality sterile water.

## Table 2
## qPCR primer sequences and conditions

| Gene | Primer name | Primer sequence (5′ to 3′) | Amplicon size (bp) | Annealing temp. (°C) |
|---|---|---|---|---|
| mtDNA tRNA$^{Leu(UUR)}$ | tRNA F3212 | CACCCAAGAACAGGGTTTGT | 107 | 62 |
| | tRNA R3319 | TGGCCATGGGTATGTTGTTA | | |
| nDNA β2-microglobulin | ß2M F594 | TGCTGTCTCCATGTTTGATGTATCT | 86 | 62 |
| | ß2M R679 | TCTCTGCTCCCCACCTCTAAGT | | |
| Alternate primers: mtDNA 16S rRNA | mtF3163 | GCCTTCCCCCGTAAATGATA | 97 | 62 |
| | mtR3260 | TTATGCGATTACCGGGCTCT | | |

7. iTaq SYBR Green Supermix with ROX (Bio-Rad).

8. 96-Well, low-profile qPCR compatible plates (Axygen PCR-96-LP-AB-C) (see Note 1).

9. Adhesive film for real-time qPCR (Axygen) (see Note 1).

**2.2. Post-PCR**

1. ABI Prism 7900HT sequence detector system (Applied Biosystems).

2. Sequence Detection System (SDS) data analysis software (version 2.2).

# 3. Methods

**3.1. qPCR Primer Design**

1. Identify regions where nuclear and mtDNA polymorphisms are rarely found (http://www.mitomap.org/MITOMAP/HumanMitoSeq). If the region is highly polymorphic, nucleotide mismatches at the qPCR primer site in the template DNA sequence may occur, which will affect the efficiency of qPCR (see Note 2).

2. Check the specificity of the primers by performing a BLAST search (http://www.ncbi.nlm.nih.gov/blast) to ensure that the primer sequences are unique for the template sequence.

3. The forward and reverse primers for qPCR should have similar melting temperature (Tm). Design the Tm for qPCR primers around 62–72°C based on the formula, $4(G+C)+2(A+T)$. The annealing temperature is determined by the lowest Tm of a primer set. The size of the amplicon is kept at around 60–150 bp.

**3.2. qPCR Setup**

1. Quantify concentration of DNA samples using NanoDrop ND-1000 UV–Vis spectrophotometer. Serially dilute the DNA samples using DNA hydration solution, beginning with a dilution of 2 ng/µl from the stock DNA and further diluting down to 0.1 ng/µl or less, depending on the tissue type (see Note 3).

2. qPCR mixture: Each sample is amplified with tRNA$^{\text{Leu(UUR)}}$ and nuclear ß-2-microglobulin (ß2M) primer pairs (Table 2) in separate wells in triplicate in a 96-well plate (see Note 4). In each reaction add 1 µl of 5 µM forward primer, 1 µl of 5 µM reverse primer (primer final concentration is 0.5 µM), 5 µl 2× SYBR SuperMix (iTaq SYBR Green Supermix with ROX, Bio-Rad), and 3 µl of 0.1 ng/µl template DNA (from muscle or liver) for a total volume of 10 µl (see Note 5).

3. The qPCR is set up for plating on a 96-well qPCR plate and ran on a Fast plate block in the ABI Prism 7900HT sequence detector system. The cycle parameters are illustrated in Fig. 1.

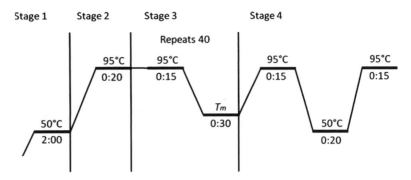

Fig. 1. Cycle parameters utilized for amplification by qPCR.

4. Follow the manufacturer's instructions for the ABI Prism 7900HT sequence detector system by using the standard format with all 96 wells selected. The program takes approximately 1 h and 15 min to complete. When done, the resulting data will show the $C_T$ values of each individual well.

5. The resulting data can be exported in .txt format to be analyzed, and the used 96-well plate containing PCR product can be disposed of properly.

***3.3. Analysis of Data***

1. Each triplicate $C_T$ value for a DNA sample must be within a deviation of no more than $0.5\,C_T$ unit from each other, if not, any outliers should be deleted and an average taken from the remaining duplicate $C_T$ values. If all the $C_T$ values of the triplicate differ from each other in greater than $0.5\,C_T$ unit, the triplicate measurements should be repeated. The large deviation is most likely due to pipetting errors or bubble formation.

2. The SDS data analysis software will generate a spreadsheet with the $C_T$ value for each qPCR. To calculate the mtDNA content, find the difference in $C_T$ values between the tRNA$^{\text{Leu(UUR)}}$ and ß2M genes (average $C_T$ value for the triplicate reactions), termed $\Delta C_T$.

$$(\text{ß2M average } C_T) - (\text{tRNA}^{\text{Leu(UUR)}} \text{ average } C_T) = \Delta C_T$$

$2 \times 2(\Delta C_T) = \text{mtDNA content}$.
Relative mtDNA content = mtDNA patient/mtDNA control.

3. MtDNA content can vary among different tissues of an individual, and is dependent on a person's age. To generate an age-matched control, take DNA samples from ten different age-matched nonsymptomatic individuals who do not carry mtDNA point mutations or deletions, and are within one standard deviation of the age-matched mean, and mix the DNA in equal amounts. The mtDNA content in age- and tissue-matched controls should always be established first. In Table 3,

**Table 3**
**qPCR results for three samples with increased mtDNA, normal control, and mtDNA depletion**

| Sample | Age | B2M 1 | B2M 2 | B2M 3 | Average | tRNA 1 | tRNA 2 | tRNA 3 | Average | $\Delta C_\tau$ | Total | Ctr | Range | % |
|--------|-----|-------|-------|-------|---------|--------|--------|--------|---------|---------|-------|-----|-------|---|
| 1 Muscle | 8.8 | 27.78 | 27.72 | 27.92 | 27.81 | 15.44 | 15.48 | 15.49 | 15.47 | −12.33 | 10,333 | 4,814 | 3,194–6,198 | 215 |
| 2 Muscle | 17.0 | 27.68 | 27.53 | | 27.60 | 18.30 | 18.35 | 18.56 | 18.4 | −9.199 | 1,176.1 | 4,933 | 3,059–6,462 | 24 |
| 3 Liver | | 27.71 | 28.01 | 27.87 | 27.87 | 17.66 | 17.69 | 17.68 | 17.68 | −10.18 | 2,329.8 | 3,134 | 2,052–3,646 | 74 |
| Muscle con. | 7–12 | 28.46 | 28.65 | 28.36 | 28.49 | 17.03 | 17.11 | 17.19 | 17.11 | −11.38 | 5,333.0 | 4,814 | 3,194–6,198 | 111 |
| Liver con. | | 28.81 | | 28.54 | 28.70 | 18.00 | 18.10 | 18.20 | 18.10 | −10.57 | 3,047.2 | 3,134 | 2,052–3,646 | 97 |

we calculated mtDNA content for three samples with increased mtDNA (muscle sample 1), normal mtDNA (liver sample), and mtDNA depletion (muscle sample 2).

# 4. Notes

1. The 96-well plate must be compatible with the ABI Prism 7900HT sequence detector system which requires Fast Optical PCR plates. The single-piece 96-well polypropylene plates have consistent well-to-well thickness for PCR process uniformity. The plates should be sealed with optically clear adhesive film to prevent cross contamination and sample evaporation.

2. Highly polymorphic mtDNA regions may also contain common deletion. The primers should be chosen from regions where mtDNA deletions rarely occur. The mitochondrial tRNA$^{Leu\,(UUR)}$ gene region is rarely deleted and contains only a few rare single nucleotide polymorphisms (mtSNP) (22). Similar to the mitochondrial tRNA$^{Leu\,(UUR)}$ gene region, the D-loop region rarely contains deletions but is more susceptible to mtSNP, potentially leading to primer mismatch. Primer mismatch may cause false-positive results showing apparent mtDNA depletion. Therefore, all depletions should be confirmed by using the alternative primers. SNPs at the primer site may be confirmed by sequencing.

3. Dilute muscle and liver DNA samples to a concentration of 0.1 ng/μl, and blood samples around 1 ng/μl, concentrations that appear to be optimal under our conditions. It is our experience that copy numbers generated from blood specimens have low specificity for the diagnosis of mtDNA depletion syndrome and therefore do not recommend the use of blood DNA for clinical diagnosis.

4. In each experimental run, include a tissue- and age-matched pooled control (a pool of ten matched samples) to ensure that the controls are always within the control range. Include a no template control (NTC).

5. The SYBR green dye is used to detect the target DNA concentration by preferentially binding to double-stranded DNA PCR products, creating a complex that absorbs blue light ($\lambda_{max} = 488$ nm) and emits green light ($\lambda_{max} = 522$ nm); this allows the targeted and amplified DNA sequence to be detected and quantified. SYBR Green is light sensitive, so once the master mixes are made, all proceeding steps should be performed in a low-light setting. When plating, it is also important to expunge the pipette slowly to help prevent bubbles in the reaction: bubbles during the PCR cycles can adversely affect the spectrophotometer reading of the SYBR Green and yield variable results from the PCR.

# References

1. Spinazzola, A., and Zeviani, M. (2007) Disorders of nuclear-mitochondrial intergenomic communication, *Biosci Rep 27*, 39–51.

2. Antonicka, H., Mattman, A., Carlson, C. G., Glerum, D. M., Hoffbuhr, K. C., Leary, S. C., Kennaway, N. G., and Shoubridge, E. A. (2003) Mutations in COX15 produce a defect in the mitochondrial heme biosynthetic pathway, causing early-onset fatal hypertrophic cardiomyopathy, *Am J Hum Genet 72*, 101–114.

3. Tay, S. K., Shanske, S., Kaplan, P., and DiMauro, S. (2004) Association of mutations in SCO2, a cytochrome c oxidase assembly gene, with early fetal lethality, *Arch Neurol 61*, 950–952.

4. Spinazzola, A., and Zeviani, M. (2009) Disorders from perturbations of nuclear-mitochondrial intergenomic cross-talk, *J Intern Med 265*, 174–192.

5. Dimmock, D. P., Zhang, Q., Dionisi-Vici, C., Carrozzo, R., Shieh, J., Tang, L. Y., Truong, C., Schmitt, E., Sifry-Platt, M., Lucioli, S., Santorelli, F. M., Ficicioglu, C. H., Rodriguez, M., Wierenga, K., Enns, G. M., Longo, N., Lipson, M. H., Vallance, H., Craigen, W. J., Scaglia, F., and Wong, L. J. (2008) Clinical and molecular features of mitochondrial DNA depletion due to mutations in deoxyguanosine kinase, *Hum Mutat 29*, 330–331.

6. Hakonen, A. H., Isohanni, P., Paetau, A., Herva, R., Suomalainen, A., and Lonnqvist, T. (2007) Recessive Twinkle mutations in early onset encephalopathy with mtDNA depletion, *Brain 130*, 3032–3040.

7. Sarzi, E., Goffart, S., Serre, V., Chretien, D., Slama, A., Munnich, A., Spelbrink, J. N., and Rotig, A. (2007) Twinkle helicase (PEO1) gene mutation causes mitochondrial DNA depletion, *Ann Neurol 62*, 579–587.

8. Wong, L. J., Brunetti-Pierri, N., Zhang, Q., Yazigi, N., Bove, K. E., Dahms, B. B., Puchowicz, M. A., Gonzalez-Gomez, I., Schmitt, E. S., Truong, C. K., Hoppel, C. L., Chou, P. C., Wang, J., Baldwin, E. E., Adams, D., Leslie, N., Boles, R. G., Kerr, D. S., and Craigen, W. J. (2007) Mutations in the MPV17 gene are responsible for rapidly progressive liver failure in infancy, *Hepatology 46*, 1218–1227.

9. Wong, L. J., Dimmock, D., Geraghty, M. T., Quan, R., Lichter-Konecki, U., Wang, J., Brundage, E. K., Scaglia, F., and Chinault, A. C. (2008) Utility of oligonucleotide array-based comparative genomic hybridization for detection of target gene deletions, *Clin Chem 54*, 1141–1148.

10. Carrozzo, R., Dionisi-Vici, C., Steuerwald, U., Lucioli, S., Deodato, F., Di Giandomenico, S., Bertini, E., Franke, B., Kluijtmans, L. A., Meschini, M. C., Rizzo, C., Piemonte, F., Rodenburg, R., Santer, R., Santorelli, F. M., van Rooij, A., Vermunt-de Koning, D., Morava, E., and Wevers, R. A. (2007) SUCLA2 mutations are associated with mild methylmalonic aciduria, Leigh-like encephalomyopathy, dystonia and deafness, *Brain 130*, 862–874.

11. Elpeleg, O., Miller, C., Hershkovitz, E., Bitner-Glindzicz, M., Bondi-Rubinstein, G., Rahman, S., Pagnamenta, A., Eshhar, S., and Saada, A. (2005) Deficiency of the ADP-forming succinyl-CoA synthase activity is associated with encephalomyopathy and mitochondrial DNA depletion, *Am J Hum Genet 76*, 1081–1086.

12. Galbiati, S., Bordoni, A., Papadimitriou, D., Toscano, A., Rodolico, C., Katsarou, E., Sciacco, M., Garufi, A., Prelle, A., Aguennouz, M., Bonsignore, M., Crimi, M., Martinuzzi, A., Bresolin, N., Papadimitriou, A., and Comi, G. P. (2006) New mutations in TK2 gene associated with mitochondrial DNA depletion, *Pediatr Neurol 34*, 177–185.

13. Ostergaard, E., Hansen, F. J., Sorensen, N., Duno, M., Vissing, J., Larsen, P. L., Faeroe, O., Thorgrimsson, S., Wibrand, F., Christensen, E., and Schwartz, M. (2007) Mitochondrial encephalomyopathy with elevated methylmalonic acid is caused by SUCLA2 mutations, *Brain 130*, 853–861.

14. Wong, L. J., Naviaux, R. K., Brunetti-Pierri, N., Zhang, Q., Schmitt, E. S., Truong, C., Milone, M., Cohen, B. H., Wical, B., Ganesh, J., Basinger, A. A., Burton, B. K., Swoboda, K., Gilbert, D. L., Vanderver, A., Saneto, R. P., Maranda, B., Arnold, G., Abdenur, J. E., Waters, P. J., and Copeland, W. C. (2008) Molecular and clinical genetics of mitochondrial diseases due to POLG mutations, *Hum Mutat 29*, E150–E172.

15. Ostergaard, E., Christensen, E., Kristensen, E., Mogensen, B., Duno, M., Shoubridge, E. A., and Wibrand, F. (2007) Deficiency of the alpha subunit of succinate-coenzyme A ligase causes fatal infantile lactic acidosis with mitochondrial DNA depletion, *Am J Hum Genet 81*, 383–387.

16. Nishino, I., Spinazzola, A., and Hirano, M. (1999) Thymidine phosphorylase gene mutations in MNGIE, a human mitochondrial disorder, *Science 283*, 689–692.

17. Bourdon, A., Minai, L., Serre, V., Jais, J. P., Sarzi, E., Aubert, S., Chretien, D., de Lonlay, P., Paquis-Flucklinger, V., Arakawa, H., Nakamura, Y., Munnich, A., and Rotig, A. (2007) Mutation of RRM2B, encoding p53-controlled ribonucleotide reductase (p53R2), causes severe mitochondrial DNA depletion, *Nat Genet 39*, 776–780.

18. Shaibani, A., Shchelochkov, O. A., Zhang, S., Katsonis, P., Lichtarge, O., Wong, L. J., and Shinawi, M. (2009) Mitochondrial neurogastrointestinal encephalopathy due to mutations in RRM2B, *Arch Neurol 66*, 1028–1032.

19. Lee, N. C., Dimmock, D., Hwu, W. L., Tang, L. Y., Huang, W. C., Chinault, A. C., and Wong, L. J. (2009) Simultaneous detection of mitochondrial DNA depletion and single-exon deletion in the deoxyguanosine gene using array-based comparative genomic hybridisation, *Arch Dis Child 94*, 55–58.

20. Shanske, S., and Wong, L. J. (2004) Molecular analysis for mitochondrial DNA disorders, *Mitochondrion 4*, 403–415.

21. Venegas, V., Wang, J., Dimmock, D., and Wong, L. J. Real-time quantitative PCR analysis of mitochondrial DNA content, *Curr Protoc Hum Genet Chapter 19*, Unit 19 17.

22. Bai, R. K., and Wong, L. J. (2005) Simultaneous detection and quantification of mitochondrial DNA deletion(s), depletion, and over-replication in patients with mitochondrial disease, *J Mol Diagn 7*, 613–622.

# Chapter 23

# Determination of the Clinical Significance of an Unclassified Variant

## Victor Wei Zhang and Jing Wang

## Abstract

After completion of Human Genome Project (HGP) in 2003, as well as the new technology development in genomic research, the most accurate genetics blueprint of human is available. Researchers started to dissect and understand the genetic map of the human species. As a consequence, analyses of novel or unclassified genetic variations become increasingly important in translational medicine. One of the medical specialties in modern medicine is clinical genetics, which is overseen by the American Board of Medical Genetics (ABMG). In 2008, ABMG published a guideline for interpretation of new variants using ACMG Standards and Guidelines (Richards et al. Genet Med 10:294–300, 2008). In this chapter, we provide updated procedures of evaluating different databases, computational tools, and structural analysis methods that we currently utilize to assist in clinical interpretation.

**Key words:** Genome Database, Computation, Structure Analysis, Amino acid conservation, VOUS, Unclassified variants

## 1. Introduction

After the completion of Human Genome Project (HGP) in 2003 and the technological innovation in genomics research, the most accurate genetics blueprint of human is available. The HGP provides a detailed human haploid genetics map as reference sequence with more than three billion base pair in length. For each individual, his/her individual genetic makeup is unique as compared to the reference map or other individuals. The understanding of these types of genetic variations is essential in elucidating evolution of human species. Moreover, in clinical settings, it has been realized that a single such variant may be the determinant factor in disease development.

Lee-Jun C. Wong (ed.), *Mitochondrial Disorders: Biochemical and Molecular Analysis*, Methods in Molecular Biology, vol. 837, DOI 10.1007/978-1-61779-504-6_23, © Springer Science+Business Media, LLC 2012

Any novel or unclassified sequence variations need to be evaluated correctly. The interpretation will have direct clinical impact on the management of patient care. Thus, accurate analyses of these novel or unclassified genetic variations become increasingly important in translational medicine.

One of the medical specialties in modern medicine is clinical genetics, which is overseen by the American Board of Medical Genetics (ABMG). The interpretation of molecular testing results requires a board-certified medical geneticist with a broad range of genetics knowledge, clinical experience, and molecular laboratory trainings to ensure that accurate diagnostic information of complex molecular results are conveyed to the patient's family. It is of great importance to have qualified molecular laboratory directors review and interpret each molecular testing result to help physicians make accurate diagnosis for patient care. The molecular testing result is indispensable information in making clinical correlation for diagnosis, prognosis, treatment, prenatal assessment, and evaluation of family members.

In 2008, ABMG published a guideline for the interpretation of new variants using ACMG Standards and Guidelines (1). In this chapter, we provide updated procedures of the determination of a variant of unknown significance (VOUS) using various available databases, computational tools, and structural analytical methods.

## 2. Materials

The procedures are in silico analyses that do not require materials or reagents for wet-bench experiments. The materials are mostly publicly available software on the Internet.

### 2.1. Internet Access and General Databases

1. Human Genome Mutation Database (http://www.hgmd.cf.ac.uk) for published literatures. Academic users can register at this web site to obtain a password to access a version of this database with limited data entry, free-of-charge. Professional license is required for a fee-based registration in order to have full access of the database.

2. Single-Nucleotide Polymorphism Database (dbSNP: http://www.ncbi.nlm.nih.gov/projects/SNP) is a free NCBI archive for genetic variations.

3. PubMed (http://www.ncbi.nlm.nih.gov/PubMed/): an NCBI database for archiving MEDLINE database of references on life sciences.

4. Online Mendelian Inheritance in Man (OMIM, http://www.ncbi.nlm.nih.gov/omim): an NCBI database that catalogs the known human diseases with genetic causes.

5. Google: a powerful search engine.

**2.2. Disease-Specific Databases for Nuclear Genes-Related Mitochondrial Disorders**

1. For POLG-specific mutation database, http://tools.niehs.nih.gov/polg/.
2. For OPA1: http://lbbma.univ-angers.fr/lbbma.php?id=9.

**2.3. Databases Specific for Mitochondrial DNA (mtDNA) Variants**

1. MitoMap: http://www.mitomap.org/MITOMAP.
2. mtDB: http://www.mtdb.igp.uu.se/.
3. For tRNA variants: http://mamit-trna.u-strasbg.fr/Summary.asp.

**2.4. Computational Algorithms for the Prediction of Pathogenicity of Missense Variants**

1. SIFT (Sorting Intolerant From Tolerant): http://sift.jcvi.org.
2. PolyPhen (Polymorphism Phenotyping): http://genetics.bwh.harvard.edu/pph (This Web site is no longer being maintained). (see Note 1).
3. PolyPhen-2: http://genetics.bwh.harvard.edu/pph2 (currently being used).

**2.5. Computational Algorithms for the Prediction of Pathogenicity of Splice Site Alterations**

1. ESE Finder 2.0: http://rulai.cshl.edu/tools/ESE2.
2. NetGene2: http://www.cbs.dtu.dk/services/NetGene2.
3. BDGP Splice Site Predictor: http://www.fruitfly.org/seq_tools/other.html.

# 3. Methods

**3.1. Interpretation of Missense Variants (see Note 2)**

Record the triplet codon and double-check the nucleotide and amino acid positions. If accurate, proceed to the following search. For any missense variants, search disease-related Web sites and publications in the order listed below. This procedure will provide a complete survey of a particular missense variant, and it is beneficial to go through these steps to gather available information and literature to help interpret the molecular finding.

*3.1.1. Human Genome Mutation Database*

1. Check Human Genome Mutation Database for published literatures (password required) (2), go to Web site (https://portal.biobase-international.com/hgmd/pro/start.php).
2. After log-in into HGMD, click right column on "GENE," type in gene name in search terms for "Exact gene symbol only," and click on "enter." A summary window appears showing the number of entries for each category of mutations, e.g., missense/nonsense, splicing, regulator, small deletions, small insertions, small indels, gross deletions, gross insertions, complex rearrangements, or repeat variations. The total number of

mutations reported in HGMD for this specific gene is at the end of this table. Alternatively, if exact gene name is not known, use gene description, chromosomal location, OMIM ID, etc., as search terms to get the necessary information.

3. After clicking the gene name, it will link to a detailed page for categories of mutations. Click on the correct category, e.g., "missense mutations," and check if the variant of interest is in the list.

4. If the mutations can be found in this category, usually a few references with the first author's name, publication date, and journal information are provided. Retrieve and review these literatures thoroughly with the qualified laboratory directors (see Note 3). If the mutations/changes cannot be found in all of the categories in HGMD, proceed to the next step.

5. Check the reported literatures and make sure that sufficient functional evidence supporting pathogenicity of the mutation, e.g., mRNA study, protein functional assay, or in vivo investigation, is provided. If not, the pathogenicity of the variant is still in question.

6. Besides HGMD and the literatures cited, additional information also may be provided in the "extra information" column, e.g., "G" for genomic coordinates, "DM" for disease-causing mutation, "FP" for functional polymorphism, etc.

*3.1.2. Single-Nucleotide Polymorphisms*

1. Search dbSNP by going to Web site (http://www.ncbi.nlm.nih.gov/projects/SNP/).

2. Enter gene name, click "Go," and select the organism to be *Homo sapiens.*

3. Click "View more variation on this gene" in the GeneView section (GeneView via analysis of contig annotation) and select options "Include clinically associated" and "in gene region," then click "Go." A page of complete single-nucleotide polymorphism (SNP) entries associated with this gene will show up. Check to see if this particular variant is listed as a polymorphism.

4. Alternatively, if the sequence containing the nucleotide changes, the sequence can be used as a query to search dbSNP database by "BLASTN" program (http://www.ncbi.nlm.nih.gov/SNP/snpblastByChr.html).

5. If the change can be located in dbSNP, record the frequency. We generally consider the change to be benign if the average heterozygosity $\geq 5\%$. This is conservative estimation compared to the general definition of polymorphism as 1%.

6. In general, benign variants are those not recorded in HGMD but found in dbSNP database with an average heterozygosity

of no less than 5%. A missense variant may be considered as a candidate for a mutation if it has an entry in HGMD or is marked as "clinically associated" in dbSNP. If the variant is recorded in dbSNP, but not in HGMD, and with an average heterozygosity of <5%, it is generally considered as having insufficient information to be defined as benign or deleterious. If this is the case, this variant is of unknown significance (VOUS).

*3.1.3. Other Databases or Resources in the Public Domain*

1. PubMed database, a public recourse for biomedical literatures, is provided by US National Library of Medicine (NLM). To search literatures in PubMed, go to PubMed Web site: http://www.ncbi.nlm.nih.gov/PubMed/, enter gene or disease name or mutation name to the query field, and click "search." Published papers on the searched items will appear as a list. Glance through the list and click on the link to read about the publications.

2. OMIM in NCBI Web page is a database for human genetic disorders. Go to http://www.ncbi.nlm.nih.gov/omim and enter the name of the gene or disease or mutation to the query field. A list of related gene/disease will appear. OMIM database provides more detailed information regarding the gene and related disease. For example, an asterisk (*) before the name of the gene indicates a known gene, a number sign (#) represents phenotype description with all identified loci of a disorder, and a plus sign (+) means a known gene and phenotype. For other information, refer to OMIM Frequently Asked Questions (FAQs) at (http://www.ncbi.nlm.nih.gov/Omim/omimfaq.html).

3. Google is a powerful search engine. Type in gene name, amino acid, or nucleotide change and click "search" or "go" for any publications regarding the variant of interest. Google scholar is also very helpful in finding related literatures, but it is less comprehensive than PubMed.

4. There are several disease-specific databases that can be further searched for publications and additional information, including expression analysis of a particular mutant protein. For example: POLG (http://tools.niehs.nih.gov/polg/) (3) and OPA1 (http://lbbma.univ-angers.fr/lbbma.php?id=9) (4). It should be straightforward to obtain related information if the variant is known.

5. If the variant has been previously reported, record the citation of the publication and print out references for review by qualified laboratory director. If the variant has never been reported in any of the above-mentioned databases, proceed to Subheading 3.1.4 for unclassified variants.

*3.1.4. Novel Variants
of Unknown Significance*

If the missense variants have not been reported in any of the databases, in silico analyses are performed in the following order:

1. Protein sequence conservation analysis is the first step to infer functional importance of an amino acid residue at a specific position. Protein sequence containing the amino acid of interest will be used as a query for analysis at NCBI protein blast Web site (http://blast.ncbi.nlm.nih.gov/Blast.cgi). Click "protein blast" under the basic blast, input the protein sequence query, and select reference proteins (refseq_protein) for search set. Click "BLAST" of the blue button. A list of orthologue proteins in different species will appear in the result page. Select and retrieve orthologue protein from different organisms throughout the evolution. We recommend the selection of the following species, if available: *H. sapiens, Bos taurus, Mus musculus/Rattus norvegicus, Gallus gallus, Xenopus laevis, Danio rerio, Drosophila melanogaster, Strongylocentrotus droebachiensis, Caenorhabditis elegan*, and *Saccharomyces cerevisiae*. Save these protein sequences, and go to http://www.ebi.ac.uk/Tools/msa/clustalw2/ Web site for multiple sequence alignment. Follow the instruction in the Web site to input these protein sequences in FASTA format, select alignment type "slow," and click "submit." When the result page appears, click "show colors" to have conserved residues colored and print the result page or save the page for future reference. Mark the amino acid position of interest (Fig. 1).

2. SIFT (Sorts Intolerant From Tolerant substitutions) is a program to predict the likely effect of the nonsynonymous substitution on protein function (5). At the SIFT Web site (http://sift.jcvi.org/), click "SIFT sequence." A protein sequence query window will appear. Input protein sequence in FASTA format and click "submit query" button. After a short period of time, a prediction page that shows the tolerant and intolerant substitutions at all positions of the protein sequence will show up. Locate the amino acid substitution of interest.

3. PolyPhen-2 (Polymorphism Phenotyping v2) is another tool to predict effect of nonsynonymous substitution based on the structure and function of a human protein (6). Access PolyPhen-2 program at http://genetics.bwh.harvard.edu/pph2/.

```
                                         c.113C>T (p.T38I)
ACADL_Human     APRQLPAARCSHSGGEE-----RLETPSAKKLTD 46
ACADL_Pig       APRLPTASRCSHSGGEE-----RLESPSAKKLTD 46
ACADL_Mouse     APRPLPSARCSHSGAEA-----RLETPSAKKLTD 46
ACADL_Chicken   AFASQPSPAPAEQHGTK-----RLEPSSAKRLTD 47
ACADL_Frog      ASAVSVPNRYQHSHPKSSDERPRLETSQAKSLMD 60
ACADL_Zabrafish AVAVSRMQHAELQHGETAAP-FRPETSMAKTLMD 58
```

Fig. 1. Analysis of amino acid conservation by sequence alignment. Alignment of ACADL protein sequences from multiple species shows moderate conservation of threonine at position 38.

Input protein sequence in FASTA format, the position of the amino acid variant, the wild-type amino acid, and the nonsynonymous variant. Click "submit query." On the top of the result page, it shows the query information and the description of the protein. The first part of the result is the prediction/confidence score and the associated categories, such as damaging, benign, or unknown. A sequence alignment also highlights the position within the wild-type sequence. Sometimes the 3D visualization window is also provided. Save and/or print out the results.

4. For a special situation, if the nonsynonymous variant is caused by nucleotide change close to the beginning or the end of an exon, additional analyses for possible effect on splicing must be performed. See Subheading 3.3 below.

***3.2. Interpretation of Deletions/ Duplications (see Note 4)***

1. If the observed deletion/duplication has not been previously reported, but results in frameshift and truncation of the protein, according to ACMG guidelines, the deletion/duplication is classified as a deleterious mutation (1).

2. If the deletion/duplication is novel, but it results in in-frame amino acid insertion or deletion, then it is classified as an unknown variant. Protein alignment should also be checked (see Subheading 3.1.4).

3. The following Web sites are to be searched in this order: HGMD, PubMed, OMIM, and Google as described in Subheading 3.1. The size of deletion/duplication detected by sequencing analysis is relatively small (usually less than 100 bp) compared to that detected by cytogenetic study.

***3.3. Interpretation of Intronic and Close-to-Exon–Intron-Boundary Variants***

1. The 5′ splice donor site contains an almost invariant sequence GU at the 5′ end of the intron. The 3′ splice acceptor site terminates the intron with an almost invariant AG sequence. If a nucleotide substitution changes one of these invariant splice site sequences, it is considered to be a deleterious mutation according to ACMG guidelines (1).

2. ESE Finder 2.0 is an algorithm for the prediction of how a sequence change in an exon-interfering exonic splicing enhancer (ESE) (7). For more information, please see the review paper on this topic (8). At the ESE Finder Web site, http://rulai.cshl.edu/tools/ESE2, enter the wild-type and the mutant sequences of interest in text format or FASTA format, and click "Send." The results of ESE Finder analysis will appear shortly. Print out the results for both wild-type and mutant sequences. If a variant is predicted to either abolish the original splice site or create a new splice site, the variant is considered likely to be deleterious.

3. NetGene2: a neural network-based prediction of splice sites for human transcript (9). Go to http://www.cbs.dtu.dk/services/NetGene2/. Input the wild-type sequence spanning the exon–intron boundary, and click "send file." The result page will appear with donor site and acceptor splice site predictions above the threshold. It also contains splice predictions for complement strand, which may not be useful for the sequence of interest. Input the mutant sequence spanning the exon–intron boundary, and click "send file." Print out the prediction results from NetGene2 for both wild-type and mutant sequences. If a variant is predicted to either abolish the original splice site or create a new splice site, the variant is considered likely to be deleterious.

4. The splice site predictor at Berkeley Drosophila Genome Project (BDGP) is based on neural network recognition of donor/acceptor splice site in a DNA sequence (10). The development of this algorithm is applied to both human and *D. melanogaster* genes. Go to the BDGP Web site: http://www.fruitfly.org/seq_tools/splice.html, and select the organism to be human. Input both wild-type and mutant sequences spanning the exon–intron boundary, and click "Submit." Print out the prediction results. The prediction also produces a confidence score based on the algorithm. If a variant is predicted to either abolish the original splice site or create a new splice site, the variant is considered likely to be deleterious.

5. A novel variant in the vicinity of exon–intron boundary may disrupt the mRNA splicing. If these prediction algorithms produce consistent results for an altered splice site, the variant is predicted likely to be deleterious. It should be noted that each algorithm has its sensitivity and specificity for the prediction of splice site changes. Some splice changes may not be detected by certain computational methods. However, they may still likely to interfere with the mRNA splicing.

### 3.4. Interpretation of Mitochondrial DNA (mtDNA) Variants (see Note 5)

1. For all mtDNA variants, both MitoMap and mtDB are used. MitoMap is a very useful database that contains human mitochondrial DNA variations from both published and unpublished sources. All entries are curated carefully by MitoMap, and server is currently maintained at The Children's Hospital of Philadelphia. Go to the MitoMap Web site at http://www.mitomap.org/MITOMAP, and under the text "MITOMAP Quick Reference" to search for point mutations, click "here" button, which will lead to allele search page. The input can be a single-nucleotide position or a range (<100 bps). Usually three categories will show up: (1) sequence polymorphisms, (2) mutations, and (3) mtDNA somatic mutations.

If the nucleotide position of interest can be identified in one of these categories, go to reference pages to get the related publications. If not, then record "NOT found in MITOMAP."

2. Human Mitochondrial Genome Database (mtDB) is another resource with comprehensive documentation of database of complete human mitochondrial variations. It contains mitochondrial DNA sequences from individuals who were healthy at the time of ascertainment, and the frequency information is very helpful in the evaluation of the nature of an mtDNA variant. Some variants may appear to be rare due to ethnic underrepresentation in the database (11). Go to mtDB Web site: http://www.genpat.uu.se/mtDB, and click "Search." A "Search mtDB" page will appear. Input the position of a single nucleotide (or up to 10 positions) and click "Do it." The result page will show the allele frequency at that position. Record the nucleotide change and the frequencies.

3. If the mtDNA variant occurs in the mitochondrial tRNA genes, the Mamit-tRNA database should be checked. The Mamit-tRNA database contains mammalian mitochondrial tRNAs with an emphasis on the structural characteristics of these tRNAs. It documents extensively on point mutations in mitochondrial tRNA genes related to human mitochondrial disorders (12). Go to Mamit-tRNA Web site at http://mamit-trna.u-strasbg.fr/human.asp, and click on the tRNA gene of interest. This will lead to the result page with schematic 2D cloverleaf representation of tRNA and a list of mutations, the associated disease phenotypes, and corresponding references.

4. For missense variants, follow the procedures detailed in Subheading 3.1.

## 4. Notes

1. PolyPhen-2 is a new version of the PolyPhen tool for annotating coding nonsynonymous SNPs. PolyPhen-2 contains enhanced multiple sequence alignments and new algorithms for variant classification that were not used in PolyPhen. Thus, there are possible discordant predictions between PolyPhen and PolyPhen-2.

2. Here we describe how these procedures are applied to define the pathogenic significance of a novel missense variant. A heterozygous novel variant, c.2987G>A (p.R996Q), in the POLG1 gene was detected by sequencing. POLG-related disorders are a group of diseases with overlapping clinical symptoms. DNA polymerase gamma is the only DNA polymerase

found in animal cell mitochondria. It bears sole responsibility for mitochondrial DNA biosynthesis. Mutations in *POLG1*, the catalytic subunit of polymerase gamma, can cause autosomal dominant or autosomal recessive disorders. By searching HGMD and dbSNP as the first step, the c.2987G>A (p.R996Q) in the *POLG1* gene was not found. Followed by PubMed, OMIM, and Google search, it was not found in any of the literatures searched. Protein alignment, PolyPhen-2, and SIFT analyses were performed (see Fig. 2). The result showed that arginine at amino acid position 996 of the POLG protein is not evolutionarily conserved. SIFT and PolyPhen-2 algorithms predict this variant to be benign. Thus, the clinical significance of this variant is unclear, and this variant is classified as a variant of unknown significance (VOUS).

3. There may be inconsistencies among the published data regarding the pathogenicity of mutations. If this is the case, procedures in Subheadings 3.1.2–3.1.4 may help to clarify the inconsistency. Occasionally, the mutation nomenclature in published literatures may not be in accordance with the updated HGVS guidelines. In this case, original gene sequence should be used to verify the nucleotide position and/or amino acid position.

4. Using a heterozygous c.2708_2711delTTAG (p.V903GfsX3) mutation detected in the *OPA1* gene as an example, the application of these procedures is described here. Mutations in the *OPA1* gene cause autosomal dominant optic atrophy (DOA), which is the most common form of hereditary optic neuropathy (13). Sequence analysis detected heterozygous c.2708_2711delTTAG (p.V903GfsX3) mutation in the *OPA1* gene. By checking the HGMD database, this mutation has been reported in OPA1 patients (14). Thus, the c.2708_2711delTTAG (p.V903GfsX3) is a reported deleterious mutation.

5. Mitochondrial disorders are a group of genetically and clinically heterogeneous diseases that may be caused by mutations in either nuclear or mitochondrial genes. Mitochondrial disorders caused by defects in mtDNA are maternally inherited. Procedures described in Subheading 3.4 are used to interpret mtDNA mutations. For example, an apparently homoplasmic novel variant, m.2765A>G (16S rRNA), and apparently homoplasmic rare variants, m.1628C>T (tRNA Val) and m.8578C>T (p.P18S, ATP6) were detected in a DNA sample by sequencing. The m.2765A>G variant has not been reported in the MitoMap database and mtDB. The m.2765A>G is located in the 16S ribosomal RNA gene, but its significance is not clear. The m.1628C>T is located in the anticodon stem of the tRNA Val. It changes a C:A mispairing to U:A pairing. The m.8578C>T variant predicts the substitution of proline with

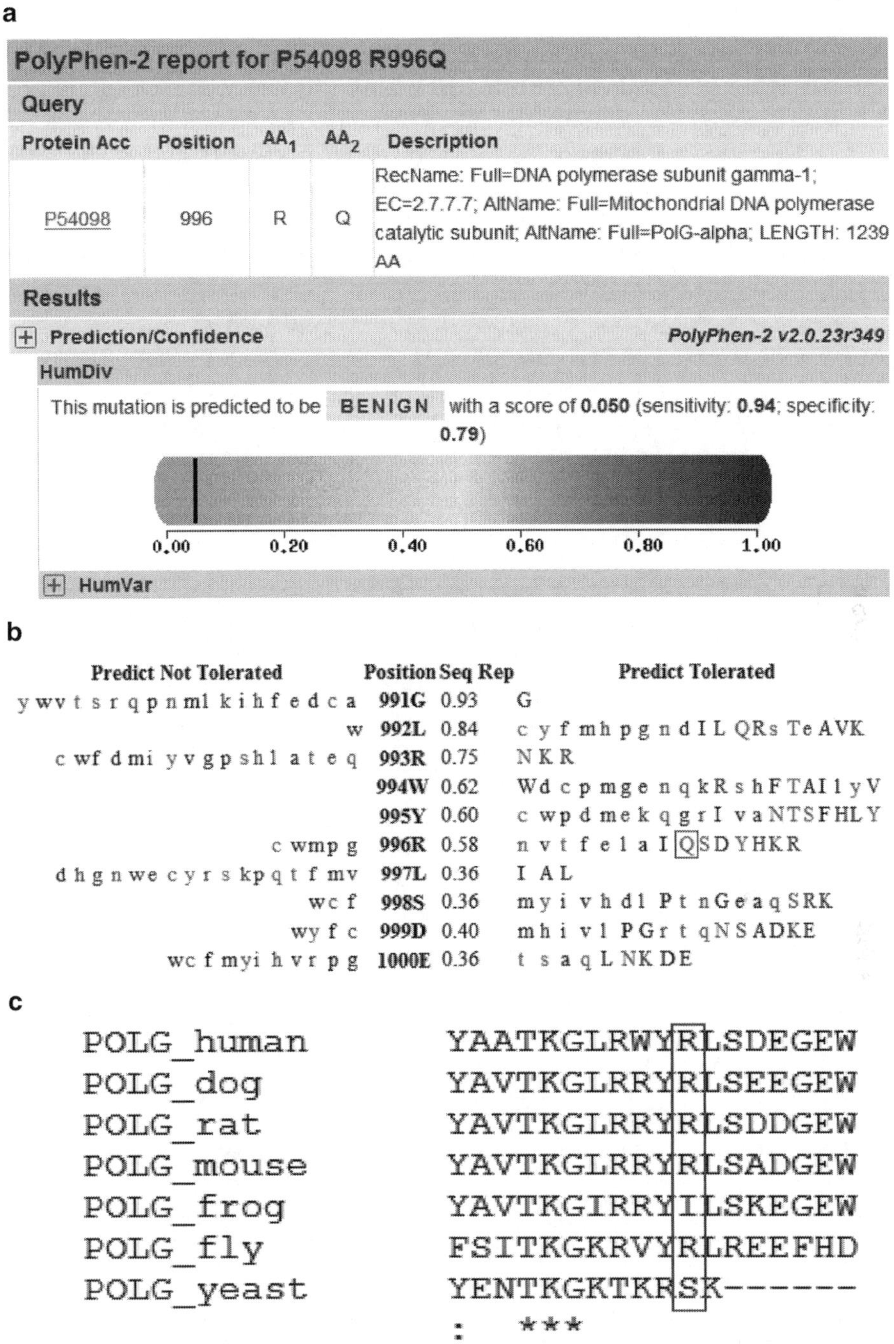

**a**

**PolyPhen-2 report for P54098 R996Q**

**Query**

| Protein Acc | Position | AA₁ | AA₂ | Description |
|---|---|---|---|---|
| P54098 | 996 | R | Q | RecName: Full=DNA polymerase subunit gamma-1; EC=2.7.7.7; AltName: Full=Mitochondrial DNA polymerase catalytic subunit; AltName: Full=PolG-alpha; LENGTH: 1239 AA |

**Results**

⊞ **Prediction/Confidence**      *PolyPhen-2 v2.0.23r349*

**HumDiv**

This mutation is predicted to be **BENIGN** with a score of **0.050** (sensitivity: **0.94**; specificity: **0.79**)

0.00    0.20    0.40    0.60    0.80    1.00

⊞ HumVar

**b**

| Predict Not Tolerated | Position | Seq | Rep | Predict Tolerated |
|---|---|---|---|---|
| y w v t s r q p n m l k i h f e d c a | 991G | 0.93 | | G |
| w | 992L | 0.84 | | c y f m h p g n d I L Q R s T e A V K |
| c w f d m i y v g p s h l a t e q | 993R | 0.75 | | N K R |
| | 994W | 0.62 | | W d c p m g e n q k R s h F T A I l y V |
| | 995Y | 0.60 | | c w p d m e k q g r I v a N T S F H L Y |
| c w m p g | 996R | 0.58 | | n v t f e l a I Q S D Y H K R |
| d h g n w e c y r s k p q t f m v | 997L | 0.36 | | I A L |
| w c f | 998S | 0.36 | | m y i v h d l P t n G e a q S R K |
| w y f c | 999D | 0.40 | | m h i v l P G r t q N S A D K E |
| w c f m y i h v r p g | 1000E | 0.36 | | t s a q L N K D E |

**c**

```
POLG_human    YAATKGLRWYRLSDEGEW
POLG_dog      YAVTKGLRRYRLSEEGEW
POLG_rat      YAVTKGLRRYRLSDDGEW
POLG_mouse    YAVTKGLRRYRLSADGEW
POLG_frog     YAVTKGIRRYILSKEGEW
POLG_fly      FSITKGKRVYRLREEFHD
POLG_yeast    YENTKGKTKRSK------
              :    ***
```

Fig. 2. PolyPhen-2 (**a**), SIFT (**b**), and protein alignment (**c**) for novel variant, c.2987G > A (p.R996Q), in POLG. Please see main text for more detailed description on interpretation of this novel variant.

serine at amino acid position 18 of the ATP6 protein. Proline at position 18 of the ATP6 protein is conserved from frog to human. Both SIFT and PolyPhen-2 algorithms predict the p.P18S in the ATP6 protein to be deleterious. The m.8578C > T (p.P18S, ATP6) variant has been reported at the frequency of 2703:1(C:T) in mtDB (http://www.mtdb.igp.uu.se/), which contains mitochondrial DNA sequences from individuals who were healthy at the time of ascertainment. Some variants may appear to be rare due to ethnic underrepresentation in the database. Thus, the clinical significance of these two variants is unclear. If these variants are also present in asymptomatic matrilineal adult relatives, then these variants, by themselves, are unlikely to be the primary cause of this individual's clinical symptoms. Targeted sequence analysis of this individual's mother will facilitate the interpretation of these two variants.

# References

1. Richards, C. S., Bale, S., Bellissimo, D. B., Das, S., Grody, W. W., Hegde, M. R., Lyon, E., Ward, B. E., and the Molecular Subcommittee of the ACMG; Laboratory Quality Assurance Committee. (2008) ACMG recommendations for standards for interpretation and reporting of sequence variations: Revisions 2007, Genetics in Medicine 10, 294–300.

2. Stenson, P., Mort, M., Ball, E., Howells, K., Phillips, A., Thomas, N., and Cooper, D. (2009) The Human Gene Mutation Database: 2008 update, Genome Med 1, 13.

3. Copeland, W. C. (2008) Inherited Mitochondrial Diseases of DNA Replication, Annual Review of Medicine 59, 131–146.

4. Ferré, M., Amati-Bonneau, P., Tourmen, Y., Malthièry, Y., and Reynier, P. (2005) eOPA1: An online database for OPA1 mutations, Human Mutation 25, 423–428.

5. Kumar, P., Henikoff, S., and Ng, P. C. (2009) Predicting the effects of coding non-synonymous variants on protein function using the SIFT algorithm, Nat. Protocols 4, 1073–1081.

6. Adzhubei, I. A., Schmidt, S., Peshkin, L., Ramensky, V. E., Gerasimova, A., Bork, P., Kondrashov, A. S., and Sunyaev, S. R. (2010) A method and server for predicting damaging missense mutations, Nat Meth 7, 248–249.

7. Cartegni, L., Wang, J., Zhu, Z., Zhang, M. Q., and Krainer, A. R. (2003) ESEfinder: a web resource to identify exonic splicing enhancers, Nucleic Acids Research 31, 3568–3571.

8. Cartegni, L., Chew, S. L., and Krainer, A. R. (2002) Listening to silence and understanding nonsense: exonic mutations that affect splicing, Nat Rev Genet 3, 285–298.

9. Brunak, S., Engelbrecht, J., and Knudsen, S. (1991) Prediction of human mRNA donor and acceptor sites from the DNA sequence, Journal of Molecular Biology 220, 49–65.

10. Reese, M. G., Eeckman, F. H., Kulp, D., and Haussler, D. (1997) Improved splice site detection in Genie, J Comput Biol 4, 311–323.

11. Ingman, M., and Gyllensten, U. (2005) mtDB: Human Mitochondrial Genome Database, a resource for population genetics and medical sciences, Nucleic Acids Research 34, D749-D751.

12. Helm, M., Brule, H., Friede, D., Giege, R., Putz, D., and Florentz, C. (2000) Search for characteristic structural features of mammalian mitochondrial tRNAs, RNA 6, 1356–1379.

13. Cohn, A. C., Toomes, C., Potter, C., Towns, K. V., Hewitt, A. W., Inglehearn, C. F., Craig, J. E., and Mackey, D. A. (2007) Autosomal Dominant Optic Atrophy: Penetrance and Expressivity in Patients With OPA1 Mutations, American Journal of Ophthalmology 143, 656–662.e651.

14. Delettre, C., Lenaers, G., Griffoin, J.-M., Gigarel, N., Lorenzo, C., Belenguer, P., Pelloquin, L., Grosgeorge, J., Turc-Carel, C., Perret, E., Astarie-Dequeker, C., Lasquellec, L., Arnaud, B., Ducommun, B., Kaplan, J., and Hamel, C. P. (2000) Nuclear gene OPA1, encoding a mitochondrial dynamin-related protein, is mutated in dominant optic atrophy, Nat Genet 26, 207–210.

# INDEX

Lee-Jun C. Wong (ed.), *Mitochondrial Disorders: Biochemical and Molecular Analysis*, Methods in Molecular Biology, vol. 837,
DOI 10.1007/978-1-61779-504-6, © Springer Science+Business Media, LLC 2012